Carbon Fibres and Their Composites

Edited by Erich Fitzer

Based on papers presented at the International Conference on Carbon Fibre Applications, São José dos Campos (SP), Brazil, 5–9 December 1983, which was jointly sponsored by the Centro Técnico Aeroespacial, the United Nations Industrial Development Organization and the United Nations Financing System for Science and Technology for Development

Springer-Verlag
Berlin Heidelberg New York Tokyo

United Nations Financing System for Science and Technology for Development (UNFSSTD)
New York

Prof. Dr. Erich Fitzer
Institut für Chemische Technik, Universität Karlsruhe
Kaiserstraße 12, D-7500 Karlsruhe

With 209 Figures

ISBN 3-540-15804-9
Springer-Verlag Berlin Heidelberg New York Tokyo
ISBN 0-387-15804-9
Springer-Verlag New York Heidelberg Berlin Tokyo

Library of Congress Cataloging-in-Publication Data. Main entry under title: Carbon fibres and their composites. „Proceedings of a UNFSSTD/UNIDO sponsored international conference in Brazil, December 1983" – Verso t.p. 1. Carbon composites–Congresses. 2. Carbon fibers–Congresses. I. Fitzer, Erich, 1921–. II. United Nations Industrial Development Organization. TA418.9.C6C313 1985 620.1'18 85-17250
ISBN 0-387-15804-9 (U.S.)

Printed in Germany

Published by arrangement with United Nations Financing System for Science and Technology for Development and sold throughout the world by Springer-Verlag GmbH and Co. KG

Typesetting and printing: Schwetzinger Verlagsdruckerei GmbH
Bookbinding: J. Schäffer OHG, Grünstadt.
2152/3140-543210

Preface

The proper choice of technology is a complex decision, particularly for developing countries, as it depends not only on local needs and conditions but also, importantly, on the national political context and, increasingly, on the international environment. This technological choice carries with it the genetic code of the nation's future development. Many developing countries which lack the needed infrastructure do not have real options; others with a reservoir of scientific and engineering skills and explicit S/T strategies, can indeed choose between alternatives.

Turning to the technologies themselves, these cover a wide spectrum: traditional technologies that are low-cost, low-energy and often better suited to meet basic needs; more sophisticated technologies which are highly knowledge-intensive and require large capital outlays for research, product design and manufacturing; and still others which depend upon a blending of modern technology with traditional methods to create products and processes more suited to local needs. Even within the group of advanced technologies, there is considerable differentiation, and those at the lower end of the product cycle are clearly within reach by the newly industrializing countries.

The UN Financing System for Science and Technology for Development has consistently advocated a pluralistic approach – a walk on many legs – so that even in the same country or region, at different locations or at different times, a variety of technical processes can co-exist, each most appropriate to the given situation. Thus, UNFSSTD is supporting Indonesia to commercialize improved solar dryers and, at the same time, Pakistan to develop a modern Silicon Technology Centre for photo voltaic applications; on the one hand, assisting Jordan to utilize local materials for low-cost housing and, on the other, helping Brazil to strengthen its capacity to develop sophisticated carbon-fibre reinforced materials for transportation and industrial applications. In the frontier technologies, the United Nations can be of help in providing access to technical information, expertise and training. But such advice should be treated with caution, and the developing country itself must, I believe, take the executing responsibility.

National development planning in Brazil emphasizes not only the development of its agricultural sector but also the promotion of energy independence with technological autonomy. Parallel to these concerns is the formation of strengthened indigenous technological capabilities in the priority sectors of the economy. The use of stronger, versatile and lighter structural materials, such as carbon fibres, are not only beneficial for industrial applications but have also importance to the agricultural sector. Carbon fibre applications can range from the construction of boats and barges for use in the Amazonas basin and the Pantanal de Mato Grosso and propellers for wind-power generation to the production of surgical instruments and artificial limbs.

The special achievement of the UNFSSTD-supported project on carbon fibres in Brazil is the linkage between R/D and actual production which it provides. While at Centro Técnico Aeroespacial (CTA) in São José dos Campos (SP) a research team developed, adapted and produced on pilot scale carbon fibre, in Camaçarí (BA) a production facility was equipped and its staff trained for the pilot production of the necessary precursor fibre. At the same time, preparations were under way to assure the industrial production and marketing of the fibres in Brazil. Recently, a major industrial complex has decided to go into production with technical assistance from CTA.

The lessons to be drawn from the Brazil experience and from other high technology projects in the developing countries are at least threefold: first, a sound base of fundamental scientific research is a prerequisite for any meaningful entry into an advanced technical field. Know-how itself has a short life-span, without the know-why to understand scientific phenomenon, and adapt the product design and the manufacturing skills that may have been acquired initially from abroad.

In dealing with advanced composite materials, for instance, the knowledge of chemists, computer specialists and materials scientists is essential to grasp the basic processes, to be followed by the skills of production engineers and managers, in order to move the process rapidly from laboratory and pilot plant, through to the market place.

Second, the process of learning-by-doing requires a major commitment to train the large variety of specialists needed to cover the whole innovation cycle. Even in those developing countries which have a large quantity of technical manpower, the quality is often deficient. Are government and industry willing to take a long-term view on developing the required technical human resources, and to make the commensurate investments in high-quality education at home and well-prepared training abroad?

Finally, advanced technology development requires a high order of entrepreneurial skills and support by venture capital. In an industrialized economy, high-tech companies are often small in size but linked to large corporations, clustered adjacent to centers of research and learning. The industrializing countries now need to create the climate in which risk-taking entrepreneurship can take root, and the policy instruments to nourish them. Here, the vast pool of third world scientists permanently settled abroad can be mobilized to assist, with transfers of both technology and capital.

South-south technical co-operation can also play a significant role in this process. Expertise from, and training opportunities in, countries at comparable levels of development can be expected to be more effective. The carbon-fibre project in Brazil is a case in point: The Chief Technical Advisor was from India. But it is unreasonable to expect that such co-operation can be based on altruism; it is best undertaken where each party sees a mutual advantage.

Again taking the São José dos Campos experience, its success was primarily due to the efforts of Brazilian technicians. But carefully-selected international expertise inputs, mobilized through UNIDO and supported by the UN Resident Coordinator's office, played a significant role. Thus, the synergy provided by various forces working together – the Brazilian Government, the international scientific community, universities, private enterprise and the United Nations – helped in the rapid acquisition of some sophisticated know-how and its prompt transfer to commercial production.

I would like to express our deep sense of gratitude to Professor Dr. Erich Fitzer of Karlsruhe University and to my colleague Dr. Niels Brandt for their contributions to this project – and for putting together the proceedings of the International Conference on Carbon Fibre Applications which brought together experts from developed and developing countries.

Rustam Lalkaka
Director, United Nations Financing System for
Science and Technology for Development

Introduction

“Advanced composites” are fibre reinforced polymers with structures similar to that of the well-known fibre glass, but composed of high stiffness fibres, the specific YOUNG’s modulus of which exceeds that of glass fibres by up to an order of magnitude. Only in such a way, one can achieve the same or even better mechanical properties with the polymers as with metallic materials. It has turned out during the last decade, that most candidates for high-modulus fibre materials consist of nearly pure carbon, although the technology with advanced composites started in the late fifties and early sixties with boron fibres, similar in stiffness and strength to today’s carbon fibres but incomparably higher in cost. Besides the strength and stiffness of these reinforcing fibres, their very low density (below 2 g/cm^3) is one decisive advantage in a broadening of the technical application of advanced composites.

Carbon fibres as reinforcing materials are relatively new. Their development started in the sixties, but the types of carbon fibres available commercially then showed only low modulus and quite low strength values as compared to those of the boron fibres already applied in US aircraft structures. At that time, the American graphite industry (UNION CARBIDE Corp.) had experience in hot working of polycrystalline graphite at temperatures above 2500 °C. Such a hot working process applied to carbon fibres brought the surprisingly high modulus values up to 750 GPa, which is about 75% of the theoretical modulus due to the high binding energy between carbon atoms with sp^2 hybridization. The breakthrough occurred by replacement of rayon-based carbon fibres by those based on fully synthetic polymers, especially on polyacrylnitrile, developed firstly by Japanese and then by British workers. This raw material enables fabrication of high-performance carbon fibres without a hot stretching process. PAN is also today the main precursor material for commercial carbon fibres.

In the beginning, advanced composites were developed only for aerospace applications because of the combination of light weight, high stiffness and high strength properties. The rapid development for this application was only briefly interrupted by the unfortunate attempt to apply carbon-fibre reinforced polymers as fan blades in the ROLLS ROYCE RB12211 aircraft engine. At that time, such an application with insufficient tests was too risky because of delaminations and catastrophic failure in the case of impact damage. Most of these difficulties from the time of immaturity are solved today in the high-performance composites, and in popular literature carbon fibres are called a miracle material. One speaks also of a revolution in material based on carbon fibre reinforced composites.

During the last decade it was learned that successful application is based not only on the material properties but also on optimum design, new high-performance production methods and special testing possibilities. First of all, the unique possibility of tailoring the structure by these materials because of their anisotropic nature, offers a completely new pioneering field for the designer. The years of flight tests in aircraft have produced confidence in this new type of material. The latest development is the vertical stabilizer for the Airbus 310 demonstrated at the aircraft show in Paris in April 1985. This structural part is the largest part made from carbon fibre reinforced composites which is in industrial series production so far, and a special production line was set up for this purpose in Hamburg recently.

Also in other fields of transportation and machinery these advanced composite materials with carbon fibres as their back bone are in vigorous further development. The newest example for the use of a fibre composite is the bogie frame developed for ultra high speed trains (350 km/h). This application is now being tested for the railway system in Germany.

Also the composites with a carbon matrix instead of a polymer matrix, the so-called carbon/carbon composites, are no longer used only for cone tips and engines in missiles, but also as a very promising replacement for asbestos in brake lining and disc materials. Another revolutionary application is the use of such carbon/carbon materials as moulds for hot forming of high melting point metals such as titanium, in the superplastic temperature range.

This book was primarily planned just as the proceedings of the "International Seminar on Carbon Fibres and their Composites" organized by the United Nations Organizations UNFSSTD and UNIDO in Brazil, December 1983, but has now been extended and updated by the authors of the various chapters and so become a modern handbook on the latest technologies. The public community interested in

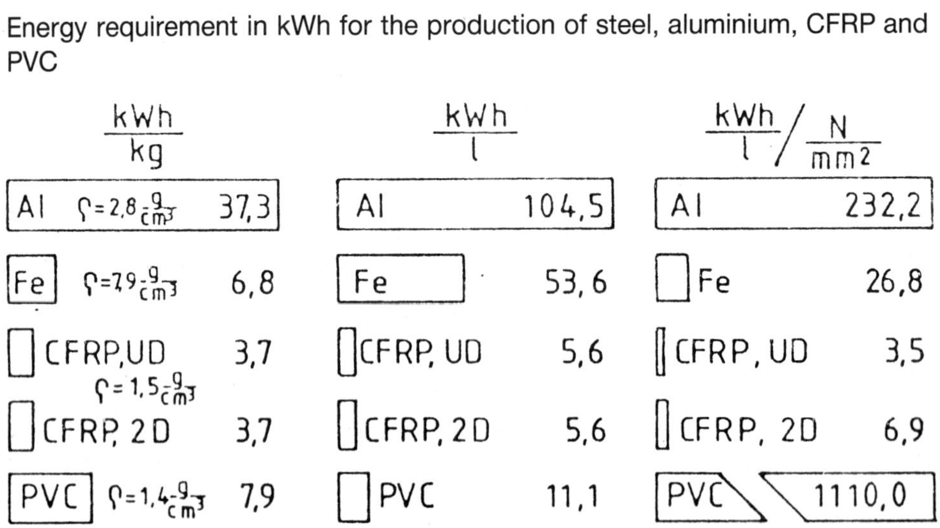

Fig. 1. Comparison of the energy requirement for the production of steel, aluminium and CFRP in PVC including raw material energy equivalence

new materials has to thank the United Nations Organization for their activities bringing such advanced technology to developing countries, and the authors of the several chapters, all internationally well-known specialists, for their enthusiastic cooperation.

The book gives a broad introduction to all fields of the new technology of advanced composites. These are development of fibres and polymer matrices, technology of fabrication processes, basic ideas of design and the new methods for testing. Finally, an overview is given of the various fields of application of such advanced composites. In more detail:

Chapter 1 treats the nature, technology and properties of the various types of carbon fibres, especially in comparison with alternative reinforcement fibres and the resulting capacity for various applications.
Chapter 2 describes the polymers, especially thermosetting matrix systems such as epoxy resins, vinyl esterresins, unsaturated polyester resins and also the new developments in high temperature resistant polyimides. In future, obviously, an increasing application of thermoplastic matrix resins can be expected. Such thermoplastic matrix systems facilitate the fabrication processes and thus will contribute to an enormous reduction in the cost of advanced composites.
Chapter 3 describes the fabrication and design methods for CFRPs. Such new material types, especially with optimized anisotropic structure, need very careful testing with sophisticated test methods.
Chapter 4 contains the most important methods for testing and describes our actual knowledge of fracture behaviour. However, more experience in this field is expected in future.
Chapter 5 and 6 are reserved for the descriptions of CFRP applications in aerospace and elsewhere. A special contribution on biomedical application is included.

We live today in a world threatened by environmental problems. In particular energy production on one hand and metallurgical technology with its high degree of air and water pollution are central points of international concern. Carbon fibre reinforced composites have the enormous advantage of a very low raw material and energy consumption compared with conventional metals, which is shown in Fig. 1. This is an additional reason why this new high technology becomes more and more interesting not only for industrialized countries but also for developing countries, especially those which are in the process of becoming industrialized countries.

In *Chapter 7* some contributions from such developing countries are presented which describe the surprising amount of progress achieved there already.

It is hoped that this book will contribute to a better understanding of these promising "advanced composites" and will initiate a further increasing interest and demand in the near future.

Prof. Dr. E. FITZER

Table of Contents

Chapter 4. Testing and Fracture Behaviour of CFRPs

Chapter 5. Application of CFRPs in Aerospace

Chapter 6. Further Applications of CFRPs (Non Aerospace)

Chapter 7. Carbon Fibres Activities in Brazil, PR China, India and South Korea

List of Contributors

Prof. Dr. D. F. Adams	College of Engineering, University of Wyoming, Laramie, Wyoming 82071/USA
Dr. H. W. Bergmann	Institut für Strukturmechanik DFVLR, Postfach 3267, 3300 Braunschweig/FRG
Dipl.-Ing. K. Brunsch	Messerschmidt-Bölkow-Blohm GmbH, Unternehmensbereich Drehflügler, Abt. D E 22, Postfach 80 11 40, 8000 München 80/FRG
Prof. Dr. R. J. Diefendorf	Rensselaer Polytechnic Institute, School of Engineering, Materials Eng. Department, Troy, N.Y. 12181/USA
Prof. Dr. E. Fitzer	Institut für Chemische Technik, Universität Karlsruhe, Kaiserstr. 12, 7500 Karlsruhe/FRG
Dr. G. W. Hastings	Head of Bio-Medical Engineering, Unit–Medical Institute, Hartshill, Stoke-on-Trent, ST4 7NY/ United Kingdom
Dr. B. J. Hayes	Technical Manager Bondes Struct. Division, CIBA-GEIGY Plastics and Additives Co., Duxford, Cambridge CB2 4QD/United Kingdom
Dr. H. Jäger	Institut für Chemische Technik, Universität Karlsruhe, Kaiserstr. 12, 7500 Karlsruhe/FRG
Dr. I. L. Kalnin	Research Associate, Dep. 13, Celanese Research Company, 86 Morris Avenue, Summit, N.J. 07901/USA
Dr. C. Y. Kim	Head of Polymer Materials Laboratory, Korea Advanced Institute of Science and Technology, P.O. Box 131, Dongdaemun, Seoul, Korea 130

Dr. T. Nagabhushanam	Department of Chemistry, Polymer Science Program, University of Lowell, Lowell, Mass. 01854/USA
Dr. B. Rhee	Department of Chemical Engineering, Chungnam National University/Korea
Dr. H. D. Stenzenberger	Technochemie GmbH, Verf. Technik, Abt. Neue Technologie, Gutenbergstr. 2, 6901 Dossenheim/FRG
Prof. Dr. H. Zeng	Deputy Director of Research Institute of Polymers, Zhongshan (Sun Yatsen) University, Guangzhou/PR of China

Chapter 1
Fibres for Reinforcement of Advanced Composites

Technical Status and Future Prospects of Carbon Fibres and their Application in Composites with Polymer Matrix (CFRPs)

E. FITZER

The structure of carbon fibres and its dependence on the presursor material gives an understanding of the properties of today's commercial fibre types. The possibilities of further improvement are considered from theoretical viewpoint as well as starting from the properties of PAN based fibres of the new generation just in development. The problem of correct terminology of this promising new material is raised. Economic considerations are presented, comprising a review on present technology, the world demand and the price situation. Finally, the various fields for application of carbon fibres – such as aerospace, transportation, sport, machinery and apparatus, cryogenic technology, as well as medicine – are discussed briefly.

1 The Nature of Carbon Fibres and its Dependence on the Precursor Material

Carbon fibres are very thin fibres, much thinner than human hairs (Fig. 1). They have diameters between 6 and 10 μm, in American units a quarter of a mill. They are also thinner than boron fibres which correspond in thickness to that of human hairs and which were used as first high modulus fibres for reinforcement of polymers in the so-called "advanced composites". Glass and polyaramide fibres have comparable thin diameters and can be used in combination with carbon fibres as hybride reinforcement in various forms, such as staple fibres, yarns, rowings or fabrics.

Carbon fibres consist of 99.9% of chemically pure carbon. They are not chemical compounds or alloys such as conventional structural materials, for instance steel or aluminium, which always contain small amounts of alloying elements.

These black thin manmade fibres, sometimes incorrectly named as "graphite" fibres – inbedded in polymers – have initiated a "revolution in materials". Even popular magazines do report on it as "miracle material" [1, 2]. "Advanced composites" contain up to 60 v/o carbon fibres. If compared with fibre glass, advanced composites are superior in light weight and high stiffness but have similar strength. Due to this combination, first time polymer materials became a competition for metals in structural application as shown in Fig. 2 [3, 4, 5]. Advanced composites have further advantageous properties, such as extremely low thermal expansion, better fatigue resistance and high corrosion resistance (Fig. 3).

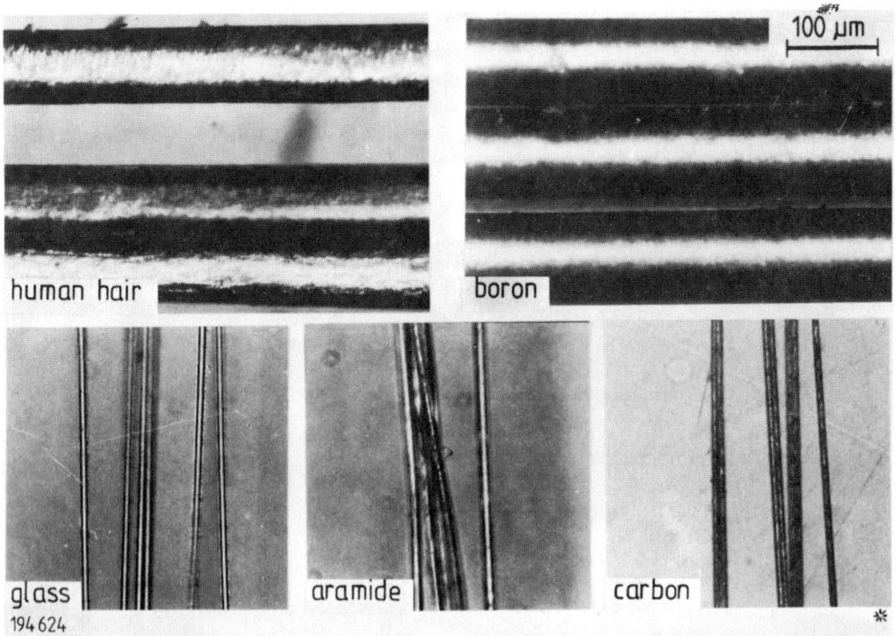

Fig. 1. Micrographs of reinforcement fibres, compared with human hairs

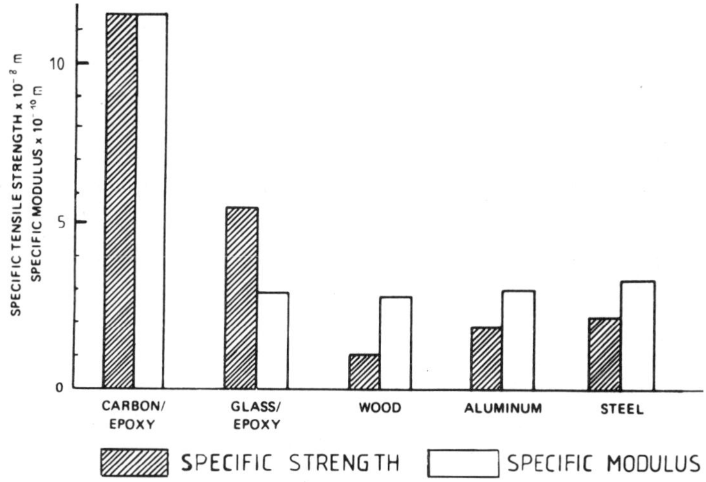

Fig. 2. Specific strength and modulus of CFRP in comparison with conventional materials [3]

The key properties for reinforcement fibres used today in composites with polymer matrix for reinforcement – the tensile strength and the YOUNG's modulus – are compiled in Fig. 4. All glass fibres, especially R- and S-types which have a higher Al_2O_3-content than E-types, show high strength (above 2000 to 3500 MN/m^2) but low YOUNG's modulus (below 100 GN/m^2). Polyaramide fibres (KEVLAR® from DuPONT or TWARON® from ENKA) are organic fibres with improved YOUNG's modulus up to 150 GN/m^2.

Only the family of carbon fibres covers a range of YOUNG's modulus values between 200 and 700 GN/m^2 and strength data from 2000 to 4000 MN/m^2 or even 5000 MN/m^2 in experimental types. Within this product family we distinguish between "PAN based" and "pitch based" carbon fibres.

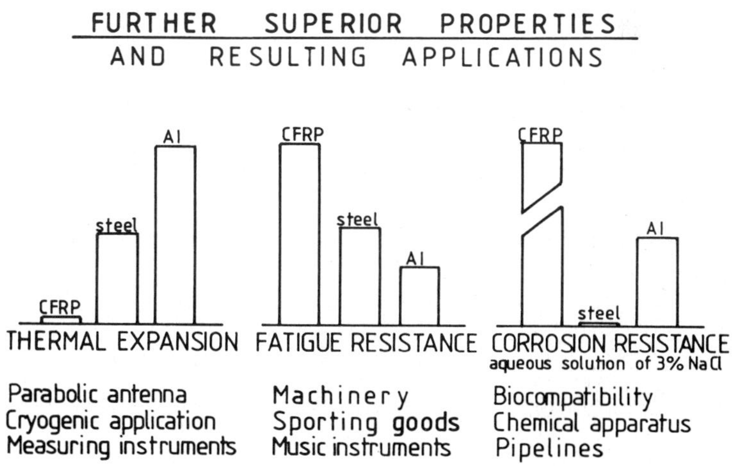

Fig. 3. Further advantages properties of advanced composites (CFRP) in contrast to conventional materials [4, 5]

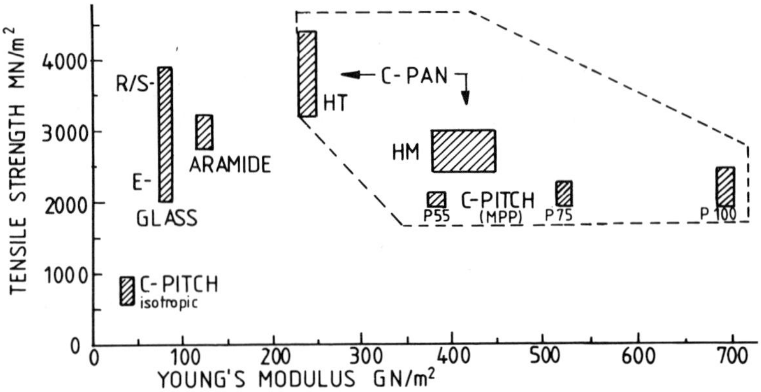

Fig. 4. Tensile strength and YOUNG's modulus of the various fibre types applied for reinforcement of composites

The more important group comprises the PAN based carbon fibres with the high tensile strength fibre type (showing only medium high YOUNG's modulus between 200 and 300 GN/m^2) and the so-called high modulus fibre type with YOUNG's modulus values above 400 GN/m^2 (but lower strength values). Extra qualities such as super high tensile strength (SHT), sometimes called new generation, are also PAN based.

Within the pitch based carbon fibres only the "mesophase pitch (MPP) based types" which reach very high modulus values but lower strength of around 2000 MN/m^2 are used as reinforcement fibres, whereas the isotropic pitch based fibres (e.g. KUREHA® carbon fibres) have mainly applications as insulation materials, as filler and for some other purposes.

Today's and future broad application of carbon fibres as structural material is mainly based on the high tensile type with much higher strain to failure, but only an intermediate modulus (IM), if compared with the high modulus types and especially with the MP pitch based ones. High modulus fibre types are used today for very special applications only, where such high modulus in spite of low strain to failure is effectively needed, for instance in space structures.

Recognizing these surprising high mechanical properties of carbon fibres, we have to consider three questions:

i) What is the reason for the extremely high YOUNG's modulus and the high strength?
ii) How can the broad variety of these properties in the family of carbon fibres be explained?
iii) Can all of these modified types be produced under reproducible conditions?

1.1 The Origin of High Mechanical Properties

These mechanical properties are caused by the very strong chemical bond between the carbon atoms within the graphitic layers, that means in crystallographic a-direc-

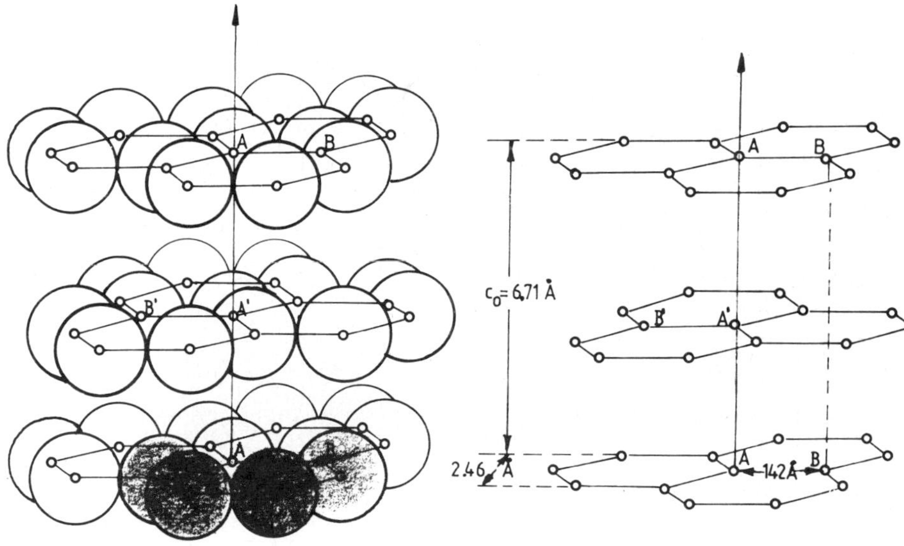

Fig. 5. The structure of the hexagonal modification of graphite [6–8]

tion. These anisotropic nonpolar, so-called σ-bonds are based on the sp^2 hybridization of the electron orbitals in this carbon structure. The structure of graphite (Fig. 5) is known almost 60 years, after it was correctly described first time independently but nearly simultaneously by both H. MARK et al. [6] and J. D. BERNAL [7]. The very dense packing of the carbon atoms within the layers can be seen from the left part of Fig. 5, whereas on the right side the usual demonstration of lattice structures with only lattice places instead of atoms has been used for better recognizing of the lattice distances.

Since about 10 years the dense packed layers became visible by high resolution transmission electron microscopy. Fig. 6 shows such an image of highly graphitized carbon in direction of the layer planes.

As known, the bond strength between the atoms controls directly the lattice elasticity of a material. The high strength of the bonds between C-atoms in the layers is due to the homopolar nature, whereas the only very week bonds between atoms of neighbouring layers is the explanation for the lower strength in this lattice direction. The high electrical and thermal conductivity of graphite in direction parallel with the layers is caused by delocalized π-electrons between the layers. Graphite single crystals, however, can not be described as a metallic conductor, but only as a semi conductor in which the conductivity bands are overlapping.

The elastic constants are known with quite high precision for graphite single crystal [9]. The most important three elastic constants, C 11, C 33 and C 44 are shown in Fig. 7. In case of complete preferred orientation of the layer planes parallel with the fibre axis the maximum theoretical value for the YOUNG's modulus would be 1060 GN/m^2, whereas the modulus perpendicular to the fibre direction can be expected as being an order of magnitude lower. The lowest value can be seen for the shear modulus C 44 with only 4,5 GN/m^2.

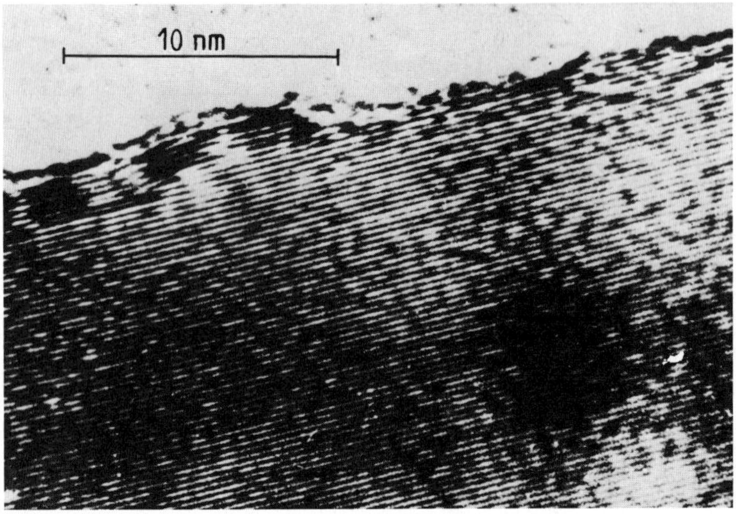

Fig. 6. High resolution TEM of highly graphitized carbon

1.2 Explanation for the Variety of C-Fibre Types

From knowledge of the elastic constants of the graphite single crystal the following consequences can be drawn for carbon fibres, which are to be used as reinforcement:

i) highest YOUNG's modulus and strength in carbon fibres will be obtained by highest degree of preferred orientation, and
ii) the low resistance against shear between the layers as typical for the ideal graphite lattice must be improved.

1.2.1 Effect of Preferred Orientation

Figure 7 shows the correlation between measured YOUNG's modulus in various carbon fibres and degree of preferred orientation, as determined by X-ray diffrac-

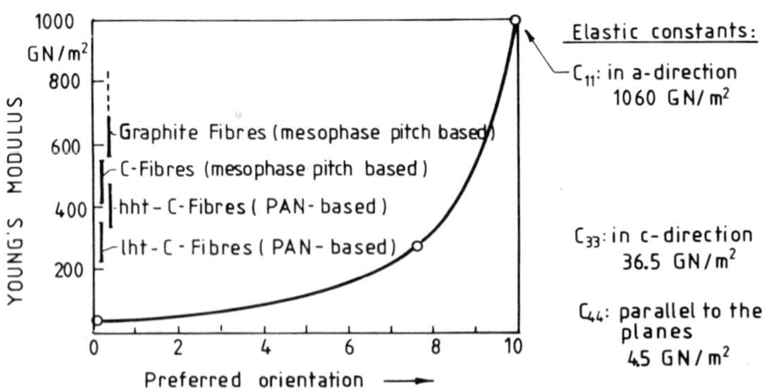

Fig. 7. Elastic constants of graphite [9] and effect of preferred orientation on YOUNG's modulus of C-fibres

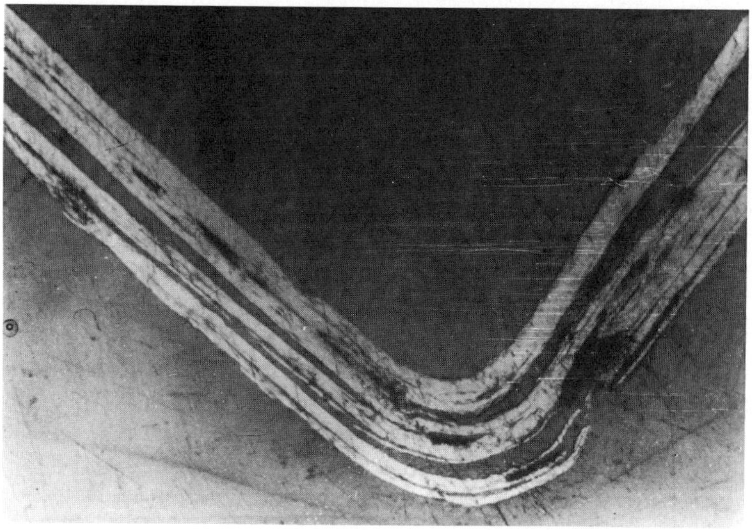

Fig. 8. Bent single crystal of graphite [10]

tion. In best cases, values up to 70% of the theoretical maximum values have been achieved already in industrial carbon fibre types. The diagram shows also, that intermediate YOUNG's modulus values (around 250 to 300 GN/m^2) as typical for HT type fibres are combined with incomplete preferred orientation.

1.2.2 Lattice Defects in C-Fibres

Figure 8 shows the deformation of a single crystal of graphite [10]. In spite of the high strength within the graphite layers, the crystal can easily be bent because of the very low shear modulus. This shear between the layers can be avoided by introduction of lattice defects.

Figure 9 shows schematically the various types of lattice defects in a graphitic structure, namely defects within the layers, defects between the layers, mostly stack-

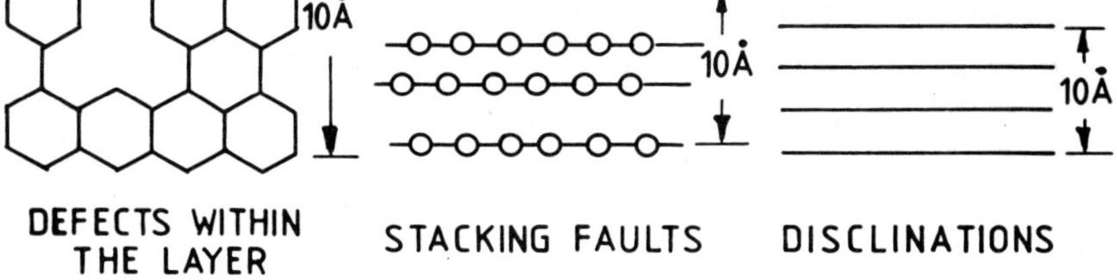

Fig. 9. Lattice defects in carbon fibres (schematically)

ing faults which cause higher lattice distance, and finally disclination of the layers. The first two defects are mostly combined, that means defects within the layer cause stacking faults and higher lattice distances. By these defects incomplete delocalization of the π-electrons is caused and shear deformation in layer direction is hindered. In case of disclinations, shear can be hindered from geometrical viewpoint. But in this case, also defects within the layer and stacking faults must be combined with the disclinations. Otherwise, disclinations can act as a crack initiator [11].

X-ray diffraction gives an indication of defect free distances within the layers. These areas of coherent diffraction behaviour vary between 20 and some hundreds Å units (2–10 nm) and depend strongly on precursor type, stabilization conditions and the final heat treatment temperature in the range between 1400 °C and 2700 °C.

Some of the lattice defects can be demonstrated by high resolution transmission electron microscopy (Fig. 10). Preferred orientation and disclinations are mainly introduced by the polymer fibre itself, that means by the degree of textile prestretching applied after spinning, and by the tensile stress conditions during stabilization. Stacking faults and defects within the layers are mainly healed by heat treatment up to graphitization temperature. However, in PAN based fibres real graphitization has never been found in contrary to highly heat treated mesophase pitch based C-fibres. That means, annealing of most stacking faults and complete delocalization of π-electrons can not be achieved (see Fig. 12 and 15).

It is understandable, that the progressing growth and improved orientation of the layers will increase the YOUNG's modulus in fibre direction.

In former years (Fig. 11) [12a, b, c] a decrease in strength was observed simultaneously. In newest C-fibre types it was shown, however, that the strength after heat treatment at 1300 °C will not be reduced during further high temperature heat

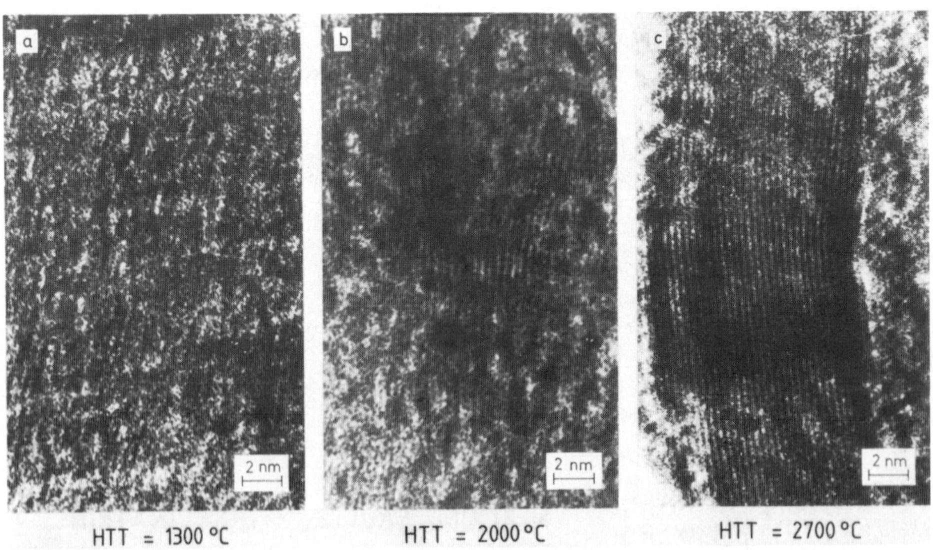

Fig. 10. High resolution TEM of PAN based carbon fibres after various final heat treatment

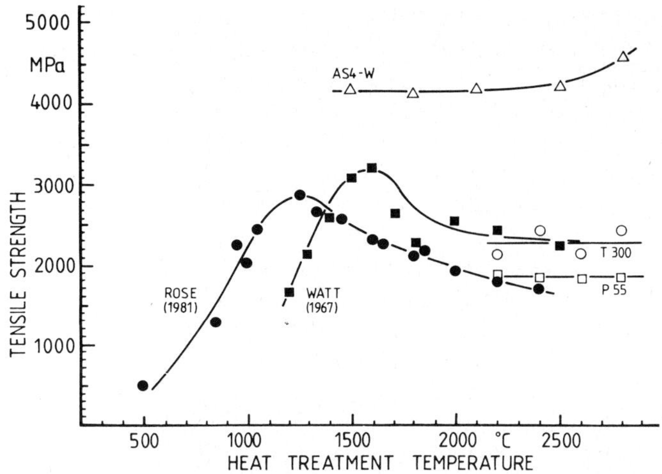

Fig. 11. Influence of final heat treatment temperature on strength of C-fibres [12 a–c]

treatment up to 2700 °C. This behaviour is explained by the use of very pure polymer precursor materials which result in homogeneous and pure carbon fibres. The lattice defects in the carbon fibres are initiated and controlled by the defect structure of the polymer under the precondition that no liquid state will be passed during carbonization. PAN, a thermoplastic polymer, has to be transformed therefore into a non-meltable ladder polymer by cyclization and cross-linking oxidation before carbonization (stabilization treatment). Carbon, resulting from such a cross-linked polymer is called "polymer carbon". The mesophase pitch based fibres show always best perfection of the graphitic structure as shown in Fig. 12. Therefore highest tendency for

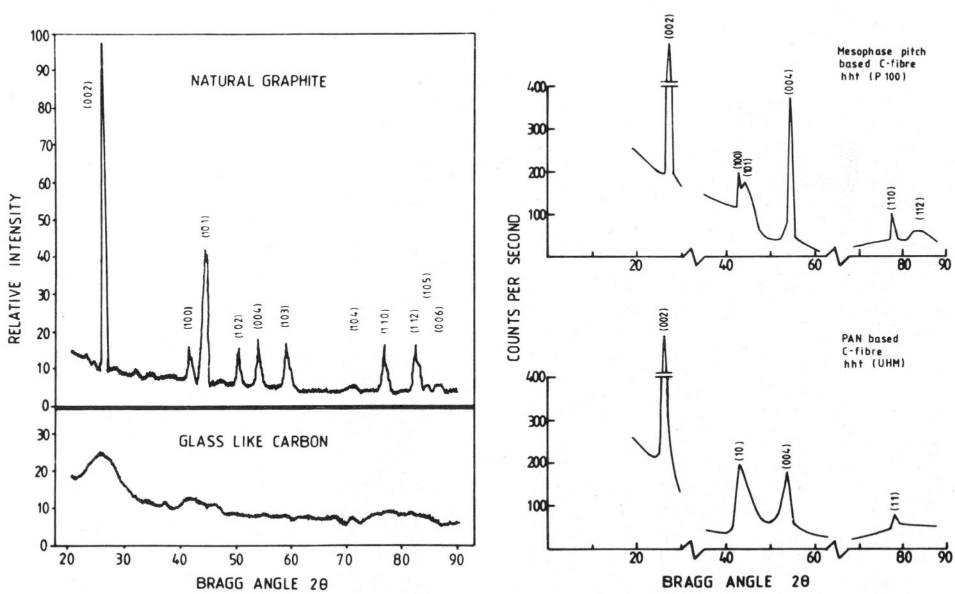

Fig. 12. X-ray diffraction profile of a graphite single crystal, mesophase based carbon fibre P 100 (partly graphitized), PAN based UHM carbon fibre and glasslike carbon, last two representing polymer carbon in spite of heat treatment at graphitization temperature

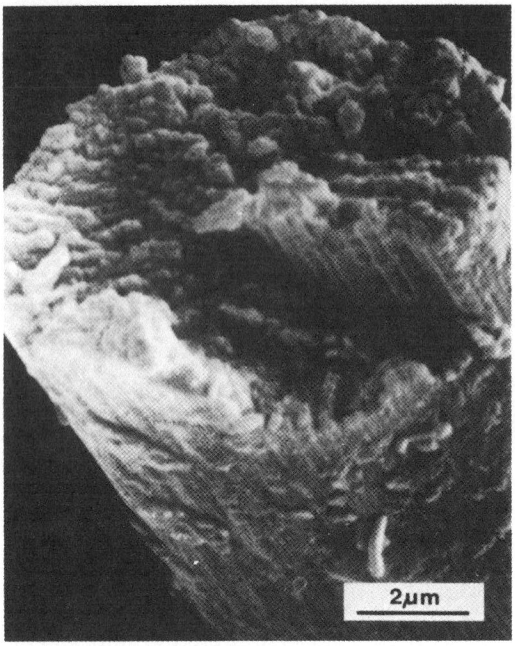

Fig. 13. SEM image of fracture surface of mesophase pitch based carbon fibres showing internal sheet like structure [11]

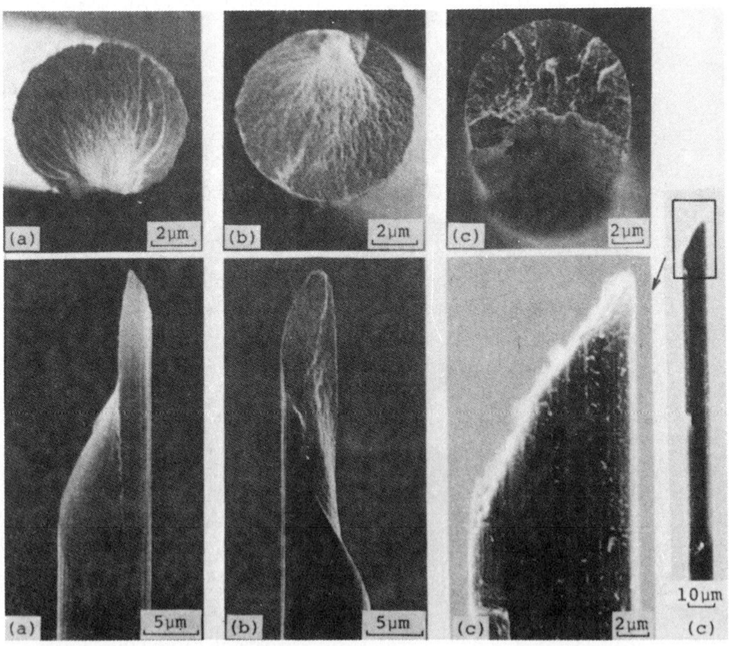

Fig. 14 a–c. Torsional fracture surfaces of carbon fibres [13]. **a** PAN based low heat treated (lht), **b** PAN based highly heat treated (hht), **c** Mesophase pitch (MPP) based (hht)

graphitization resulting in formation of sheet like structural elements is observed. An example is demonstrated in Fig. 13 representing the fracture surface by REM. It is assumed that the fracture is initiated by the best graphitic structural planes [11] as described later in Chap. 2.3.1.

Another indication of shear sensitivity because of beginning graphitization is shown in Fig. 14 by means of the torsional fracture surfaces [13]. The fracture surface of high modulus fibres (middle of the figure) is increased if compared with that of the high tensile carbon fibres (left hand side). The fracture surface of mesophase pitch based carbon fibres (right hand side) is extended by an order of magnitude because of the low resistance against shear attaque.

From scientific viewpoint this result is not surprising because mesophase pitch is the ideal precursor for best graphitizing carbons. All coke raw materials for synthetic graphite are made in this way. The graphitization behaviour can be recognized from decrease of the average layer distance $\bar{c}/2$, tending to 0.335 nm, the value of single crystal graphite (Fig. 15). The structural defects within the resulting carbon

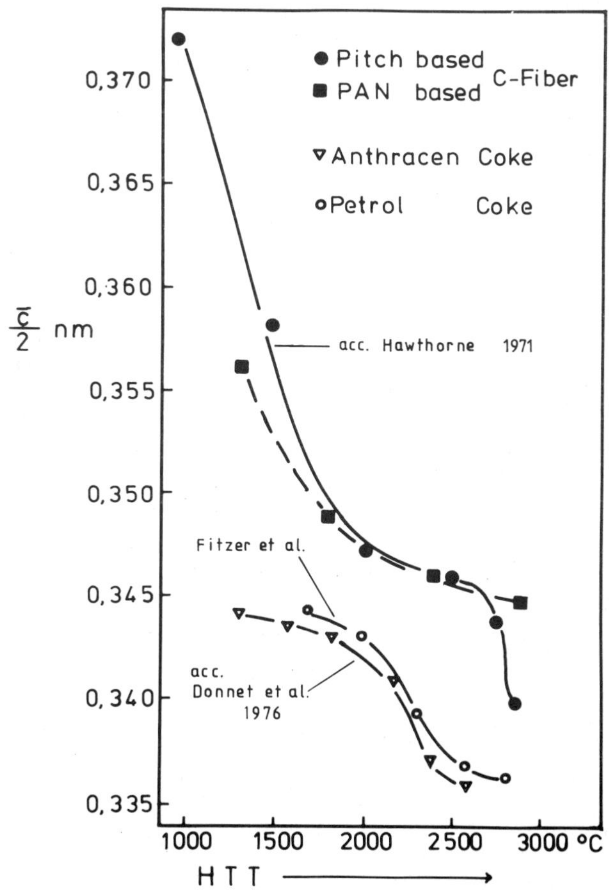

Fig. 15. Decrease of layer distance by heat treatment in carbon fibres and in graphitizing petroleum and anthracen coke [12 c, 14 a–c]

needed for high quality C-fibres must be introduced during oxidation treatment of the pitch fibre, similar as during stabilization treatment of PAN. The defects caused in the resulting carbon fibre are indicated by higher lattice distances as shown in Fig. 15 [12c, 14a, b, c].

Contrary to the ladder or even step ladder structure of oxidized PAN, oxidized pitch mesophase will retain sheet like polyaromatic structures with less degree of defects and thus join the tendency of graphitizing carbons under extreme heat treatment, as indicated by enhanced decrease of layer distance. The task to form non graphitizing carbon from a pitch mesophase is an intrinsic contradiction.

1.3 Performability of High Quality C-Fibre Production

In industrial production it is relatively easy to achieve reproducible YOUNG's moduli because this property is controlled by the prestretching conditions of the polymer fibres and the final heat treatment temperature of the carbon fibres which both parameters can be realized in reproducible manner.

It is much more difficult to achieve reproducible strength due to its sensitivity to flaws, notches and other structural defects in the carbon fibre as well as in the polymer precursor. As mentioned before, very pure and homogeneous polymer fibres have been recognized as important precondition for achievement of reproducible strength of the carbon fibres. It will be shown later, that this precondition is realized in industrial production not only in case of special low filament PAN yarn but also with multifilament tows (see Fig. 29).

Such demand is obviously more difficult to fullfill with natural or semi-natural raw materials such as in case of pitch precursor fibres from petroleum origin. As shown in Fig. 16 [15] such pitches consist of many hundred of different species, varying

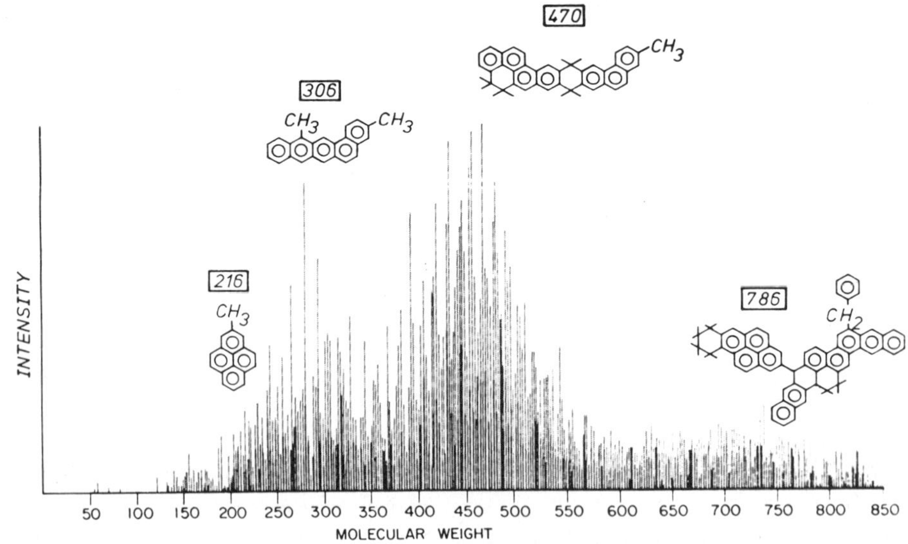

Fig. 16. Molecular weight distribution of a typical petroleum pitch [15]

upon crude composition and process conditions in the refinery. The chemical composition, the molecular structures and the molecular weight distribution of all these pitch components will control mesophase formation and mesophase behaviour during spinning and subsequent stabilization and thus the microstructure of the resulting carbon fibre. These problems coming from the raw material will increase additionally the difficulties in obtaining reproducible fibre properties in industrial production.

2 Properties of Today's Commercial Carbon Fibres and Considerations on Further Improvement

2.1 Characterization and Nomenclature of the Various Fibre Types

It was shown before, that not only mechanical properties of the carbon fibres – these are first of all YOUNG's modulus and tensile strength – but also the fracture behaviour depends on the precursor material and the final heat treatment. For more complete characterization of the various types of C-fibres an information about the way of fabrication is needed. The nomenclature for today's still incomplete classification varies in different countries because of traditional reasons:

The British manufacturers who have developed as first high quality carbon fibres on PAN basis used for highly heat treated fibres (heat treated at graphitization temperatures) the term "type I" fibres. They understand under "type II" such carbon fibres which have been heat treated around 1400 °C maximum only.

The American industry prefers the term "carbon fibres" for the fibres according to British terminology "type II" and the term "graphite fibres" for the highly heat treated fibres corresponding to British nomenclature "type I". Unfortunately, American producers use the term "graphite fibres" in combination with composites even if only low heat treated PAN based carbon fibres are used for reinforcement (see for instance "graphite epoxy"). From scientific viewpoint such a description is wrong and misleading.

As recommended by IUPAC [16] the term "graphite" can only be used if carbon material with crystalline order in c-direction is considered. This order is recognized in X-ray diffraction diagrams by modulation of the two-dimensional lines 10 and 11 into three-dimensional ones (see Fig. 12).

This state of graphitization can also be characterized by mean layer distances evidently below 0.34 nm (see Fig. 15). Such crystalline order is not achieved in the PAN based carbon fibres, and not at all in low heat treated ones. Therefore, the term "graphite fibre" should not be used for this industrially most important carbon fibre type. The term "graphite fibre" can be considered with restriction for highly heat treated mesophase pitch based carbon fibres because some crystalline order in c-direction can be observed (see Fig. 12 and 15), especially if stabilization treatment (oxidation before carbonization) was incomplete.

There exists another fibre type recently developed by Japanese workers [17] which deserves the use of the term "graphite fibres" from scientific viewpoint. These are the fibres grown by gas phase deposition on fine dispersed catalysts (CVD-fibres). They are effectively graphitizable, and if highly heat treated, they exhibit excellent graphite structure. According to the theoretical considerations discussed above, they are completely unsuitable for reinforcement purposes, but can gain interest for electrical application, especially because of their suitability to form intercalation compounds [18].

The variety of carbon fibres can also be characterized by the fibre properties: "high tensile" type (HT) and "high modulus" type (HM). So far as PAN is used as precursor high tensile type corresponds to a final heat treatment temperature below 1400 °C (low heat treated) and high modulus fibres to those of at about 1800–2700 °C. This original today still widely used classification does not define the newer PAN based HT types with high strain. These new types are therefore described as new generation of high strain HT fibres or as IM fibres, indicating intermediate modulus. From fabrication viewpoint, they are obviously heat treated at low or intermediate temperatures i.e. below 1400 °C (iht) or above 1400 °C (iht) but below 2000 °C.

The description by means of properties HT and HM became also difficult for mesophase pitch based carbon fibres. Starting from such a precursor, a high strength carbon fibre has not been achieved so far but only a high modulus type even if heat treatment has not exceeded temperatures of 1800 °C (intermediate heat treated), whereas heat treated at graphitization temperatures results in ultra high modulus fibres (UHM).

The following new *terminology* is in discussion by IUPAC [19]:

> "Because of the different properties depending on the used precursor, the precursor should be mentioned in the description of the carbon fibres, that is: polyacrylonitrile (PAN) or mesophase pitch (MPP) based carbon fibres. As nearly all carbon fibres used for reinforcement of polymers are PAN based, it seems unnecessary in most cases to mention this origin (PAN). However, in case of MP pitch based fibres, it should be always mentioned.
> The heat treatment can be described by small indices letters additional to the term PAN or MPP based carbon fibres such as
> lht – meaning low heat treated
> (maximum 1400 °C)
> iht – meaning intermediate heat treated
> starting from 1300 °C, but below 2000 °C)
> hht – meaning highly heat treated
> (at graphitization temperatures, that means above 2000 °C)"

The new generation of high strain HT fibres belongs obviously to "lht" and "iht". These fibres can be classified as SHT (super high tensile), or HTHS (high tensile – high strain) and IM (intermediate modulus) respectively.

As highly heat treated mesophase pitch based carbon fibres will graphitize partially (see Fig. 12 and 15) one can use the term MPP based *graphite* fibres. The carbon fibres made from isotropic pitch (non mesophase based pitch) as used for thermal insulation or as filler material are rarely used for reinforcement purposes. They are non graphitizable and can be described as isotropic pitch (isoP) based carbon fibres.

2.2 Mechanical Properties of Today's Carbon Fibres

Figure 17 shows the stress/strain behaviour of the various fibre types. Within the PAN based as well as the mesophase pitch based carbon fibre groups the YOUNG's moduli reach maximum values if highest heat treatment temperatures have been applied. These HM types are indicated by the heat treatment temperatures in the left parts of the respective property areas. The right parts in these property areas correspond to the lowest applied heat treatment temperatures. The strength decrease between HT and HM types in all former and some still today's PAN based fibres is indicated by the arrow in left direction.

The most important carbon fibre grade for application in structural elements such as in aerospace, machinery, sporting goods and so on, is the PAN based lht type, heat treated at 1400 °C. Commercially available qualities show strength values up to 4000 MN/m^2, types in development promise further increased strength. This tendency is indicated by the arrow in right up direction in Fig. 17. The most important advantage of this fibre type for technical application is its high strain to failure. The efforts of all fibre producing companies are directed to develop products with even higher strain to failure by means of further increase of the tensile strength, but preserving the same intermediate modulus (250–300 GN/m^2).

The disadvantage of strain to failure below 1% as in PAN based high modulus types and especially below 0.5% as in all types of mesophase pitch based carbon and graphite fibres can only be tolerated for applications where the high YOUNG's modulus is absolutely needed, for instance in stiffness controlled parts of CFRPs in large

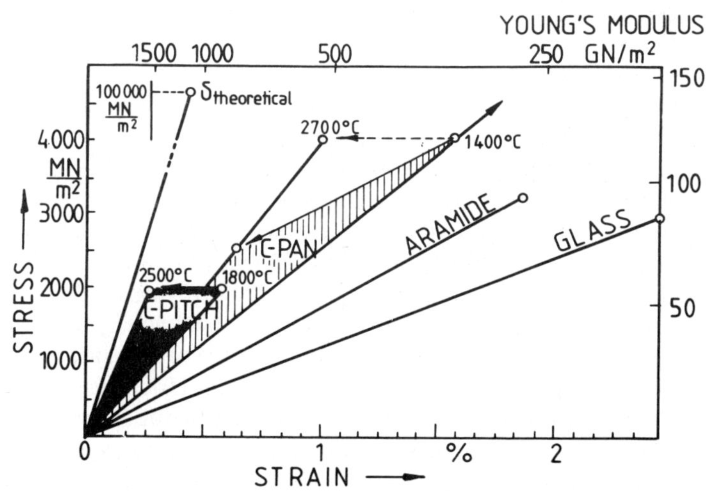

Fig. 17. Stress strain behaviour of the various reinforcement fibres

space structures, but never in strength control structural parts. In carbon/carbon composites (these are fibre reinforced all carbon composites with carbon matrix instead of polymer matrix), these high modulus types and even more the pitch based carbon and graphite fibre types are most advantageous [20 a, b].

2.3 *Considerations on Mechanical Properties Which can be Expected in Future*

2.3.1 *Theoretical Maximum Values for the Mechanical Properties*

The theoretical maximum YOUNG's modulus of a carbon fibre is given by the elastic constant C 11 of the graphite single crystal with 1060 GN/m^2 (GPa), see Fig. 7. Its hypothetical stress-strain behaviour is included in Fig. 17. This value could be realized only in case of perfect preferred orientation of defect free graphitic layers. YOUNG's moduli which correspond to 70% of this theoretical maximum value have been achieved in commercial ultrahigh modulus carbon fibres.

There exist also theoretical considerations on maximum strength of a material, assuming simultaneous fracture of all bounds in the considered cross section. This needed maximum fracture energy must be equivalent with the energy obtained by formation of the new surfaces. This theoretical strength σ_{theor} is given by the YOUNG's modulus E, the surface energy γ and the lattice distance a, according to the relation

$$\sigma_{theor} \sim \sqrt{\frac{E \cdot \gamma}{a}} \quad \text{with} \quad \sigma_{theor} \sim \frac{E}{10} \text{ for graphite.}$$

With the theoretical maximum YOUNG's modulus a theoretical maximum tensile strength of 100 GPa (10 000 MN/m^2) would result for carbon fibres.

Because of the anisotropic nature of the chemical bounds and the low shear modulus in graphitic carbon, theoretical strength considerations must be limited to the part of layers oriented in fibre direction, as expressed by the YOUNG's modulus measured experimental in fibres. A theoretical maximum strength of 70 GPa (70 000 MN/m^2), for the high modulus type and such one of 25 GPa (25 000 MN/m^2) for the high tensile high strain types can be calculated.

In practical measurements so far only 3% are achieved in case of HM fibres with a measured strength of 2,1 GPa (2 100 MN/m^2) but already 20% for the best HT types with a measured strength of 5 GPa (5 000 MN/m^2).

For isotropic structures, this lower practical strength is explained by defects of critical minimum size as notches and flows. For the graphite structure such critical defect length would be in the order of 50 nm (500 Å). By low angle scattering and TEM measurements the size of the voids in the direction in which fracture proceeds, the transverse direction, is only in the order of a few nm and thus very much less than the critical flow size. Additionally, fracture was observed very often by scanning microscopy not to start from largest but from smaller flows. Therefore, simple

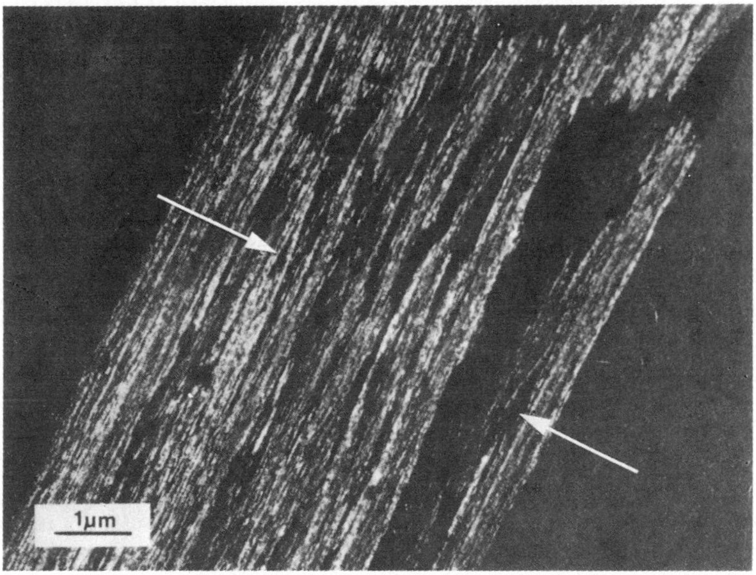

Fig. 18. Dark-field (002) TEM image of longitudinal section of pitch based carbon fibre showing misoriented crystallites (marked by arrows) [11]

considerations on critical flow sizes as contributing to strength limitations can be ruled out [11]. Considerations on misoriented layers as source of cracks because of the extremely low shear modulus C 44 according to REYNOLDS-SHARP mechanism of fracture [21] result in more realisitic predictions of ultimate tensile strength. In such calculations, the crystallite size $\overline{L}_c$ – which is an order less in PAN based carbon fibres than the critical flow size according to GRIFFITH – as well as the misorientation angle become decisive parameters in strength limitation. Under some realistic assumptions, an upper strength limit of 7,5 GPa (7 500 MN/m^2), which implies a strain to failure of 1,9%, is discussed as probable. PAN based HM fibres will be limited to an ultimative tensile strength of around 5 GPa (5 000 MN/m^2) due to their structural composition of large crystallites some of which will be misoriented sufficiently to cause failure by the REYNOLDS-SHARP mechanism of fracture (see Fig. 18) [11]. There do not exist precalculations of ultimate strength on MP pitch based carbon fibres. Because of the enhanced $\overline{L}_c$ growth (graphitization tendency) a lower strength limit can be expected.

All these calculations are based on the low shear modulus C 44 as known from ideal graphite crystals. Imperfect lattice structures will increase the shear modulus which will favour the ultimate strength data of the PAN based fibres and especially the HT types. Quantitative data on the increase of C 44 by the defect structure are not available, however.

In Fig. 19 the various structural models of HT and HM type carbon fibres are presented, which are based on analysis of X-ray and electron diffraction as well as TEM and intercalation studies [24]. This newest model shows the staples of graphitic layer planes, which form tubes but not strips or bands as formely assumed. Struc-

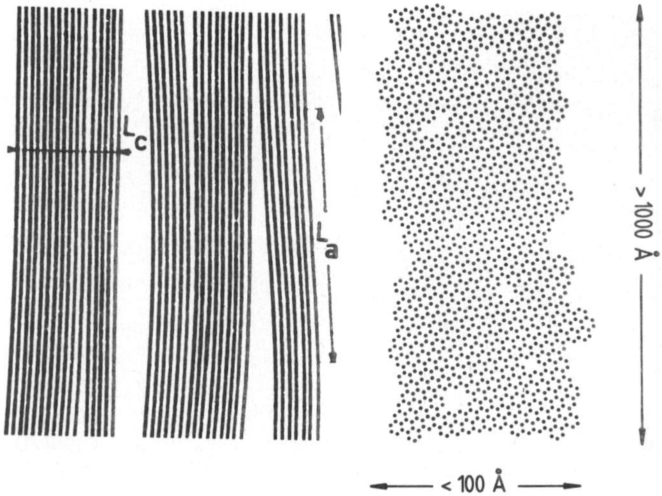

Fig. 19 a–c. The various structural models of C-fibres. **a** 2-dimensional RULAND model derived from stretch graphitized rayon based fibres [22]

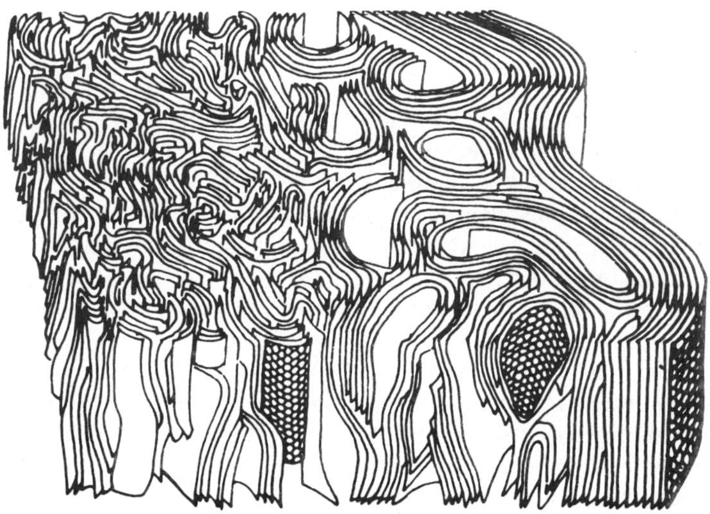

Fig. 19 b. 3-dimensional JOHNSON-model resulting from careful TEM observations of PAN based HT-fibres [23],

tural perfection of the layer itself and flattening of the curvature of these tube structures is caused by heat treatment at graphitization temperature. There do not exist comparable models of MPP based HM fibres, but further improvement of the sheet like layer structure is indicated by all quantitative measurements but misoriented areas with increased staple hights do remain as source for cracks (see Fig. 18) [11].

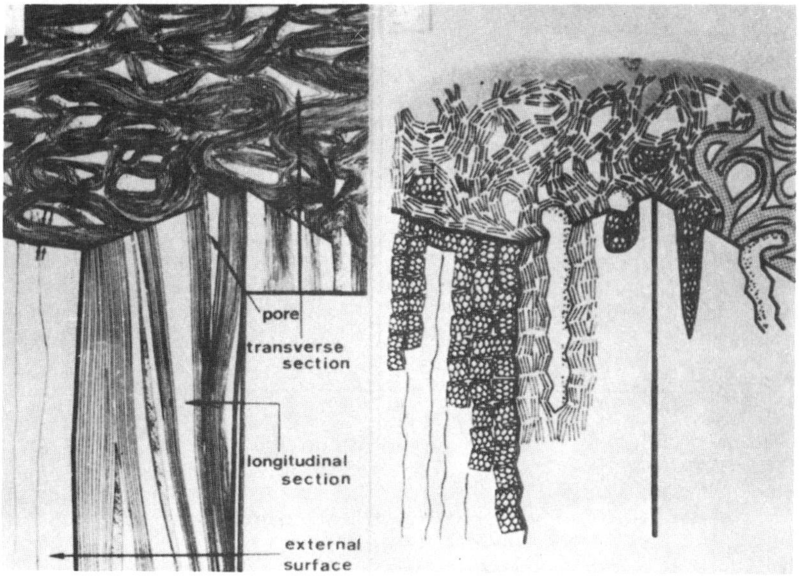

Fig. 19c. newest OBERLIN model for HM- and HT-fibres [24]

2.3.2 Influence of Test Methods

The ultimate strength of a material with brittle fracture behaviour depends on the defects of minimum size as shown above. Not the nature of such defect, but rather the probability of its presence in the considered volume will control the measurable strength of material. Therefore, we can expect an increase of effective fibres strength with decreased length of the test sample. For the description of the probability for the presence of critical sources for cracks, the WEIBULL statistic is found to be suitable also for carbon fibres.

Figure 20 shows some values of average tensile strength as function of gauche length in monofilament testing [25a] and with strand samples [25b]. This dependence on testing conditions requires standardization of the testing methods for characterization of carbon fibres. Mostly, a gauche length of 30 mm is used for the single filament method, and a gauche length of 120 mm for the strands.

These two different methods are explained in Fig. 21. The strand method uses samples in form of fibre bundles which are fixed by epoxy resin in exact position of prefered orientation. The cross section is measured by weight, length, density and number of monofilaments of the fibre bundle. This method gives directly average strength values of all monofilaments within a yarn.

The monofilament method is restricted to the fracture behaviour of one single filament. This method gives more specific results. An exact determination of the cross section of the fibre is necessary. Mostly, the cross section of a bundle is determined by optical microscopic method and used as basis for calculation of a monofilament. The single filament method is more time consuming. The minimum gauche length for the single filament method is experimentally limited with about

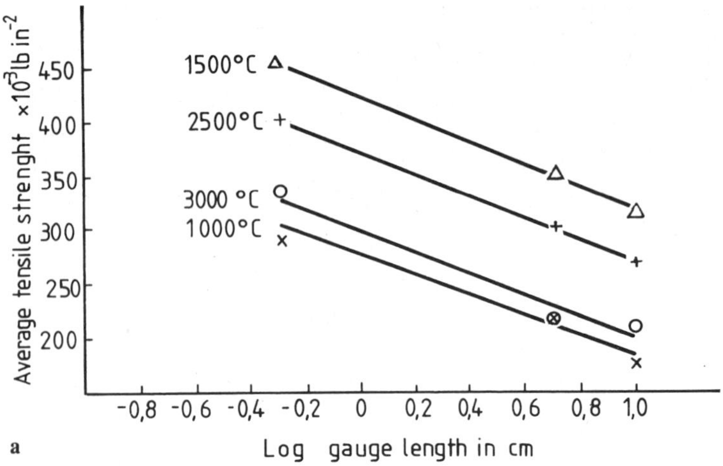

a

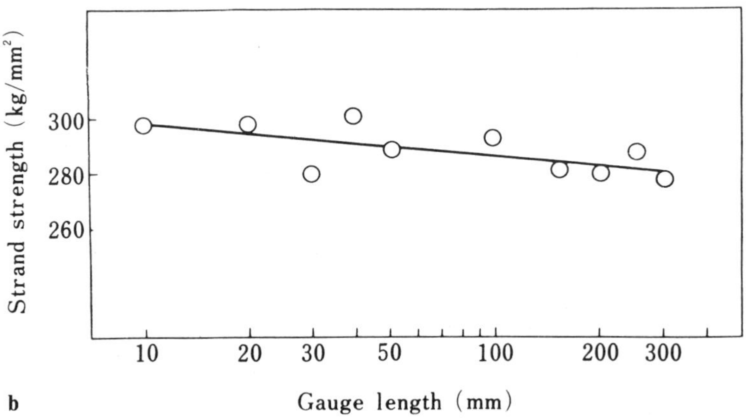

b

Fig. 20 a, b. Average tensile strength of carbon fibres as function of gauche length; **a** for monofilament measurements [25 a]; **b** for strand measurement [25 b]

10 mm. For smaller gauche lengths the knot test must be applied in which the tensile strength is calculated from the diameter of the loop when fracture occurs. Practical gauche length of 1 mm and even less can be realized by this method (Fig. 22).

For technical application mostly the strand method is used, because it gives more realistic values for strength of the fibres within the composites. Obviously, the epoxy resin inhibits crack initiation at some critical defects in the fibre surface, whereas in monofilament tests such a defect acts effectively as crack initiator.

The stress strain curve based on own measurements as shown in Fig. 23 confirms the above conclusions. The diagram represents the elastic behaviour of isolated fibres measured by the monofilament method, that of a unidirectionally reinforced composite with 60 v/% of fibres and that of the pure matrix without fibres. Dotted lines reflect the behaviour of the fibres and the composites with non surface treated carbon fibres, the full lines that of the surface treated ones after wet oxidation. By

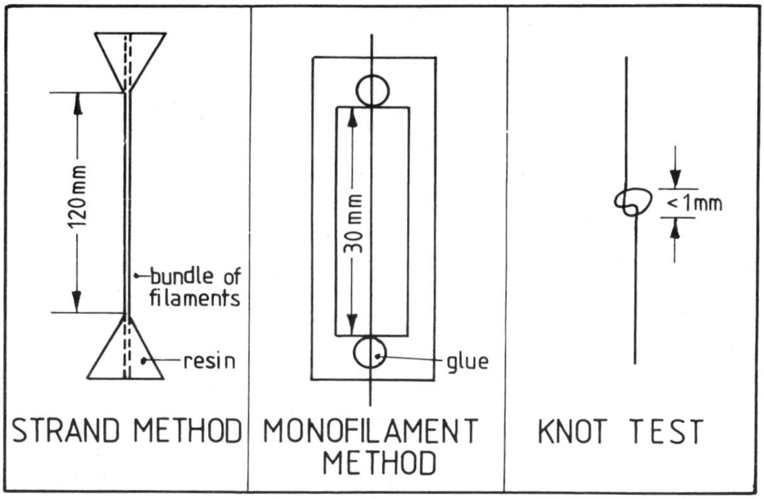

Fig. 21. Methods used for tensile strength test of carbon fibres

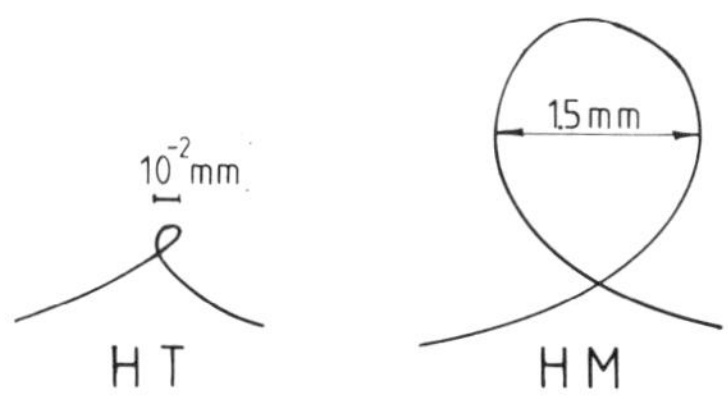

Fig. 22. Minimum diameter in knot tests for HT type (left) and HM type (right) carbon fibres

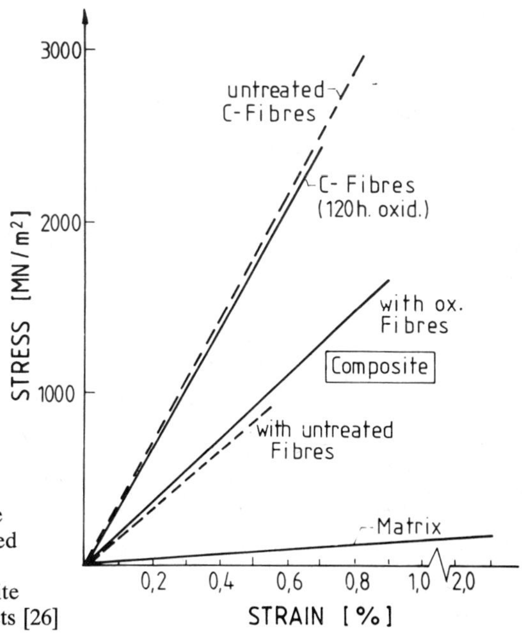

Fig. 23. Stress strain behaviour of carbon fibres, pure matrix material and composite with non surface treated and surface treated carbon fibres. The surface damages by treatment are compensated in the composite by the matrix, which inhibits surface defects [26]

the applied surface oxidation the strength of the fibre was reduced, but the strength of the composite is now found to be increased. This is a clear indication, that the defects introduced additionally by surface oxidation are compensated by the epoxy matrix [26].

2.3.3 The New Generation of PAN Based lht and iht C-Fibres

Before we draw a conclusion from these theoretical considerations we should take a look at table Fig. 24 with the compilation of mechanical properties of some of today's commercial fibres, available at least in small quantities. From the viewpoint of design and application of advanced composites, the PAN based "lht" and "iht" carbon fibre type is most important. Fibres with strength values from 4000 to 5000 MN/m^2 are grades of the so-called "new generation" of high tensile types. The intermediate level for YOUNG's modulus of 250 to 280 GN/m^2 will guarantee a strain to failure up to 2%. There is a realistic chance for industrial manufacture of this type in large quantities. So far, these values can be considered as practically upper quality limit in near future.

Also from the viewpoint of compressive strength the new generation of PAN based lht carbon fibres is superior. It is well known, that all reinforcement fibres with sheet like structure such as polyaramide fibres, but also the "carbon fibres hht" and especially the mesophase pitch based carbon and graphite fibres exhibit a very low compressive strength in a polymer composite. This disadvantageous property is improved with the intermediate modulus/high strength (IM-HT) fibre types.

2.3.4 The Impact Behaviour of C-Fibre Reinforced Composites

An additional advantage of the new generation of PAN based lht C-fibres can be seen in the improved impact strength of the composites. It was considered always, that impact resistance of a composite with brittle fibres can be improved if pure

PRODUCTION PARAMETERS		PRODUCER	TYPE	TENSILE STRENGTH MN/m^2	YOUNG'S MODULUS GN/m^2	STRAIN TO FAILURE %
PAN-based	lht	Toray	Torayca T300	3000	250	1,2
	lht	Toray	Torayca T800	5000	250	2,0
	hht	Toray	Torayca M40	2400–3000	400	0,60 – 0,75
	lht	Toho	Besfight ST1	3600	240	1,5
	lht	Toho	Besfight ST2	4080	240	1,7
	lht	Toho	Besfight ST3	4320	240	1,8
	lht	Hercules	IM6	5000	280	1,8
Mesophase PITCH-based	mht	UCC	Thornel P55	2100	380	0,55
	mht	UCC	Thornel P75	2100	520	0,40
	hht	UCC	Thornel P100	2400	690	0,35

192739 State: 1985

Fig. 24. Mechanical properties of commercial carbon fibres (state mid 1985)

brittle fracture of the composite is avoided by some delamination of fibre pull-out which stops crack propagation into the fibre. The consequence would be an optimization of the interlaminar shear strength, avoiding extremely high ILSS values. The designer, however, likes high interlaminar shear strength as basis of tailorable structures.

As found recently [27] this simple correlation is valid only for high modulus fibres with a low strain to failure. In case of fibres with strain to failure above 1% a good adhesion between fibre and matrix, that is a high value of ILSS up to 120 MN/m^2 causes an improved resistance against impact. This behaviour can be explained by the elastic deformation of the fibre itself if a good adhesion is guaranteed. Such a stop of crack propagation by the energy consumption due to elastic deformation was discussed before in theoretical consideration in literature [28]. As a consequence of our results, increased interlaminar shear strength in carbon fibre reinforced polymers will increase impact resistance for these fibre types (see Fig. 25). Also this behaviour supports the exceptional position of PAN based carbon fibres of this new generation as candidate type for future broad application of these advanced composites.

The technological means for production of these new generation of PAN based carbon fibres are not discussed in literature.

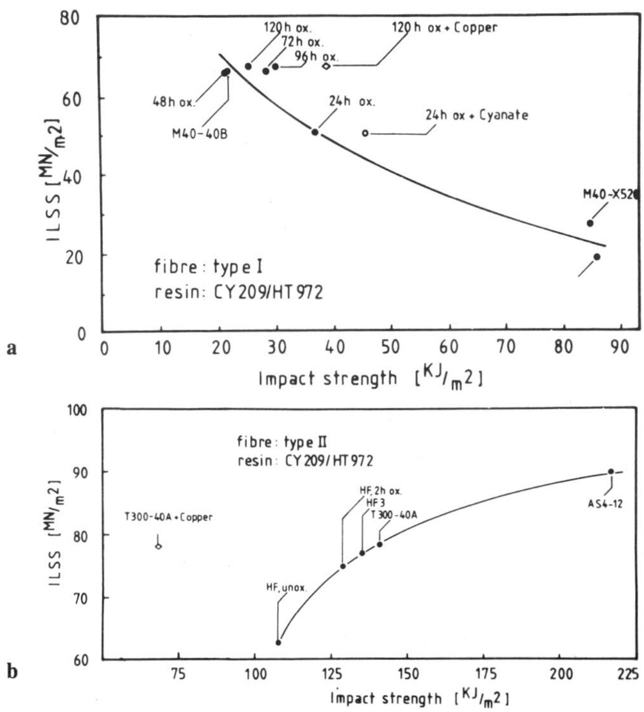

Fig. 25 a, b. Correlation of interlaminar shear strength and impact strength of fibre reinforced composites [27]. **a** PAN based highly heat treated (hht) carbon fibres; **b** PAN based low heat treated (lht) carbon fibres

The first availability of these superior quality fibres was explained by selection of special bobbins from the normal production [29]. For controlled reproducible production obviously the following parameters are decisive:

a) low probability for critical defects, that means lower diameter of fibres,
b) very pure precursor fibres,
c) controlled process conditions for spinning, textile pre-stretching and stabilization.

It should be mentioned, that in patent literature also MP pitch based C-fibres are claimed with strain-to-failure up to 1.4% [30]. Because of the known lower strength of MPP C-fibres (between 2000 and 3000 maximum) the YOUNG's modulus must be decreased to values between 100 and 200 GN/m^2 only. The results are described to be achieved by chemical pretreatment of the pitch, thus avoiding the dominant preorientation of the liquid crystals. No commercial fibres of this type are available so far.

3 Economic Considerations

High modulus carbon fibres have been introduced for the first time mid of the Sixties to replace boron fibres. The price was in the order of $ 500/kg at least, similar as that of boron fibres. There was a decade long discussion on economy of scale, but the high price level was protected additionally by a latent patent situation.

Mid of the Seventies when the price for the least expensive carbon fibre type was still above $ 100/kg, an US company started a price war with the introduction or at least the announcement of low cost mesophase pitch based carbon fibres. A price reduction to at least $ 20/kg was promised because of the very low cost of the primary raw material, that is petroleum pitch. From standpoint of reinforcement fibres, today the only final result of this scientifically and technologically most interesting initiative is a radical price breaking of PAN based carbon fibres, down to the level around $ 35/kg, whereas the price for the pitch based types was doubled during last years again. It seems that the claimed low costs of the raw material are compensated by far by the unexpected high production costs, obviously in the process step of spinning. Additionally, the properties of both types of carbon fibres (PAN based on one hand, and mesophase pitch based on the other) are so different, as shown before, that there does not remain any alternative to replace PAN based carbon fibres by pitch based ones, even if economic reasons would justify it. But MPP based C-fibres must be considered as an extension of the broad variety of achievable properties in C-fibres.

3.1 Present Technology

The fabrication processes for carbon fibres consist of three steps (Fig. 26) [31]:

the spinning,
the stabilization, and
the carbonization.

Fig. 26. Process steps for carbon fibre fabrication

Stabilization is an oxidation treatment in air between 200 and 300 °C in order to transform the thermoplastic fibre into a nonmeltable precursor fibre, suitable for carbonization without changing the fibre form. This stabilization treatment needs residence times in the order of 1 h for PAN, but shorter times for pitch [32, 33]. This behaviour was claimed as a process advantage for pitch fibres. Also the spinning step appeared advantageous for pitch because melt spinning can be applied. All polymer specialists will prefer melt spinning processes, if compared with dry spinning (solution spinning) or wet spinning (solution spinning with subsequent precipitation of the fibres in a coagulation bath). Only melt spinning offers a fast production by cooling without solvent problems and time consuming washing of coagulated fibres. Melt spinning can not be applied for PAN.

Modern polymer technology has developed very economic wet spinning processes based on high multiple spinnerets up to 300 000 holes. Yarns with the same number of monofilaments can be spun in one coagulation bath simultaneously (see Fig. 27). Melt spinning as well as dry spinning are restricted to a maximum of about 2000 holes in a ring type spinneret, which causes much higher investment. The spinning rate, however, is faster, by which the advantage of multifilament spinning is partly compensated.

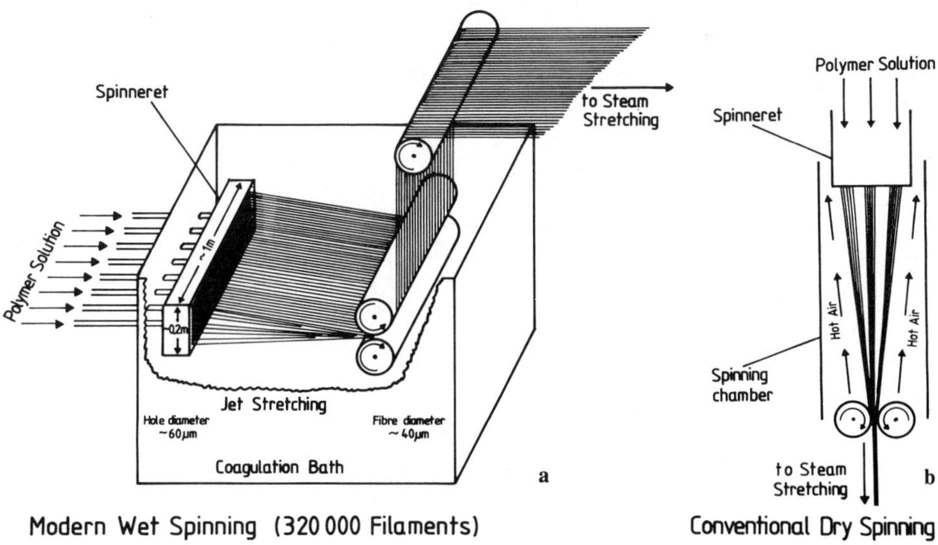

Fig. 27 a, b. Polymer spinning (schematically): **a** modern wet spinning (320 000 filaments), **b** dry spinning (2500 monofilaments)

It seems, that all other process steps regarded as more disadvantageous for PAN fibres have less influence for the overall economy of processing than this most economic spinning step. This rationalized spinning, however, demands consequently the application of a multifilament tow for the carbon fibre production.

3.2 Carbon Fibres in Form of Multifilament Tows

The idea to use heavy tow PAN precursor fibres to produce heavy tow PAN-based C-fibres is followed only by a few of carbon fibre producers (RK Textile in UK, SIGRI in Germany and GREAT LAKES in USA). As shown from the price forecast in 1982 and the effective prices in 1983 and 1985 in Fig. 28, it seems probable, that lowest price limit of heavy tow will be more than 20% below lowest price limit of low filament tows with 12,000 monofilaments.

Strength values of heavy tow as compared with low tow qualities are shown in Fig. 29 [34]. All results are obtained with the monofilament method. There do not remain severe differences in strength properties between low filament and heavy tow.

The processing suitabilities of multifilament as compared with low filament carbon fibres are shown in Fig. 30 [34]. Multi tow has best chances in application as sheet molding reinforcement, injection molding with staple fibre or rowings, pulltrusion technique as well as wet winding.

3.3 Production Capacity, World Demand and Price Development

The total production costs of a material are controlled by
1. raw material costs,
2. production costs,
3. scale of production.

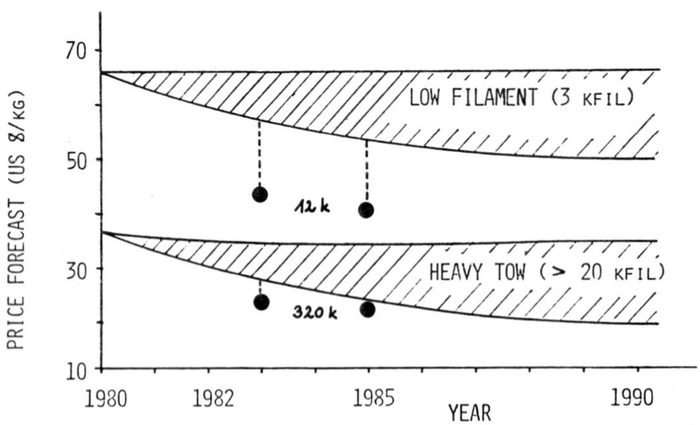

Fig. 28. Price forecast for carbon fibres: curves basis 1980, points basis 1983 [34]

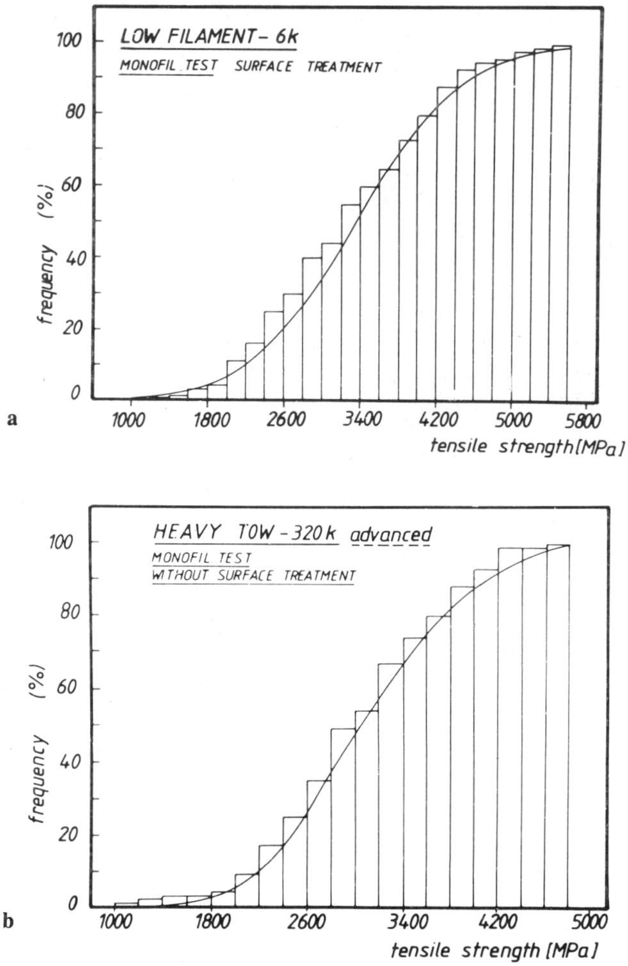

Fig. 29. Frequency of tensile strength in low filament (6000 monofilaments) and heavy tow (320 000 monofilaments) PAN based low heat treated (lht) carbon fibres [34]

As mentioned before, the problem of economy of scale was taken for justification of high prices since last 20 years. It seems, however, that the year 1982 marks a turn in production scale caused by a steeper increase of world demand (see Fig. 31 [35]).

Figure 31 shows mainly the production capacity of the TORAY group. There are also strong increasing capacities in USA (UCC, HERCULES, COURTAULDS and others) as well as in UK (RK Textile), in France (RHONE POULENC/TORAY) and in Germany (ENKA/TOHO). There is no doubt, however, that today's large Japanese production is mostly export oriented as shown in Fig. 32 [35]. Additional advantage for the high performance qualities of the Japanese producer as pioneers is the permission of international aviation control organizations for the use of TORAY products in commercial aircrafts based on long years tests.

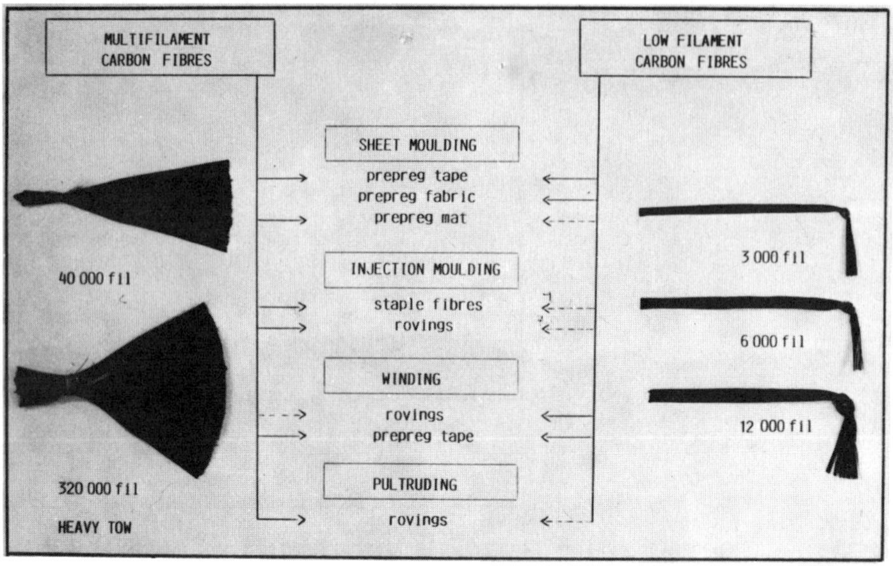

Fig. 30. Processing suitability of multifilament and low filament carbon fibres [34]

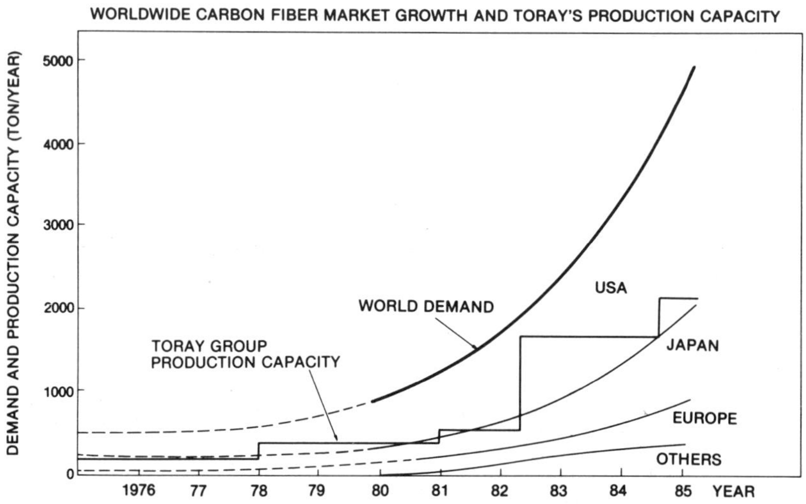

Fig. 31. World wide carbon fibre market and production capacity of the TORAY group [35]

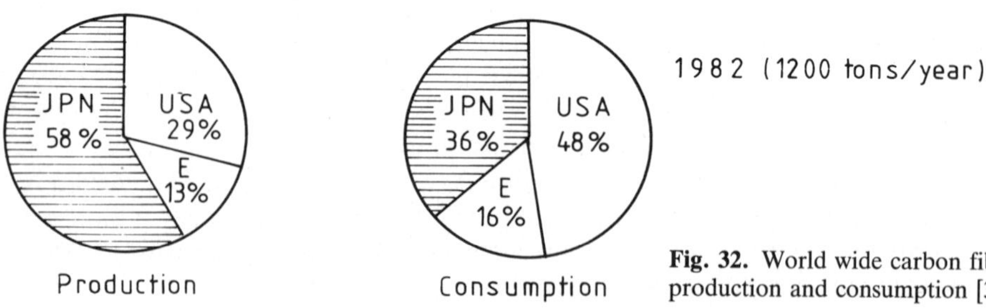

Fig. 32. World wide carbon fibre production and consumption [35]

4 The Various Fields for Application of Carbon Fibres

It is difficult to get exact newest information on the end use distribution of carbon fibres and it is even more difficult to learn about the amount needed for military applications.

Figure 33 shows an earlier publication [35], which is concerned on 1981 data only. At that time there were two groups of main application, namely aerospace in Western countries and sporting goods in Japan, which again are exported all over the world. With lower carbon fibre price and increasing development of process technology for composites, the growing demand will be directed mainly to the field of general industrial application.

The following contributions in the Chap. 5 and 6 will discuss in more detail the various fields of application. Only a brief survey can be given in this introductory paper.

4.1 Application in Aerospace

Light weight structures are preferentially needed in aerospace applications. The US space shuttle (Fig. 34 [36]) for instance consists in main parts of carbon fibre reinforced epoxy resins (cargo bay doors and booster rocket casings). But also modern commercial aircrafts use increasing amounts of carbon fibre reinforced composites instead of aluminium not only in highly industrialized countries such as US (BOEING 757 and 767, the European countries (Airbus 310) but also in the host country of the Seminar. Fig. 35 shows the progress of this development in EMBREAR commuter plane [37]. Fig. 36 [38] gives a forecast of probable replacement of conventional metals by advanced composites in commercial aircrafts until the year 2000.

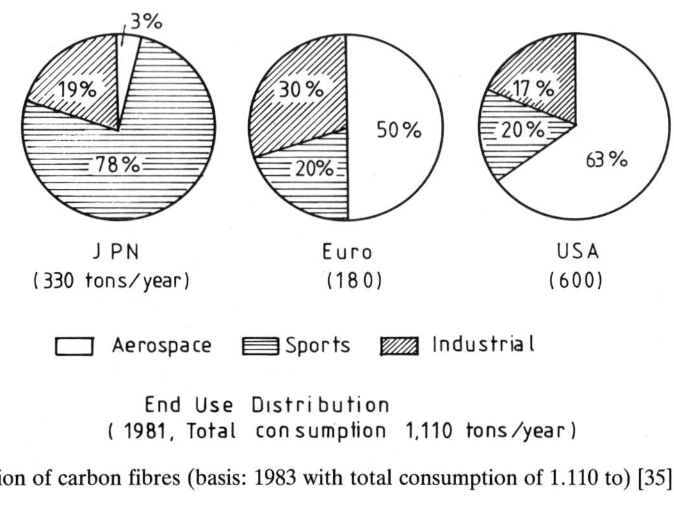

Fig. 33. End use distribution of carbon fibres (basis: 1983 with total consumption of 1.110 to) [35]

Fig. 34. US space shuttle: doors of the load room in construction at ROCHETYNE International, made of CFRP [36]

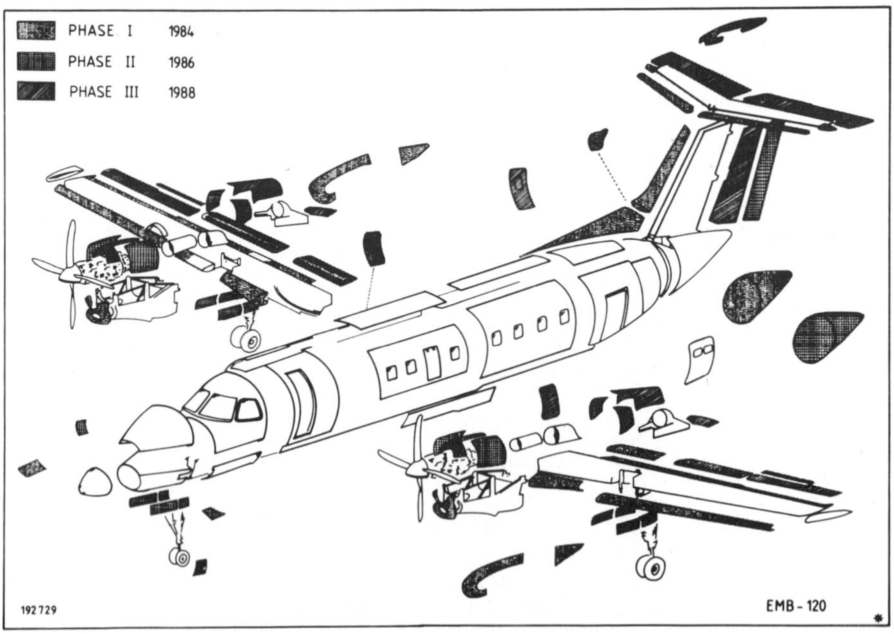

Fig. 35. Structural parts in EMBREAR's commuter plane, Brazil [37]

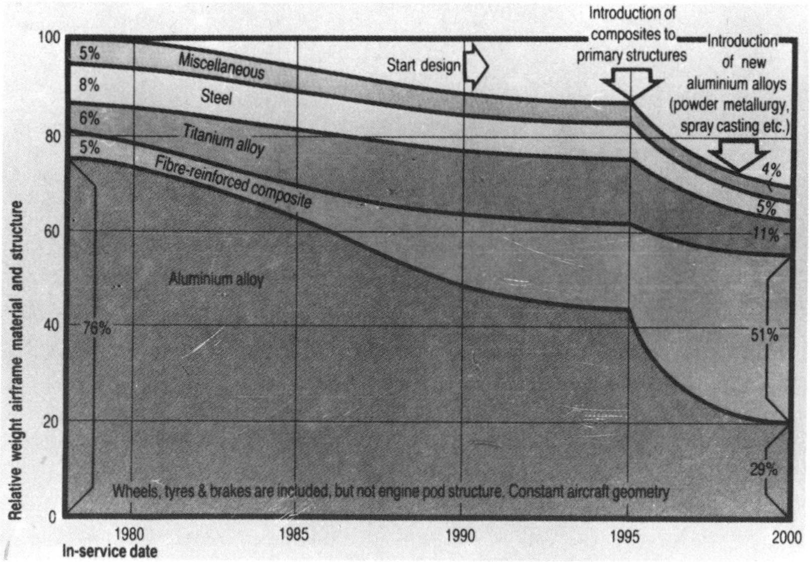

Fig. 36. Forecast for materials in commercial aircrafts industry until year 2000 [38]

There are already airplanes in development whose whole structure consists of composites instead of metals (for instance the LEAR FAN, as shown in Fig. 37). Also in inner parts of airplanes metals are replaced by advanced composites, for instance the columns supporting the floor panels or the seats for the passengers (see Fig. 38 [40]).

Fig 37. LEAR-FAN Comuter plane with a structure consisting completely of CFRP [39]

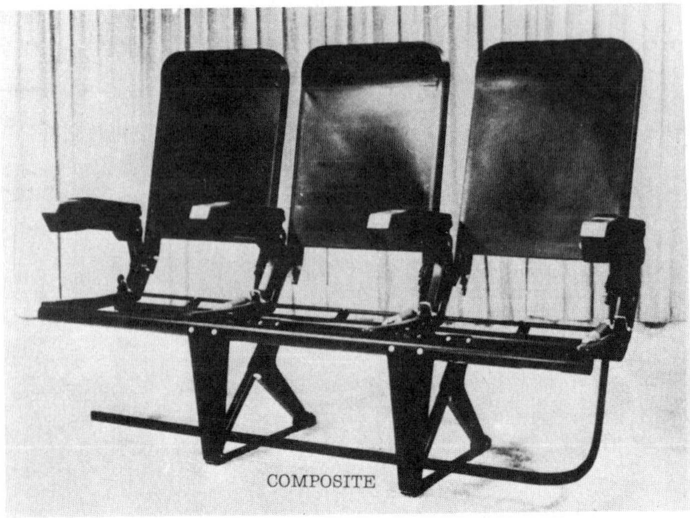

Fig. 38. Seats in airplane made from carbon fibre reinforced composites [40]

4.2 Application in Transportation

Because of the requirements for the replacement of metals in primary structures of airplanes and based on years of long critical tests, the quality of carbon fibre reinforced composites and their processing technology have been largely improved. Car industry can take advantage in using this experiences for replacement of vari-

Fig. 39. Frame of a FORD model car made from advanced composites [41]

ous metal parts in the cars. It seems today, that parts of the outer structure will be fabricated by glass fibre reinforced composites mostly. Highest chances for advanced composites are given at the moment to leaf springs, driveshafts, parts of frames and others where higher stiffness is needed. Fig. 39 [41] shows parts of a frame of a car. In racing cars, such structural elements are already in practical application (Fig. 40) [42].

One can assume, that all leading automobile companies are mostly interested to use advanced composites as part of the engines. The communication on this subject, however, is very restricted because of commercial secrets. Fig. 41 [43] shows a tentative version of a connecting rod made from carbon fibres.

Fig. 40. Formula I racing car with CFRP frame [42]

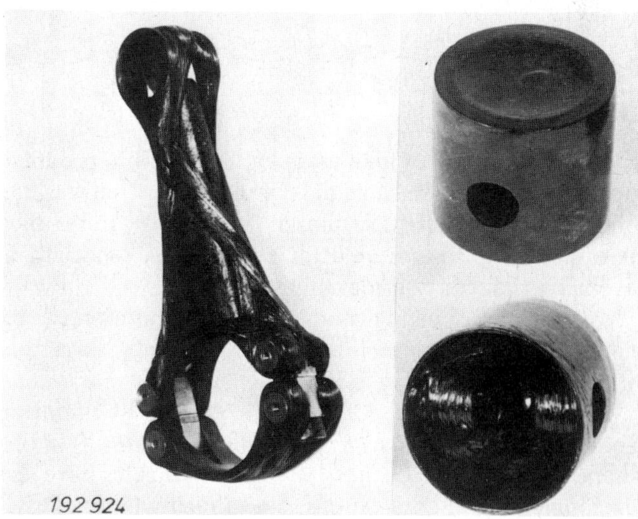

Fig. 41. Connecting rod and piston made of C-fibres (in development) [43]

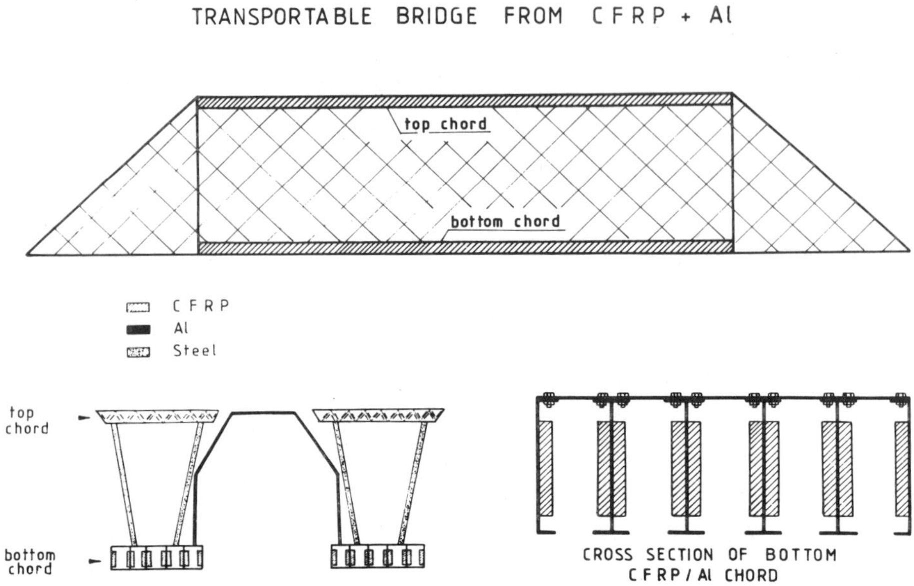

Fig. 42. Transportable bridge with loadcarrying structural elements of carbon fibre reinforced composites, made by pulltrusion [44]

A further new development for transportation purposes, a transportable bridge with 40 m length must be mentioned. It has been designed for vehicles up to 60 t weights [44]. All load-carrying structural elements were made from unidirectionally carbon fibre reinforced epoxy, thus achieving a meter weight of the bridge of only 500 kg (see Fig. 42).

4.3 Sporting Goods

Since the beginning of high modulus carbon fibre production for aircrafts and space applications, a considerable part of the production was used in extravagant sporting goods, mainly in golf sticks and in fishing rods. In the meantime, the superiority of these materials is confirmed for general use, especially as the carbon fibre prices have been reduced considerably.

Also for other sporting goods carbon fibre reinforced composites are gaining more and more importance such as in tennis racketts, ski and especially in cross-country ski sticks and so on (Fig. 43) [45].

Application of carbon fibres will gain most importance in disciplines, which depend on the capability of the equipments, such as in bicycles, rowing, sailing and others. The Olympic Games 1984 in Los Angeles have shown some new impressive developments and applications, e.g. a racing bycicle.

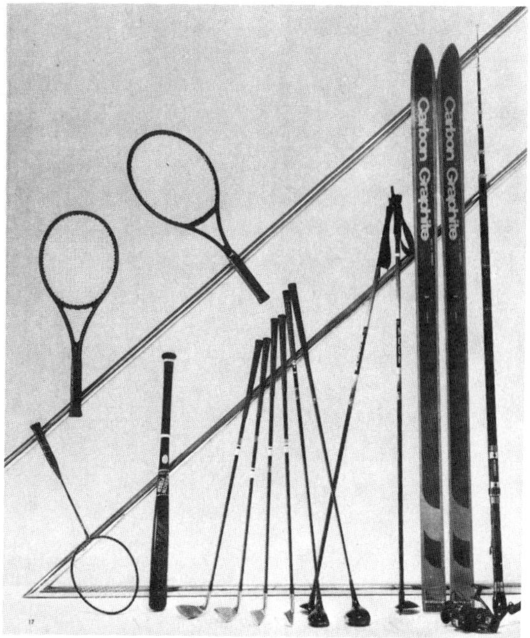

Fig. 43. Sporting goods in which C-fibre reinforcement is used in commercial scale [45]

4.4 Application in Machinery and Apparatus

The successful applications for turbine blades, compressor blades, windmill blades, ultra centrifuges, fly-wheels and in other machinery is not referred to here because of the following main lectures gives by the other experts. Only one interesting new application for structural elements with large size should be mentioned: these are light weight outer tubes for offshore oil drilling equipments as shown in Fig. 44 [46]. In this case, especially the high corrosion resistance of such advanced composites is needed. First in field application was tested by ELF in cooperation with Aerospaciale in France in 1984.

4.5 Cryogenic Application

Materials for cryogenic temperatures require some other combinations of properties than for conventional temperatures. Especially the disadvantageous embrittlement of metallic materials at low temperatures is mentioned. Carbon fibre reinforced composites are most suitable for application at cryogenic conditions because:

1. the room temperature strength properties remain unchanged, even at lowest temperatures.
2. The interlaminar shear strength in polymer matrix composites improves with lowering of the temperature. Fig. 45 [47] represents own measuring results.

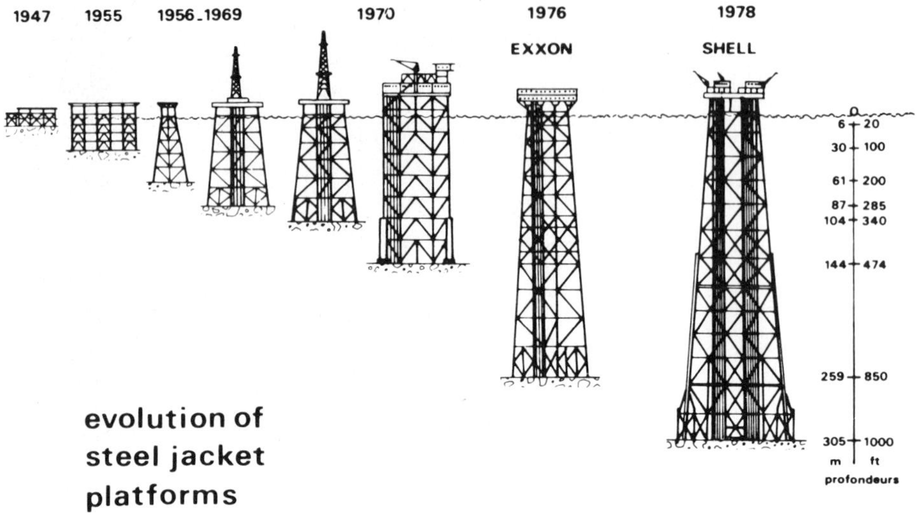

COMPOSITE TUBES/STEEL TUBES COMPARISON

(4"1/2 SATELLITE TUBES)

	STEEL	GLASS	CARBON
OUTSIDE DIAMETER (MM) .	114.3	118	118
TOTAL THICKNESS (MM) ..	7.4	7.3	7.3
SPECIFIC GRAVITY OF MATERIALS	7.8	2.1	1.55
LINEAR WEIGHT IN AIR (KG/M)	19.6	6.6	5.1
LINEAR WEIGHT IN WATER (KG/M)	17.0	4.7	3.3

Fig. 44. Application of carbon fibre reinforced tubes in off shore drilling. The light weight of carbon fibre reinforced composites becomes most important with increasing drilling depth [46]

3. Carbon fibres have a good thermal and electrical conductivity. Surprisingly, these transport properties do nearly disappear at lowest temperatures, because the contribution by electrons is completely eliminated. This behaviour is shown in Fig. 46 [48].

These properties offer promising applications in pulsed superconductive electron-magnets.

Finally, carbon fibres as substrat for superconducting niobium carbonitride layers are in development for this special application (see Fig. 47 [4–9]).

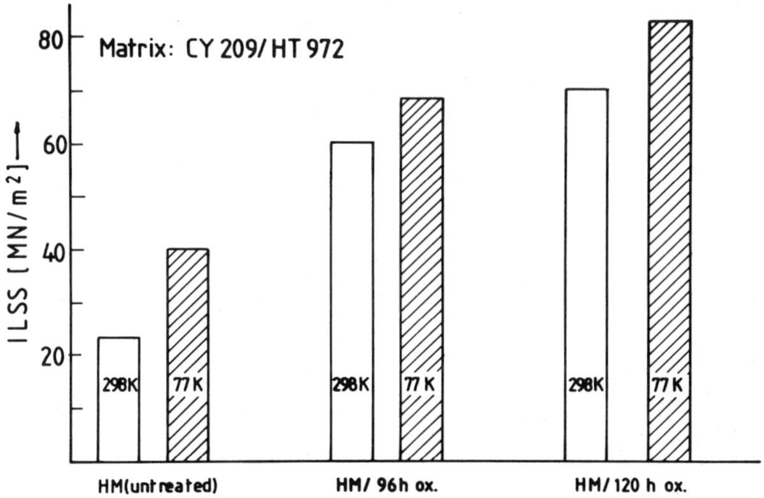

Fig. 45. Increase of interlaminar shear strength of fibre reinforced composites if measured at cryogenic temperature [47]

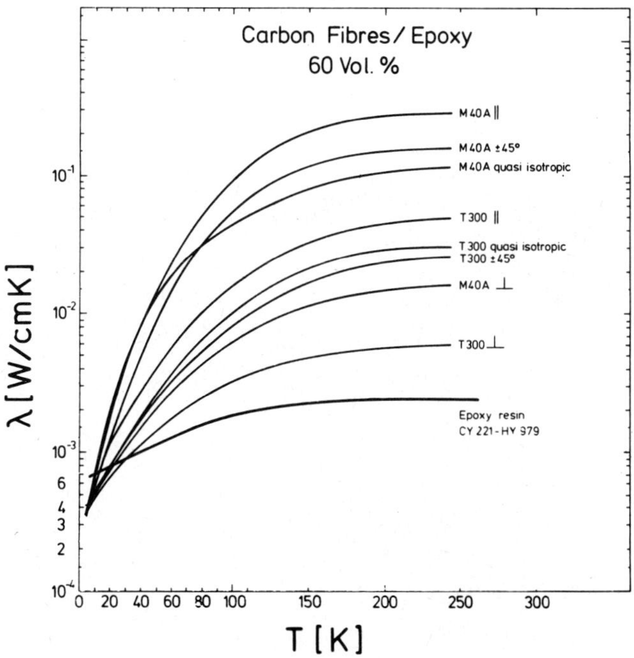

Fig. 46. Thermal conductivity of carbon fibres in comparison with polymers at lowest cryogenic temperatures [48]

Superconductors on C-Fibers for Fusion Experiments

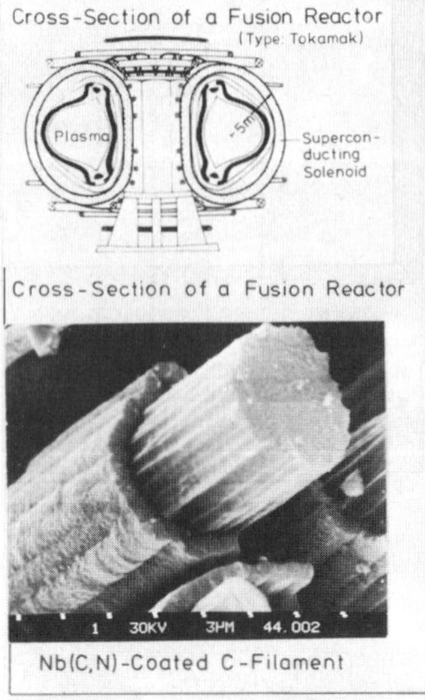

Magnetic Coils	Conventional Nb_3Sn-Wires	New Development Nb(C,N) on C-Fibers
Crit. Temp.	18 K	15–18 K
Crit. Field	22 Tesla	4 - 40 Tesla
Flexibility	Multifilamentary Wires	CVD-Coatings on C-Fibers
Tensile Strength	Not Sufficient, Combination with Steel Wires Needed	Sufficient, C-Fiber as Carrier

Fig. 47. Superconductive niobium carbon nitride coating on carbon fibres for application in superconductive ultra strong magnets [49]

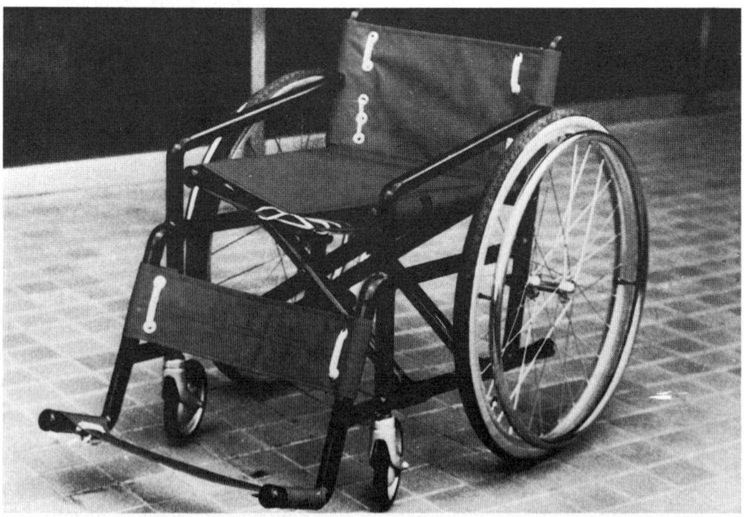

Fig. 48. Light weight wheelchair, made from carbon fibre reinforced composites [45]

4.6 Medical Application

Also this subject will be treated in a separate paper in Chapt. 6. We have two groups which have to be considered, firstly the application in apparatus, and secondly, the application as implantation material. In the first group, Fig. 48 [45] shows successful application, namely a wheelchair made from carbon fibres with the enormous advantages of the low weight. May I draw your attention to the second group especially. Fig. 49 [50] shows a the proved application of carbon fibres as ligament

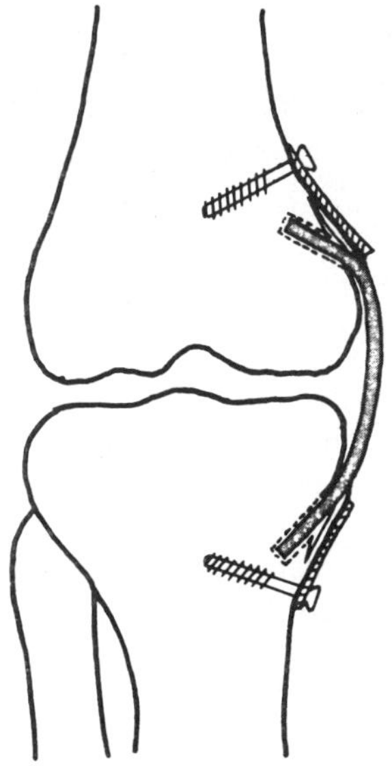

Fig. 49. Replacement of ligaments by carbon fibres [50]

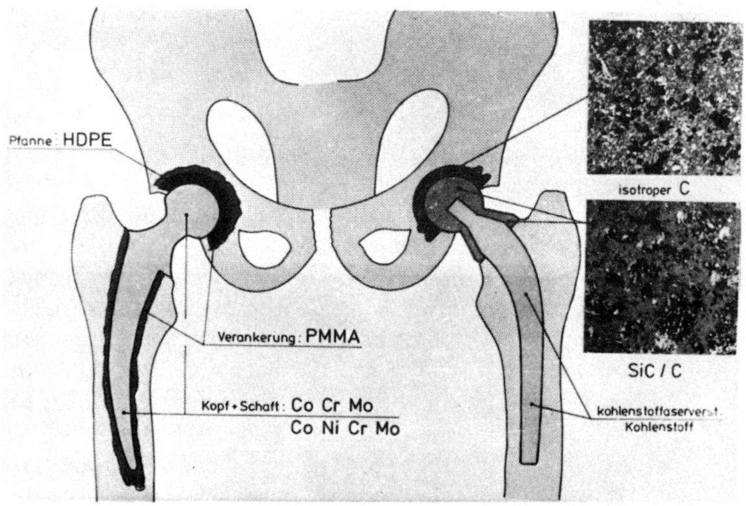

Fig. 50. Carbon fibre reinforced implantation material for hip joints [51]

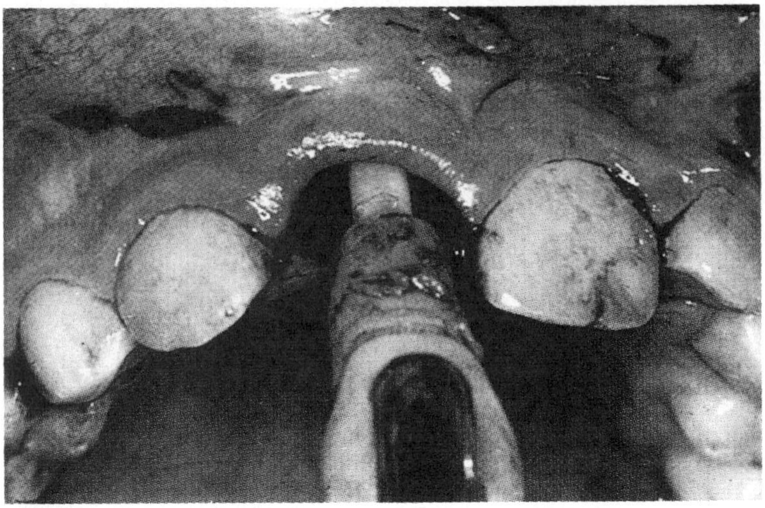

Fig. 51. Fixation of tooth roots by carbon fibre reinforced polysulfone [52]

replacement at the human knee. These operations are performed successfully with hundreds of patients in Germany already. Fig. 50 [51] shows a human hip joint replacement. This application is supported in Germany by a year-long Governmental program. It is hoped that this application will be realized in the near future.

Finally, carbon fibres are used already in fixation of tooth roots as shown in Fig. 51. In this application carbon fibre reinforced polysulfone is applied [52].

5 Final Conclusions

i) *We can conclude:* Carbon fibres as the backbone of new revolutionary "advanced composites" are available today in high performance quality. The prices have been reduced drastically already and will be reduced even more in future by large scale production and additionally by large scale application of carbon fibres in form of multifilament tow.

ii) *We must recognize, however:* Considerably high costs of today's high performance advanced composites with low filament tows are strongly influenced by the process technology, which uses prepreg/autoclave curing techniques.

iii) *We propose therefore:* For broader economic application the manufacture of composites must be simplified by use of other techniques such as pulltrusion, sheet molding or injection molding. Especially thermoplastics as matrix offer a new economic way to use this "miracle material"carbon fibres also in complicated shaped structural parts made of advanced composites by simple hot working, similar to forging of metals.

iv) *We understand selected applications:* Due to the outstanding thermostability of the carbon fibres it is understandable, that the design engineer wants to apply these still expensive carbon fibres first of all in combination with a polymer that shows outstanding properties and as further consequence for most critical application cases. They ask therefore at first for thermoplastic polymers which have highest thermal resistance and complete corrosion resistance. Even the polysulfones, which were candidate materials as heat resistant matrix for carbon fibre reinforced composites during the last years have not brought the expected break-through because they have not proved to be resistant enough against solvents and oils.

A new thermoplastic matrix material, the so-called PEEK, a polyetheretherketone developed by ICI is now in discussion. Carbon fibres combined with this matrix promise to become a candidate future material, if the processing difficulties with this high melting polymer can be overcome, and if the high costs for this special polymer can be justified by the outstanding properties of the composite. This again means the start with a combination of most advanced but expensive materials and thus an exclusive application for cases, where suitable material alternatives do not exist. It reminds us of the situation 10 years ago with C-fibre reinforced epoxies, namely their application in aerospace only, where high material costs are easily acceptable.

v) *But finally, we expect a break-through:* One should not forget, that carbon fibres, if they are available at low prices – even if they have outstanding properties – could be applied also in conventional low price thermoplastic polymers for broadest applications.

As general conclusion, one can state that there remains much development work to be done for chemists, materials scientists and engineers to start a new material age with this "revolutionary" material carbon fibre.

Acknowledgement

The helps of Mr. Heine and Mrs. Philipp are thankful acknowledged.

6 References

1. Newsweek, February (1982), p. 36
2. Financial Guardian, Tuesday, April 19 (1983), p. 23
3. Bradshaw, W. G.: Sprechsaal 113 (1980), pp. 442–450
4. Deutsch, G. C.: in 23th National SAMPE Symp. and Exhibition, Anaheim (1978), p. 34
5. Fitzer, E., Heine, M.: CHEMRAWN III, World Conference on Resource Material Conversion, The Hague (1984)

6. HASSEL, D., MARK, H.: Z. Physik, 25 (1924), p. 317
7. BERNAL, J. D.: Proc. Roy. Soc., London, A 106 (1924), p. 749
8. FITZER, E.: Contemporary Topics in Polymer Science, Wilmington, USA, plenum press, New York, 5 (1983) pp. 101–138
9. KELLY, B. T.: High Temp. – High Pressures, 5 (1973), pp. 133–144
10. COOPER, C.: Chemistry and Industry, 18 (1982), pp. 678–683
11. JOHNSON, David, J.: Chemistry and Industry, 18 (1982), pp. 692–698
12a. ROSE, P. G.: Verarbeiten und Anwenden kohlenstoffaserverstärkter Kunststoffe (VDI-Verlag) (1981), p. 5
12b. WATT, W., MORETON, R., JOHNSON, D. J.: Nature, 221 (1967), p. 690
12c. FROHS, W.: Ph.D. Thesis, University of Karlsruhe, FRG (1984) in preparation
13. SAWADA, Y., SHINDO, A.: International Symp. on Carbon, Toyohashi, Japan Ext. Abstr. 3 A 06 (1982), p. 296
14a. HAWTHORNE, H. M.: Int. Conf. on Carbon Fibres and their Composites, The Plastic Institute London (1971), paper 13
14b. DONNET, J. R., EHRBURGER, P.: Proc. 4th Int. Carbon and Graphite Conf., Soc. of Chem. Ind., London (1976), p. 384
14. c) FITZER, E., LAUDENKLOS, P., SCHWARZ, G.: Ber. der Deutschen Keram. Gesellschaft, Band 53, 6 (1976), pp. 169–174
15. LEWIS, J. C.: Plenary lecture CARBONE '84, Intern. Carbon Conference Bordeaux, France (2–6 July, 1984)
16. FITZER, E.: CARBON *20* (1982), p. 445
FITZER, E.: Pure and Applied Chemistry (IUPAC) in press
17. ENDO, M., KOYAMA, T., INAGAKI, M.: Extend. Abstr. 2 B 09, Toyohashi, Japan (1982), p. 212
18. DRESSELHAUS, M. S.: Plenary lecture CARBONE '84, Intern. Carbon Conference Bordeaux, France (2–6 July, 1984)
19. FITZER, E.: Pure and Applied Chemistry (IUPAC) (in preparation)
20a. FITZER, E., GKOGKIDIS, A.: in Proceedings Symposium on Petroleum Derived Carbons, St. Louis (Washington DC: American Chem. Society (1984) pp. 506–522
20b. FITZER, E., GKOGKIDIS, A., HEINE, M.: High Temp. – High Press., 16 (1984), p. 363–392
21. REYNOLDS, W. N., SHARP, J. V.: CARBON, 12 (1974), pp. 103–110
22. RULAND, W.: Journ. Appl. Phys., 38 (1967), p. 3585
23. BENNETT, S. C., JOHNSON, D. J.: Proc. Fifth London Carbon and Graphite Conf., vol. 1. Soc. Chem. Ind. London (1978), p. 337
24. GUIGON, M., OBERLIN, A., DESARMOT, G.: Fibre Science and Technology, 20 (1984), pp. 117–198
25a. MORETON, R.: Fibre Science and Technology (1968), p. 273
25b. TORAY Data-Sheet No. TY-031 (1984)
26. WEISS, R.: Thesis, University of Karlsruhe (1984)
27. FITZER, E., JÄGER, H., WEISS, R.: 16th Bienn. Conf. on Carbon, San Diego (1983), pp. 471–478
28. HANCOX, N. L.: Composites, March (1971), pp. 41–45
29. TREWIN, E. M.: SAMPE, London (1983), p. 14–1
30. DE 3330575 (1984)
31. FITZER, E., HEYM, M.: Chemistry and Industry, 18 (1976), p. 663–676 (Chemical Society Publ., Lechworth)
32. DOS 3132784 (1981)
33. US Pat. 4331620 (1982)
34. BÖDER, H., GÖLDEN, D., POPP, G., VOGELSANG, J.: SAMPE, London (1983), p. 12–1
35. Information (leaflet) bei TORAY Industries, Inc. (1983)
MATSUDA, H.: International Conf. Interface – Interphase in Composite Materials, Université de Liège (1983)
36. Fabrication at ROCKETYNE INTERNATIONAL, by courtesy Dr. BERGMANN (1983)
37. Private communication by EMBREAR, San José dos Campos, Brazil (1983)
38. Engineering, September 1982, p. 611
39. By courtesy Dr. BARRY J. HAYES (CIBA-GEIGY PLC, United Kingdom)
40. Information by FIBERITE

41. DHARAN, C. K. H., THOMS, J.: in 24. Nat. SAMPE Symp. and Exhibition, San Francisco (1979), pp. 1550–1566
BEARDMORE, P., HARWOOD, J. J., HORTON, E. J.: in Proc. of the 3rd Int. Conf. on Composite Materials, Paris (Pergamon Press) (1980), vol. 1, p. 11
42. Information (leaflet) by HERCULES Inc. (1983)
43. v. GELLHORN, E.: Plastverarbeiter, 35 (1984), p. 75–79
44. FÜSSINGER, R.: in Processing and Uses of Carbon Fibre Reinforced Plastics (VDI-Verlag Düsseldorf) (1981), pp. 215–227
45. Information (leaflet) by TORAY INDUSTRIES, (1983), p. 15–1
46. DELACOUR, J.: SAMPE, London (1983), vol. 2, paper 15
47. FITZER, E., HARTWIG, G., WEISS, R.: 8. ETPC (Europ. Thermophysical Properties Conf.) Baden-Baden 1982, published in High Temp. – High Press., 15 (1983), pp. 147–150
48. HARTWIG, G.: ICMC/ICEC – Conference, San Diego, USA (1981)
49. BRENNFLECK, K., DIETRICH, M., FITZER, E., KEHR, D.: Proc. of 7th Int. Conf. on CVD, Eletrochem. Soc. Inc., Princeton, N. J., (1979), pp. 300–314
50. CLAES, L., BURRI, C., NEUGEBAUER, R.: 3. Münchener Symposium für experimentelle Orthopädie, Georg Thieme Verlag, Stuttgart (1981), p. 56
51. BRÜCKMANN, H.: Thesis, University of Karlsruhe, FRG (1979)
52. KIRSCHNER, H.: Seminar "Anwendung von Kohlenstoff in der medizinischen Technik", Institut für Chem. Technik, University of Karlsruhe, FRG (1984)

Comparison of the Various New High Modulus Fibers for Reinforcement of Advanced Composites with Polymers, Metals and Ceramics as Matrix

R. J. DIEFENDORF

1 Introduction

High modulus fibers are used mainly as fibrous reinforcement in resins, metals, or ceramics to provide strength and stiffness. It is instructive to consider what factors led to the development of composites. Structural designs are as likely to be limited by the stiffness as by the strength of the construction material. Hence, engineers have always desired stiffer, stronger, less dense and lower cost structural materials. There are a number of materials in the upper center part of the periodic table, such as boron, carbon, silicon carbide and alumina, which all have significantly higher modulus/weight ratios than the common engineering metals. Theoretically, the high modulus would result in high strength if the materials were perfect. Unfortunately, most of the high modulus materials are covalently bonded and are brittle. Small flaws can produce catastrophic failure at drastically reduced stresses. (The flaws may be introduced during manufacture or during service.) Except in some unusual protected environments, primary structural elements can not be made from materials which fail catastrophically. Composite materials offer the potential of using these brittle materials in structures which will not fail catastrophically. The brittle materials are made into fibers to give redundancy and placed into a matrix. The matrix serves to transfer stress into and out of the fibers. The matrix and matrix/fiber interface must also serve to stop cracks which originate in the fibers from propagating through the solid. Hence, a localized impact may break a few fibers, but the crack can be stopped by a ductile matrix or fiber/matrix debond. Even ceramic fiber/ceramic matrix composites can be made which are relatively tough. The use of composites to attain non-catastrophic failure with brittle materials has penalties. The volume fraction of reinforcing fibers is generally 0.65 or less, and fibers only reinforce in one direction. While fibers can be oriented in three orthogonal directions, composites offer the greatest potential if fibers are oriented predominantly uniaxially or biaxially. Fortunately, many structures are loaded mainly in one direction.

Comparison of high modulus fibers must necessarily compare their performance in composites. The task is made more difficult by the diverse applications, particularly if non-ambient temperature uses are considered. This paper will be separated into two parts: firstly, the requirements for fibers which are to be used in resin, metal or ceramic matrix composites will be defined. Secondly, present fiber performances will be compared to each other and the desired composite requirements.

2 Composite Requirements

The desired properties for fibers are often ranked quite differently for resin, metal and ceramic matrix composites. Also, since metal and ceramic matrix composites commonly are exposed to elevated temperature, (if only for processing), the high temperatures characteristics of the fibers must also be considered.

2.1 Resin Matrix Composite Reinforcement

Fibers for resin matrix applications would have the highest modulus, lowest density, and highest tensile, compressive, and interfacial bond strength between resin and fiber, while still failing in a non-catastrophic mode. Non-catastrophic failure requires a balance between the values of the tensile strength of the fiber and the interfacial bond strength. Too high an interfacial bond strength for a given fiber tensile strength will produce a more brittle composite which will be flaw sensitive and generally have a lower tensile strength. Too low an interfacial bond strength will produce poorer composite shear and tensile strengths, but the composite will be less flaw sensitive. The exact value of the ratio of fiber tensile strength to interfacial bond strength should be selected for the particular application. Hence, for high damage tolerance, a slightly lower interfacial bond strength might be desirable. Other factors affect the fracture behavior. For example, higher fiber modulus, low fiber strength, higher volume fraction, brittle matrix, low test temperature, and high loading rate will all produce a more brittle fracture. Except for a few applications, fibers are produced with only one surface treatment but that is optimum for many applications. Practically, the shear strength of a composite must exceed 35 MN/m^2 (5,000 psi) to have any practical application. This requirement forces the strain-to-failure for a fiber to be at least 0.5% for any structural application for two reasons. Firstly, strains in structures often reach 0.2% (a common fatigue limit strain in metals), and local strains, as under rivets, are likely to be several times higher. Secondly, this low strain-to-failure corresponds to low fiber strength, and a brittle composite would result if the shear strength is maintained at the minimum of 35 MN/m^2. Obviously, for more critical structural applications the strain-to-failure should be several times higher. An aircraft manufacturer for example, would not use a fiber with less than a 2.0 percent strain-to-failure with a 100 MN/m^2 composite shear strength for primary structural applications such as the wing box.

Thermal expansion coefficient is also an important property for reinforcing fibers. In many cases, the thermal expansion of a composite will be lower than the structure to which it is attached. Careful design must be used or binding and warping will result with a temperature change. However, the negative thermal expansion coefficient of some fibers can be used to produce a zero thermal expansion coefficient, provided the fiber modulus is sufficiently high. Finally, stress rupture is important for sustained load applications.

One secondary consideration is the form of the reinforcement. Generally, continuous fiber in the form of tow with zero or a quarter to a half twist per centimeter is desirable. Fine diameter (<15 μm) is desirable to minimize stresses which arise from fiber high curvatures in parts. (However, a large number of fine fibers is harder to handle during fabrication than fewer large diameter (>100 μm) filaments.) Another consideration is electrical properties. For radar domes, or printed circuit boards for example, an insulating fiber is required. However, for electromagnetic shielding, or static discharge, a conductive fiber is required. Although either insulating or conductive fibers may be used in corrosion resistant composites, conductive fibers may cause corrosion of attached metal parts. Finally, other properties such as thermal conductivity or coefficient of friction may be important for certain applications.

2.2 Metal Matrix Reinforcement

The mechanical requirements for reinforcements for high temperature applications results in many of the same chemical elements and compounds being selected as presently used in resin matrix composites. Materials with high modulus and high strength generally have high melting points, and good high temperature properties. (Exceptions are the high modulus organic fibers.) Hence, most fibers developed for use with resin matrices would appear to be good candidates for reinforcement in high temperature composites.

High temperature stress rupture and creep resistance are the properties of general concern in most high temperature structural material applications. The refractoriness (i.e., melting temperature) of the fiber is often a good indication of these properties because thermally activated slip and creep become dominant at 0.5 of the melting temperature for simple crystal systems. Oxidation resistance must also be thoroughly considered in fiber selection because of the uncertainty of filament exposure to the environment or the possibility of oxygen diffusion through the matrix. These problems are particularly acute if the fiber also has a fine diameter such as the carbon filaments. Fiber diameter will also effect the choice of processing techniques and subsequent forming operations that can be used. Fine diameter (25 μm or less) multi-tow filaments such as FP-alumina and graphite are more conveniently processed by liquid infiltration while filament over 100 micrometers in diameter can be processed using powder metallurgy techniques.

Ceramic and carbon fibers could offer a greater potential as reinforcements in metal matrix composites if not precluded by stability and other limitations. The over-riding consideration in selecting a metal matrix for composites is its chemical and physical compatibility with the reinforcing fiber.

A major requirement for the reinforcements is that they be compatible with their environment. Ideally, the reinforcement would be oxidation resistant and compatible with all potential matrices. Only the oxides are truly oxidation resistant, but some compounds have adequate oxidation resistance if they form an adherent glassy coating. Compatibility with matrices is also required. Numerous fiber and matrix combinations have been proposed as potential high temperature composites based on their unique mechanical, physical and chemical properties. The record of successful development of such composites has been disappointingly low because of chemical or physical incompatibility between fiber and matrix.

Chemical compatibility must consider two factors. One factor is sufficient interaction between the fiber and the matrix to assure the strong bonds necessary for off-axis strength of continuous reinforced composites or for stress transfer between reinforcements. The other is the ability of the matrix and fiber to co-exist without mutually induced disintegration by interdiffusion and reaction. The dilemma faced in composite development is that the conditions which result in both types of compatibility are mutually exclusive. Useful composites can be produced only if a successful compromise between bonding and disintegration can be established. Two principal techniques used to achieve this result are applying a protective coating to the filament, and adding constituents to the matrix that restrict interaction.

The carbides, nitrides, and borides all react with most potential metal matrices, and the question of importance is how fast. The most kinetically stable carbides are those of niobium and hafnium, although silicon carbide is also attractive for lower temperatures. Coating, however, is not an ideal solution to compatibility problems because of two critical considerations. The first is the obvious economic penalty incurred by the addition of the coating step during composite fabrication. Second is the criticality of optimum coating thickness. Too thick a coating can result in sources of brittle failure while too thin a coating results in insufficient protection.

Mechanical compatibility of the fiber with the metallic-matrix requires consideration of the thermal expansion properties of the fiber, especially if cyclical thermal excursions are expected in service. Exact matching of thermal expansion properties of fiber and matrix is neither practical nor absolutely always necessary. However, the particular operating conditions and the performance requirement will determine the criticality of this variable. For example, a macrodimensionally stable large structure can be achieved in a graphite-copper composite, but when the surface quality must meet optical standards the thermal mismatch between fiber and matrix precludes this material for use.

Finally, corrosion couples may be generated between electrically conductive fibers and the matrix. In turbine applications the thermal conductivity of the fiber as well as the matrix can be utilized for heat dissipation and therefore must be considered when designing cooled components. In this case metal fibers are preferable to ceramic fibers. Conversely, ceramic fibers could be more advantageous for applications where heat retention or insulation is desired.

Finally, compatibility is not the only consideration. For metal matrices, no single matrix alloy will be optimum for all desired properties and the development of suitable materials invariably involves compromises for one or more of the significant properties. The question that remains to be settled is whether sufficient properties can be retained after such a trade-off to render the alloy useful and economical for composite application. While lower melting point metals, such as aluminum, can form useful composites with certain fibers, it is not apparent that suitable systems exist for higher melting point matrices.

2.3 Ceramic Matrix Composite Reinforcements

Brittle materials, while often having superior high temperature properties, are limited by inadequate fracture toughness at near ambient temperature. The problem then is to increase fracture toughness while not sacrificing strength. In ceramic matrix composites the emphasis is on toughening at lower temperature, $<1000\,°C$, so that they may survive heating or cooling. This is in contrast to metal matrix composites where the emphasis is on improving high temperature behavior.

The general principles for toughening ceramic composites are similar to those for other types of composites:

1. to increase the local driving forces necessary to propagate cracks to failure;
2. to locally increase the amount of mechanical energy consumed per unit area of propagation of any crack; or
3. to decrease the local strain by cracking which reduces the stress concentration.

However, the mechanisms by which these principles are realized in ceramic matrix composites may be the same or very different from other composites. The differences in mechanisms manifest themselves in two important ways. First, in most other composite systems, applying the first two principles above would generally increase the strength of the body. However, in ceramic composites this is not necessarily the case. In some, toughening may be obtained at the *sacrifice* of some strength rather than increasing it, due to the introduction of larger effective flaws and/or lower Young's moduli. In other ceramic composites, strength may either be unaffected or may be increased. An important difference between ceramic and other composites is the temperature regime in which toughening is emphasized.

While ceramics can be toughened by incorporating fibers having a higher Young's modulus than the ceramic matrix, e.g. as illustrated by the work on use of ceramic fibers in glass or crystallized glass matrices, its applicability to ceramics is much more limited. The limitation arises because of Young's moduli of most matrices and fibers of interest are similar and the descriptive terms of matrix and reinforcement lose much of their meaning except geometrically.

Some benefits accrue at higher temperature, if fibers have a higher modulus and/or a lower rate of decrease of their modulus with temperature. Hence, some modulus transfer of load may also occur at temperatures reasonably in excess of 1000 °C. In addition to improved toughness at ambient temperatures, ceramic matrix composites will require a combination of properties to accommodate particular high temperature applications including: creep rupture strength, oxidation/hot corrosion resistance, and good mechanical and thermal fatigue characteristics. The challenge is to design a ceramic composite with a balance of properties superior to competitive materials. The development of composites faces additional problems which include chemical and mechanical compatibility between matrix and reinforcement, and anisotropy of properties. Chemical instabilities, such as interdiffusion at elevated temperatures, may degrade the properties of the reinforcements or interface, yet some reaction may be required to provide adequate stress transfer between the reinforcement matrix interface. The compatibility problems of reinforcements with ceramic matrices can be severe. The oxides begin to react with the elements, carbides, nitrides, and borides in the temperature range of interest, but would be attractive combinations for lower temperature use. The nitrides and the element carbon are also marginal in the desired temperature range. The various carbides, nitrides, and borides appear to be mutually compatible from a reactivity view in the lower end of the temperature range of interest.

High temperature atmospheric oxidation can be a problem with ceramic matrix composites. Carbon fiber/carbon matrix composites, while having good mechanical properties, must be completely protected from oxidizing gases. The problem is particularly difficult as the anisotropy of the composite (local or global) makes application of protective surface coatings very difficult. The coefficient of thermal expansion can not be matched to the *different* coefficients of thermal expansion of a carbon/carbon composite in its different directions. Although various designs similar to those used in early glass to metal seals may minimize problems, the surface coating is usually in tension at ambient temperature and can act as a crack initiator to the whole composite. Microcracking of the protective coatings are common. At temperatures above the application temperature of the coating, the coating will be in compression and may present an effective barrier, even though microcracked, to oxidation for short times. The leading edges of the space shuttle rely on this behavior.

An important consideration of brittle reinforcement/brittle matrix systems is residual stresses arising from mismatch in coefficient of thermal expansion upon fabrication or use. (In addition, stresses generated from processing may also be high.) These stresses can provide beneficial prestresses for applications, but generally they result in deleterious internal stresses. Interestingly, sometimes these stresses are so high that extensive microcracking occurs within the composite and this can produce a tough material. Carbon fiber/carbon matrix composites are a good example.

In summary, carbon, oxides, carbides, nitrides, and borides all appear attractive for ceramic matrix composite applications. Several ternary oxides (mullite) provide improved creep resistance and could be desirable for reinforcements. A potential important advantage of developing oxide ceramic matrix composites would be the inherent oxidation resistance. In general, the distinction between matrix and reinforcement in ceramic systems is less well defined. The major reason for using composite technology in ceramic systems is to improve toughness.

3 Reinforcement Fibers

Fibers are available for a fair range of materials over a wide range of diameters. However, it should be realized that all fibers that might be desired are not presently available in appropriate diameters. Generally, some processes produce fine diameters (<25 μm) only, while others can only produce large diameter fibers economically. Similarly, some processes are sufficiently versatile to produce almost any of the desired materials, while others are specific to one. Fine diameter fibers are desirable for:

1. high strength,
2. large surface area for bonding, and
3. the ability to be bent in a sharp radius of curvature.

Fine diameter fibers do appear to be desirable, and even required, for ceramic matrix composites.

Fiber/matrix compatibility studies for high temperature composites appear to indicate that relatively large amounts of reaction will usually occur in many potential systems. The implication is that large diameter fibers would be required for these systems whether diffusion barriers are used or not. The result is that the fiber reinforcement should have a minimum diameter of 15 μm, and preferably larger for many composite systems. However, present processes designed to produce fine diameter fibers at low cost may not be attractive for many of the high temperature composites, since the processes are not easily modified to produce large diameters. Processes which produce large diameter fibers, such as chemical vapor deposition, are very versatile but unfortunately expensive.

3.1 Fiber Processes

Generally, reinforcement fibers must have very high strength to be useful, since limited volume fractions and plying of uniaxial layers drop the potential strength rapidly. While high strength reinforcement could be single crystalline, the high strength is usually obtained by a very fine or elongated grain microstructure. The problem is to find and develop processes which will produce the very fine grain sizes desired.

3.2 Glass Melt Spinning

In the sense that a glass is the limit of a fine grained microstructure, glass drawing is a useful process when systems with suitable viscosities exist (Fig. 1). Silica is the basis for high strength glass fibers, although the composition is usually modified by the additions of boron, aluminum, magnesium calcium oxides. Generally, the modulus of glass fibers is mainly determined by the silicate network, although S type glasses have a higher specific modulus than E glass because of a slightly higher modulus and lower density (Table 1). Higher moduli have been obtained by adding berylium oxide to the glass, but the improvement is partially negated by a higher density. In any case, the toxicity and cost of berylium glasses have generated little interest in the slightly higher modulus. Most elements and binary compounds with high modulus have low viscosity and would require extremely high quench rates to produce a glassy structure. Even with the very complex crystal structure of boron, only very limited success has been achieved in obtaining short lengths of boron filament from quenching a melt. Hydrodynamic instabilities from low viscosity and crystallization both caused problems with boron. More complex melts, particularly

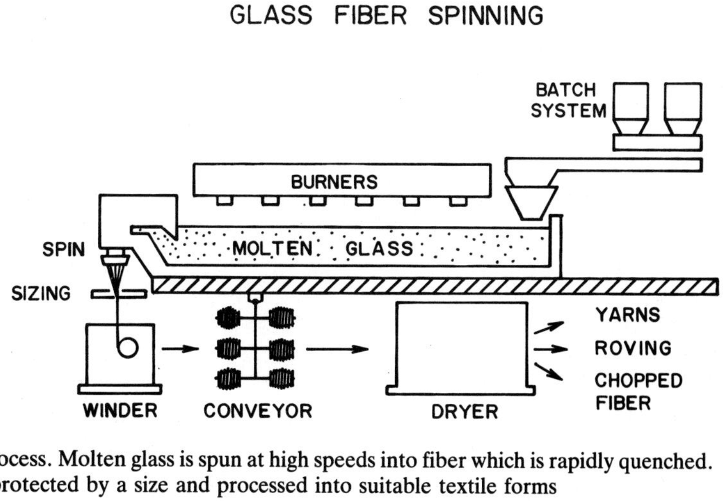

Fig. 1. Glass fiber process. Molten glass is spun at high speeds into fiber which is rapidly quenched. The fiber surface is protected by a size and processed into suitable textile forms

eutectics, might allow some useful systems to be discovered, but it should be realized that melt containment may be difficult. Furthermore, devitrification of the glasses may limit usage. No suitable systems to give high moduli glasses that might be melt spinnable have been identified.

The other processes that produce fine grained high strength fibers all are carried out well below the recrystallization temperature.

Table 1. Fiber Properties

	Modulus GPA	MSI	Strength GPA	KSI	Strain %	Density gm/c.c.
Glass						
E-HTS	69	10	3.5	510	3.5	2.55
S-2	82.7	12	4.5	660	4.5	2.4
CVD Filament						
Silicon Carbide						
tungsten substrate	418	61	3.1	450	0.8	3.3
carbon substrate	418	61	3.9	575	0.9	3.1
Boron	397	58	3.4	500	0.9	2.6
Carbon Fiber						
PAN Precursor						
Low Modulus						
(AS, T-300, XA, Celion, etc.)	220–240	32–35	3.0–3.3	435–480	1.3–1.4	1.75
High Modulus (HMS)	330–350	48–51	2.3–2.6	339–375	0.7	1.91
Typical new grades						
Union Carbide						
T-300A	230	33	3.6	520		
T-700	230	33	4.5	660		
T-800	267	39	5.3	780	2.0	
T-40	294	43	4.2	620		
Hercules						
AS4	233	34	3.6	520	1.5	
AS6	246	36	4.1	600	1.65	
IM6	288	42	4.2	620	1.6	
Pitch Precursor						
Union Carbide						
P-55S	380	55	2.1	300	0.5	2.02
P-75S	500	75	2.1	300	0.4	
P-100S	685	100	2.2	325	0.3	
P-120S	822	120	2.7	400	0.3	
Inorganic Fiber						
Silicon Carbide	200	29	2.7	400	1.3	2.8
Alumina						
duPont (FP)	340–375	50–55	1.7	250	0.4	3.7
Sumitomo	210	31	1.8	260	0.8	3.2
Boron Nitride	310	45	2.1	300	0.8	2.1
Organic Fiber						
Kevlar 49 (PPT)	124	18	3.6	525	2.5	1.44
PBT	304	44	2.7	400	0.9	

3.3 Chemical Vapor Deposition

Chemical vapor deposition is the formation of a solid by the decomposition or reduction of one or more gaseous molecules upon a heated substrate. Fine grained materials may be formed at one-third or less of the melting point of the material for simple crystal structures and somewhat higher for more complex crystal structures. The crystallite size also depends on the rate of deposition and back etching, and the surface mobility. For fiber production, the deposition geometry shown in Fig. 2 is typical. Tungsten or carbon filament is run through a chamber where the filament is resistance and RF heated and suitable gases decompose on the heated wire. Although the linear deposition rate is quite high by plating process standards, the contact time required to deposit the necessary coating thickness is tens of seconds or more. Hence, for any appreciable production rate, many reactors must be run in parallel. The cost for the substrate and deposition gases, and the capital equipment for the reactors and gas handling is quite high. Hence, the filament tends to be quite expensive. However, CVD can be used to deposit almost any type of large diameter fiber or coating. A problem arises from the kinetic nature of the deposition process. Compounds may be deposited off stoichiometry or even with second phases, and the composition may vary within the fiber. While these variations may give improved room temperature properties, high temperature properties may be severely degraded or vice-versa. (An example is SiC deposited with excess silicon.) CVD is used at present to produce boron and silicon carbide filament (Table 1).

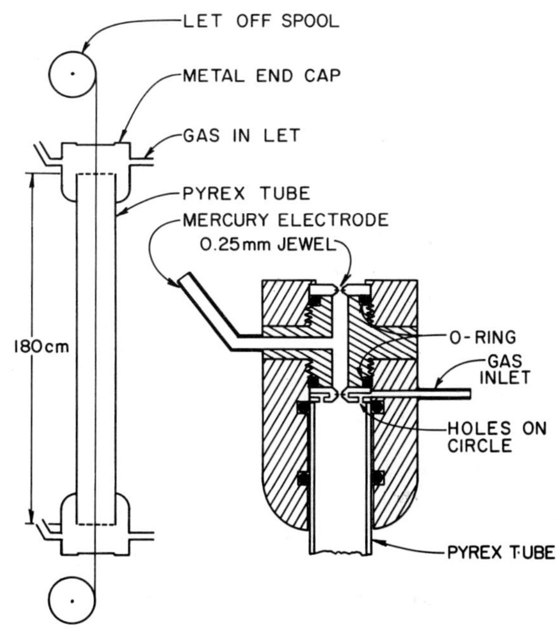

Fig. 2. Chemical vapor deposition of filament. A conductive substrate is passed through a cylindrical chamber where reactive gases deposit the desired material on the heated substrate

3.4 Precursor Polymeric Fiber Decomposition

Carbon, silicon carbide and alumina fibers can be made by careful pyrolysis of a polymeric or gel precursor fiber. The use of a polymeric precursor allows the fiber to be spun using the procedures, modified as required, of the synthetic fiber industry. The result is generally a bundle of fine diameter fibers which upon a controlled heat-treatment decompose to the product fiber. The most well known examples are the carbon fibers whose processes are shown in Fig. 3. Most carbon fiber is produced from a special polyacrylonitrile (PAN) copolymer fiber. In this process, the PAN is stretched to align the polymer chains parallel to the fiber axis, and these linear backbones are converted into ribbons of continuous hexagonal rings of carbon and nitrogen by an oxidation. These ribbons have a sufficiently high glass transition temperature that part of the orientation is retained after a carbonization cycle to 1000 °C. Most of elements other than carbon are removed in the carbonization cycle. While the orientation of these ribbons are generally parallel to the fiber axis, this alignment can be improved by heat-treating to higher temperatures which increases the modulus of the fiber. Recent improvements in lower modulus PAN precursor carbon fibers promise two percent tensile strain to failure by year's end. Unfortunately, similar improvements in tensile strength have not been made in the high modulus PAN precursor fibers.

The second major process for producing carbon fibers uses liquid crystalline fractions of coal tar or petroleum pitch as precursors. The sheet-like aromatic hydrocarbons are easily oriented parallel to the fiber axis upon melt spinning of these liquid

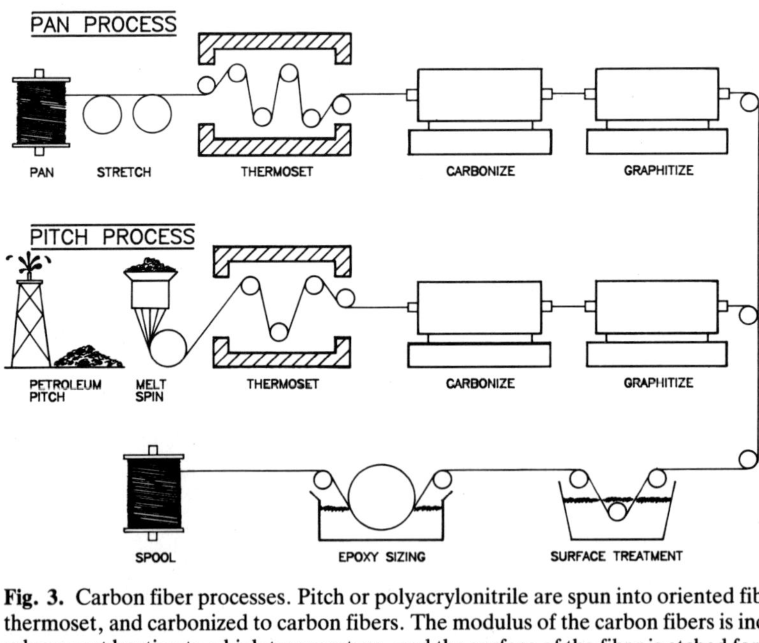

Fig. 3. Carbon fiber processes. Pitch or polyacrylonitrile are spun into oriented fibers, are thermoset, and carbonized to carbon fibers. The modulus of the carbon fibers is increased by subsequent heating to a high temperature, and the surface of the fiber is etched for improved bonding to the resin matrix

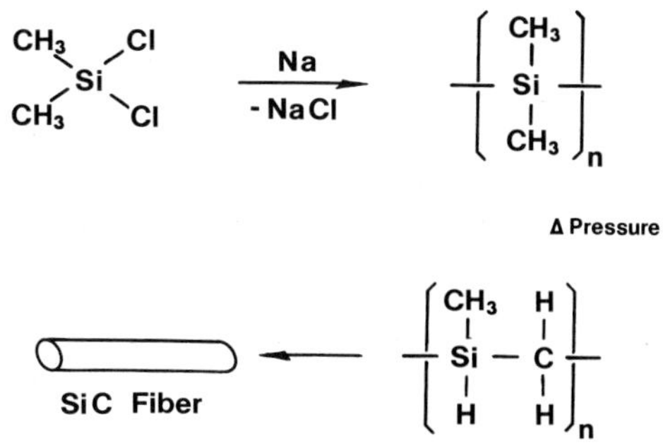

Fig. 4. Silicon carbide fiber process. Dichlorodimethylsilane is polymerized and then thermally treated to form a polycarbosilane polymer which can be spun into a fiber. The polycarbosilane is then pyrolysed to a silicon carbide fiber

crystalline materials. The pitch fibers are made infusible by a thermosetting oxidation step and then carbonized. Just as the ribbons in PAN can be more perfectly aligned by higher temperature heat treatment causes these sheets to undergo rearrangement to a more aligned and higher modulus fiber. Fiber properties for both precursors are reported in Table 1.

Silicon carbide fiber has been developed by Yajima and Nippon Carbon Co. in Japan (Fig. 4). As the polymer has a Si/C ratio of one-half, excess carbon is present in the fiber and this often leads to variable properties. The values are shown in Table 1. Dow Chemical is developing SiC fiber in the U.S. The preliminary mechanical properties are reported to be about equal to Yajima's better values.

Although some work was performed in the 1960's in the U.S. to make inorganic fibers via a polymeric route, only alumina fiber has reached a production stage. The properties for these fibers are reported in Table 1. While the strength is only of the order of 200,000 PSI at room temperature, it maintains most of its short time tensile strength to 1000 °C. However, at high temperatures required for turbine service applications, alumina would not have the necessary creep resistance. Also, compatibility problems limit its use in high temperature metal matrices, but it is attractive for lower temperature use. It also may be of interest in ceramic matrix systems.

3.5 Host Conversion

In one embodiment of this process, a precursor fiber is converted by reaction with a gas phase species to form the new fiber (Fig. 5). Carborundum Company (Division of SOHIO) has made both boron and silicon carbide filament by reacting a rayon precursor carbon fiber with boron or silicon chlorides at elevated temperature. In the case of boron carbide, the room temperature modulus was 410 GPA (60×10^6 PSI), while the strength was 1.2 GPA (180×10^3 PSI). Obviously, this process could be used to produce other carbide fibers, or coatings on carbon fibers with relative ease.

In an extension of this concept, Carborundum has converted B_2O_3 glass fiber to either boron nitride or boron carbide fiber. The resulting boron nitride fiber generally has about a 27 GPA (4×10^6 PSI) modulus and 0.34 GPA (50×10^3 PSI) strength which is too low for most reinforcement applications. Strain annealing this fiber at high temperature increases the mechanical properties to those shown in

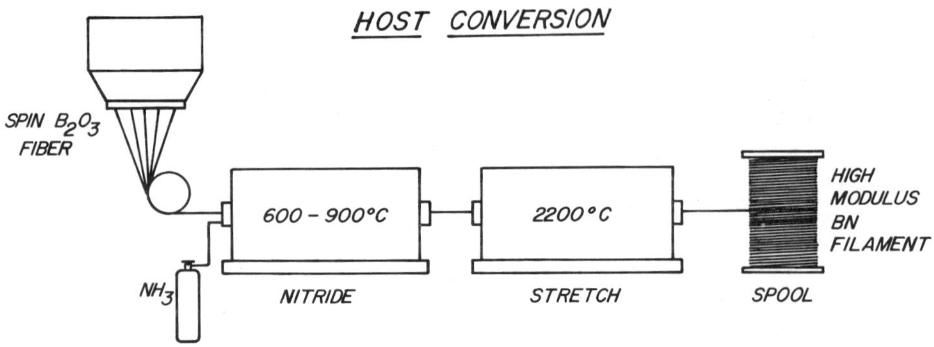

Fig. 5. Boron nitride filament. Boric oxide fiber is spun into a fiber, converted to the nitride by heating in ammonia, and stretched at high temperature to produce high modulus boron nitride fiber

Table 1. Individual fibers have tested as high as 480–550 GPA (70–80 × 10^6 PSI) in modulus and 2.4 GPA (350 × 10^3 PSI) in strength. A big advantage of boron nitride filament is its excellent dielectric properties. However, its behavior in metal matrices is very similar to high modulus carbon fibers, e.g. it is hard to wet, and is reactive.

A process somewhat different in concept is the use of a rayon fiber to absorb metallic nitrate salt and then to carefully pyrolyse to a salt laden carbon fiber. The carbon is then oxidized and the remaining oxide carefully sintered. Several oxide fibers were prepared by Union Carbide Corporation in this manner, the most notable being stabilized zirconia. Fiber strengths were in the 0.7–1.4 GPA (100–200 × 10^3 PSI) range. Although mainly used for insulation, some ceramic matrix composites were made from these fibers, and the strengths were as high as 0.35 GPA (50 × 10^3 GPI). Unfortunately, the fibers are no longer available commercially.

3.6 Organic Fibers

The ultrahigh modulus of graphite results from the strong sp^2 bonding in the basal plane. High modulus organic fibers are possible if a portion of the multifused aromatic rings in graphite can be incorporated into a polymer. Highly oriented para-polyphenylene or polyacene ladder polymers provide attractive specific moduli. The problem is how to get these intractable polymers into a highly oriented fiber form. A major breakthrough came with the development of Kevlar® aramid fiber (Fig. 6). The enhanced solubility due to good solvents and the amide linkage in polyphenyleneterephthalamide (PPT) allowed formation of an easily oriented liquid crystalline polymer solution. This solution could be spun into a highly oriented fiber, whose properties could be further enhanced by a high temperature crystallization heat-treatment. Properties are listed in Table 1. As might be expected from a rigid rod polymer structure with relatively weak bonding between adjacent polymer molecules, these rods might be expected to buckle at relatively low compressive stresses. Also, the offset which is introduced by the amide linkage produces a

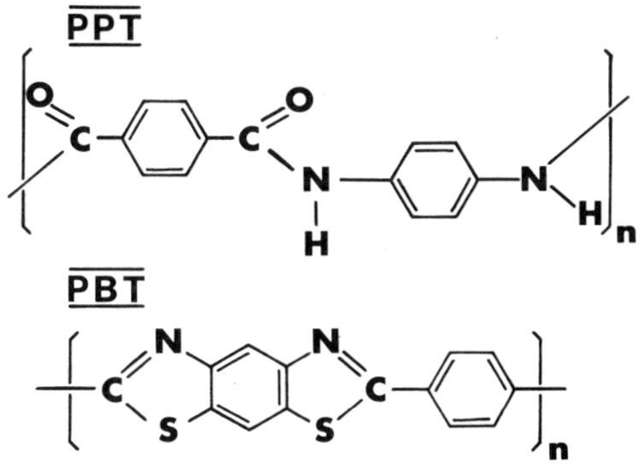

Fig. 6. High modulus organic fibers. PPT: polyphenyleneterephthalamide. PBT: polybenzothiazole

slightly kinked structure along the rod (Fig. 6a). This might be expected to further reduce the load required for buckling as well as decrease the modulus. A truly straight rigid rod polymer such as polybenzothiazole (PBT) might be expected to provide a higher modulus which it does (Fig. 6/Table 1). However, the compressive failure load is the same as for PPT, although the failure mechanism is quite different.

Linear chain type polymers would also give high modulus if the chains were fully extended parallel to the fiber axis. For example, fully extended polyethylene has a theoretical modulus of 250 GPA with a density slightly less than unity. Values of modulus as high as about 2/3 the theoretical have been observed experimentally, which is twice that of aluminum. Compressive strength of fully extended polyethylene parallel to the fiber axis have not been measured experimentally, but can be estimated from the elastic constants to be about 180 MPA. This low value which can be expected for a flexible chain in compression would limit the structural use of fully extended polyethylene severely, as would the low melting point of the polymer. Other linear chain polymers with higher chain-to-chain interaction would improve compressive and transverse properties, but would be harder to process. Recent patents have been published on these other polymers.

3.7 Whiskers

After GALT and HERRING's discovery of high strength tin whiskers, much research on whisker growth and properties was performed in the 50's and 60's. Many problems exist in growing whiskers in large quantity with good quality at an economically viable price. Because of this, interest in whisker technology has wanned, except for SILAG and TOKAI SiC whiskers which are made from rice hulls or similar by-products. The main application of these whiskers has been to aluminum and other low-melting metallic matrices with relatively good results. SiC whiskers would be too reactive for higher melting metallic matrices, except with prohibitively thick (relatively) protective coatings. However, the application of these whiskers to ceramic systems, even just for model systems, appears attractive. The fine dimensions of the whiskers could allow for additional toughening mechanisms to apply.

4 Comparison of Fibers

Many of the fibers described in the previous sections are experimental fibers which are not produced in quantity. Comparison of fibers which are primarily used for resin matrices will be limited to glass, Kevlar 49, carbon and boron. Similarly, only boron, silicon carbide, and alumina fibers will be compared with respect to application to metal matrix composites.

4.1 Resin Matrix Composites

The properties of composites depend not only on fiber and matrix properties and the interfacial bond between them, but on the fabrication and test techniques as well. Furthermore, mechanical test data are usually obtained from relatively small well made test specimens. For composites that are not flaw sensitive, the test results from small specimens may adequately describe larger scale hardware. However, for more brittle composites, the application of small specimen test data to a large structure will overestimate the strength of the structure, for the structure is more likely to have strength reducing flaws. The property data is for small test specimens and only for uniaxial laminates (Table 2). For the newer carbon fibers, the tensile strength of the composite is increased although not as much as might be expected from the fiber

Table 2. Typical Fiber Reinforced Composite Properties

Resin Matrix	E-Glass	S-2 Glass	Boron	AS, T-300 Celion Type Carbon	HMS Type Carbon	Kevlar
0° Tensile Modulus (GPA)	47	54	210	150	210	76
0° Tensile Strength (GPA)	1.15	1.31	1.73	1.6	1.12	1.38
90° Tensile Modulus (GPA)		25	23	6.9	8.5	5.5
90° Tensile Strength (GPA)		0.46	0.06	0.041	0.037	0.3
0° Compressive Modulus (GPA)	47	54	28	150	210	76
0° Compressive Strength (GPA)	0.54	0.69	3.1	1.6	0.99	0.276
Short Beam Shear Strength (GPA)	0.105	0.90	0.103	0.131	0.072	0.04–0.08
Density (gm/c.c.)	2.18	2.13	2.0	1.55	1.6	1.38
Impact Strength (ft. lb.)		280	40	100	30	150
Thermal Coefficient of Expansion $\times 10^6$ 1 °C		3	4		−2	−6

Aluminum Matrix	Sumitomo Al_2O_3	duPont (FP) Al_2O_3	Nippon SiC	Carbon	Boron
0° Tensile Modulus (GPA)	150	200	100–110	140	230
0° Tensile Strength (GPA)	0.86	0.6	0.8	0.7	1.5
90° Tensile Modulus (GPA)		106			180
90° Tensile Strength (GPA)		0.09	0.08	0.02	0.28

strength. Similarly, it is not apparent whether the improved composite strengths observed in test specimens will be carried over to structures. Although the composite tensile strength for these newer carbon fiber is higher, no increase in compressive strength has been observed. It may be that the fiber itself is failing internally just as Kevlar and the higher modulus carbon fibers do. (The very high modulus pitch fibers fail at less than 1/2 the value of the tensile strength in compression.)

Perusal of Table 2 shows that boron and high modulus carbon fiber are most desirable for high stiffness applications. Boron filament also has good tensile and compressive strength properties. The major disadvantages of boron filament are cost, machining and limited allowable laminate curvature. Lower modulus carbon fibers have "balanced properties" and moderate impact strength. Cost is also moderate. Kevlar has good tensile strength, moderate stiffness, is tough, but is relatively poor in compression. Cost is moderate. For specific strength (strength/density) applications, Kevlar and the new carbon fibers are attractive. Glass is hard to beat for pure strength limited applications because of low cost, strength and toughness. Finally, very high modulus carbon fibers have a sufficiently negative coefficient of thermal expansion that a zero thermal expansion coefficient laminate can be produced. (Kevlar also has a negative thermal expansion coefficient but its modulus is too low to produce a zero thermal expansion coefficient.) Obviously, depending on the relative importance of the design and cost requirements, a different high modulus fiber would be selected.

4.2 Metal Matrix Composites

The data is more fragmentary for metal matrix composites, especially because of the lack of high temperature data. Data are shown here for the reinforcement of aluminum at room temperature (Table 2). All fibers give a good increase to the stiffness and strength of aluminum parallel to the fiber axis. Good transverse properties are observed with boron and alumina fiber, but not with carbon. Also, corrosion in the aluminum/graphite may be a problem in some environments.

5 Conclusions

A wide range of fibers is available and more will be developed in the near future. The widely varying properties of the fibers and differing requirements for applications means that no one fiber will completely dominate the others, at least in the near future. Each fiber with its unique characteristics tends to be optimum for particular uses.

Acknowledgment

The author would like to thank S. H. Chen for preparing the drawings and J. A. Margosian for typing the manuscript, especially since it meant the loss of their weekends.

Carbon Fibre Surfaces – Characterization, Modification and Effect on the Fracture Behaviour of Carbon Fibre – Polymer Composites

I. L. KALNIN and H. JÄGER

The types of existing fiber surface treatments, the resulting changes in the fiber surface properties, and the effects of the surface condition on the microfracture of oriented carbon fiber – polymer composites are discussed.

1 Introduction

The development of the strong and stiff carbon fiber (CF) and its use as the reinforcing element in lightweight structural parts is one of the major technological achievement of the past decade, and the production volume of CF is projected to expand as new applications arise [1]. As the usage of the various CF types increases, so does the understanding of those CF properties that are needed for a good composite performance. This performance often depends on the degree of adhesion between the fiber and the binder phase that in turn depends considerably on the state of the CF surface. Already the early CF development work in Great Britain indicated that an adhesion promoting surface treatment is necessary to realize an adequate composite strength [2]. Since then, the number of different surface treatments has increased greatly, although the understanding of their action is still far from complete.

2 Surface Treatment

The finding that surface oxidation of carbon blacks, while changing their initially hydrophobic surfaces into hydrophylic ones, greatly increases their adhesion strength to polymeric materials [3], has been found equally applicable to CF, as a consequence of which the treatments most commonly described and used are oxidative. In addition to being oxidative or non-oxidative, they can be classified further as being carried out in

1. gas phase or liquid phase;
2. at relatively low temperatures ($< 150\,°C$) or at elevated temperatures, typically $> 400\,°C$;
3. by chemical, electro-chemical, or other means, and finally,
4. whether the treatment results in an overall fiber weight gain or loss.

The oxidative gas phase treatments are typically carried out by heating the CF in air at 400–800 °C [4], although other oxygen-containing gases, such as CO, CO_2, H_2O, etc. may be used or added to the air to modify the oxidation kinetics and mechanism. The high temperature gaseous treatments will result in surface etching the extent of which must be strictly controlled, lest the filament strength be degraded by excessive pitting [5]. Treatments at lower temperatures 120–150 °C, by ozonized air [6] or at room temperature by plasma excited oxygen or ammonia [7] with a lesser tendency toward pitting are known. Likewise, oxidative liquid phase treatments using chemical or electrochemical means have been explored widely. Of the chemical ones, many aqueous solutions of oxidants, such as nitric acid, hypochlorite [8, 9], chlorate or dichromate [10] were found useful. Also, the electrochemical (anodic) surface oxidation being particularly adaptable to continuous operation has been utilized [11, 12]. Other surface treatments that improve the fiber resin adhesion by depositing more active forms of carbon have been reported, such as the highly effective, but, expensive whiskerization treatment [13], the deposition of pyrolytic carbon [14] or the grafting of adhesion promoting polymeric species [15].

To summarize, then, the main principles for effective surface treatment were well established by 1973. The most successful ones involve controlled oxidation, either from the gas or the liquid phase. These treatments have the advantages that the oxidation

1. can be carried out by means of inexpensive commodity chemicals,
2. removes the original weakly bonded layer of surface carbon, about 15–50 nm thick, and
3. reacts with the surface carbon to form several types of oxygen-containing groups, some of which are strongly chemisorbed to the surface carbon atoms and serve as the bonding sites for the polymeric binder.

3 Internal and Surface Structure

In order to better understand the interaction of the surface with the binder, one ought to be familiar with the surface structure which is basically similar to the internal structure of the CF. The early model, which represented the CF as a conglomerate of highly oriented interconnected undulating submicron size ribbons [16], has been modified substantially in view of the extensive studies of DIEFENDORF and coworkers [17], the plasma etching work of BARNET and NORR [18], and the identification of phase heterogeneity, including the possible presence of a three-dimensional graphite phase [19, 20]. According to this the high modulus CF contains at least two phases, both comprised of turbostratic graphitic ribbons, plus three-dimensional graphite inclusions when the fiber is heat treated above 2000 °C. One of the turbostratic phases consists of wider, thicker, more axially oriented ribbons, and the other one of narrower, less oriented, intertwined ones (fibrils). Both phases contain numerous microvoids, microcracks, and other lattice faults. The less oriented material is present primarily in the central portion of the fiber ("core"), while the more perfect ribbons are located on the outside ("sheath") with a mixed region in between. A current model of such a CF is shown in Fig. 1a, b. Although the surface is the most highly oriented part of the fiber, it contains exposed edge planes as well as basal planes. The sp^2 carbons comprising the high energy planes have a free bond and, therefore, will chemisorb oxygen eagerly to form a very strong carbon-oxygen complex [21] that can bond further with the binder resin. The low-energy basal planes that interact by the much weaker π bonding will not form strong bonds at the surface. In addition, the surface exhibits a more or less fibrillar microtexture, microporosity, cleavage cracks, crystallite boundaries, foreign inclusions or impurities, and fracture-inducing flaws. All of these features will affect the fiber-resin adhesion, and therefore, require characterization in order to assess their significance.

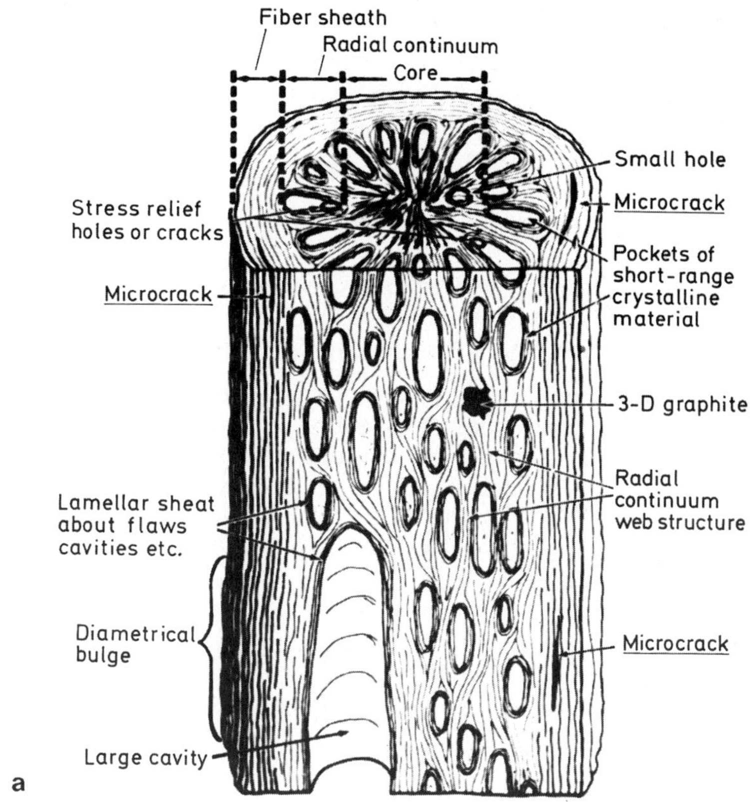

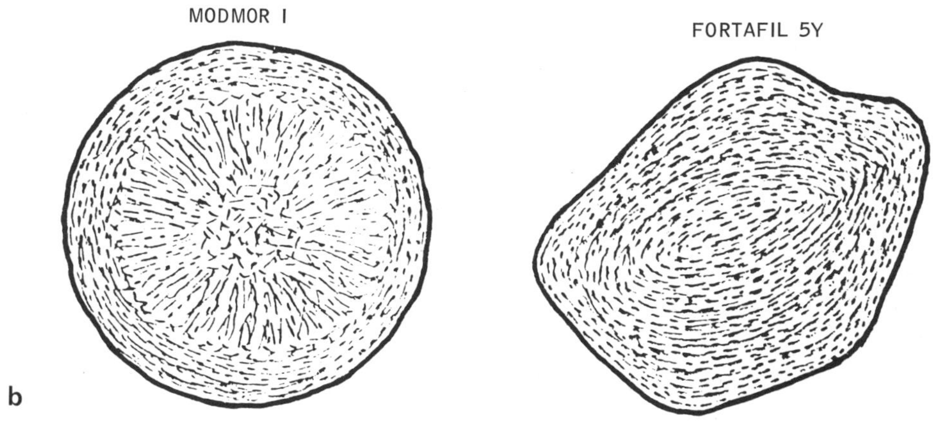

Fig. 1a. Model of axial section of a high modulus ex-PAN fibre [18]. **b** Model of cross sections of some ex-PAN fibres [17]

4 The Surface Properties of CF

A complete characterization of the CF surface would require the knowledge of:

1. the surface composition, i.e. the quantity and nature of the surface atoms and their reactivities,
2. the surface energy and its change upon sorption of the gas or liquid species of interest,
3. the specific surface area, rugosity (roughness), and accessible microporosity,
4. the surface flaws and their distribution and
5. the fiber shape, size, surface texture and anisotropy.

Unfortunately, much of the published characterization has been quite limited in scope, the pertinent information being summarized below.

After an oxidative treatment, the surface oxygen content increases slightly, to ca. 100–300 μg/g of C in case of HT-type carbon fibres and to ca. 10–40 μg/g of C for HM type C-fibres (Fig. 2) [57]. Part of the oxygen is chemisorbed very strongly onto the most energetic edge carbon atoms to give the before mentioned surface complexes, the rest forming both acidic groups, or neutral groups, most likely lactone and hydroxyl ones. These surface oxides can be determined by desorption analysis. Their composition depends on the applied oxidation process (see Table 1) [59]. As mentioned, both –C=O and –C–O– configurations have been identified by ESCA. The portion of the highest energy surface, likely to give the strongest adhesion,

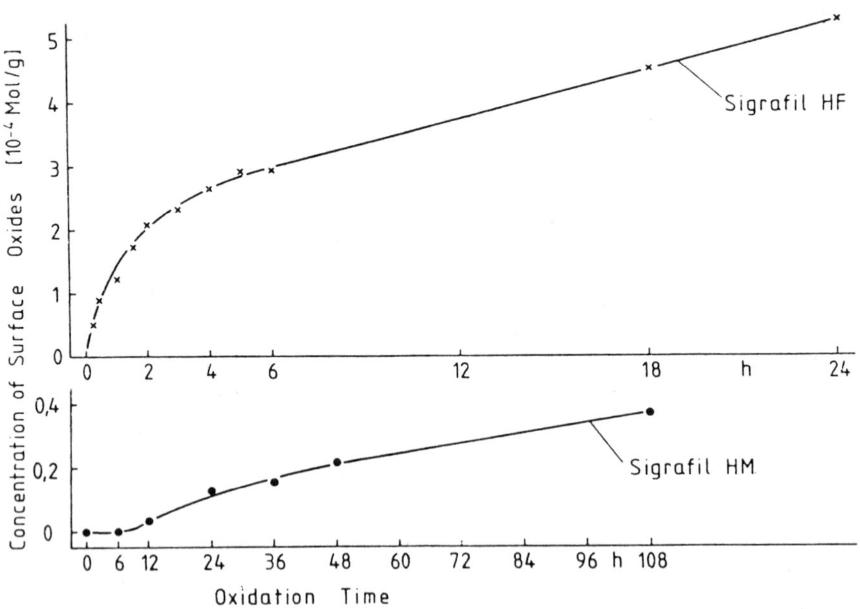

Fig. 2. Amount of surface oxides on HM- and HT-C-fibres with increasing oxidation time in 68% HNO_3-solution

Table 1. Comparison of desorption products of dry-, wet- and anodic oxidized HM-C-fibres with the untreated one

Surface Treatment	Desorption Products [μmol/g]				
	H_2O	CO_2	CO	CH_4	C_2H_6
Untreated fibre	20.6	1.7	1.1	0.5	–
Dry (air)	17.4	1.9	3.2	1.7	1.1
Wet (HNO_3)	11.9	5.5	10.9	1.0	0.3
Anodic	20.3	12.9	19.6	2.5	0.8

decreases sharply with increasing fiber modulus. Thus, for a 250 GPa modulus fiber the values are ca. 4% of the total SSA increasing to 10% after the manufacturer's surface treatment, whereas, for a similar 400 GPa modulus one, the corresponding change is from ca. 0.5 to 1–3% of the SSA [22]. Judging from the concurrent liquid adsorption studies [23], both types of oxygen-containing groups, the "active complex" and the acidic one, are capable of bonding further to polar organic molecules, such as butylamine and presumably to other epoxy hardeners. However, these groups occupy only ca. 10–12% and 20–25% of the total SSA of the high (400 GPa) and medium (250 GPa) modulus ex-PAN fibers, respectively, Table 2. For the ex-rayon CF, the percentage may be slightly larger.

It is evident that the majority of the CF surface atoms must bond to the resin through low energy interactions, such as, dispersion (London), dipole-dipole, dipole-induced dipole, electrostatic, π-bonding, hydrogen bonding, or donor acceptor interactions [27] which for CF have not been analyzed. A method based on contact angle measurements of the solids against a series of reference liquids, permitting the separation of the total surface energy into a polar and dispersive component [28] has been applied to type A (both treated and untreated) HTS, HMS, and

Table 2. Chemical characterization of CF surfaces

Fiber Trademark & type	Specific Surface Area (SSA) (m^2/g)	"Active" Surface Area[a] (% of SSA)	Acid group Content (% of SSA)	Acid group Content (μequiv./g)	Oxygen Content (μg/g)
Torayca T-300 [24]	0.55		20	13	400
Torayca M-40A	0.4		10	7	220
Rigilor AC (no S.T.) [25]				7	
Rigilor AC (S.T.)				16–20	
Grafil A-U [22]	0.48	4 ± 1			
Grafil A-S	0.56	10 ± 4			
Grafil HM-U	0.40	0.5			
Grafil HMS	0.42	2 ± 1			
Thornel 50-S [26]	1.1	7	26		
Celion GY-70	0.75				200

[a] Surface area occupied by oxygen adsorbed at 340 °C on samples after vacuum degassing at 950 °C

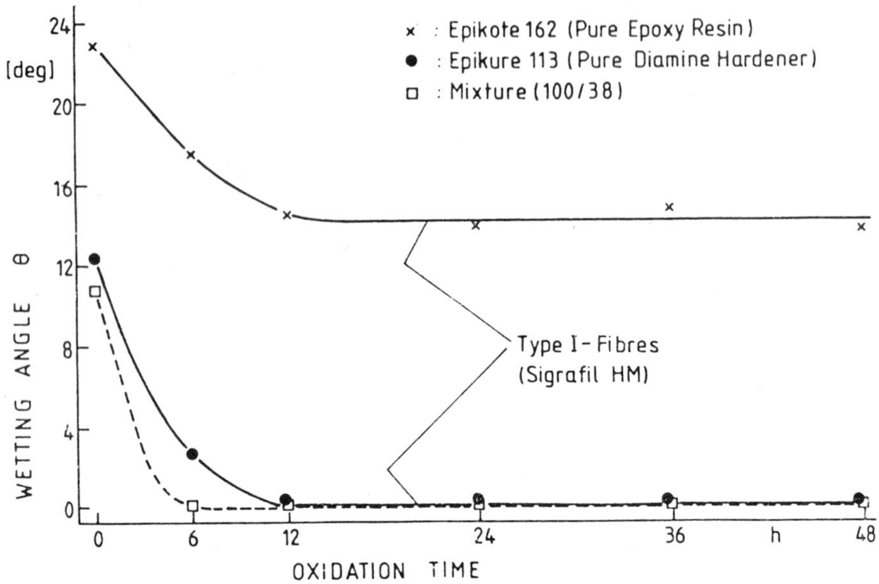

Fig. 3. Influence of oxidation time (68% HNO_3) on wetting angle of carbon fibres in various solutions of epoxy formulation components

Modmor II CF [29]. The resulting total surface energies were rather low and about the same, 53–57 mJ/M^2, with the polar, and dispersive contributions divided about evenly.

Furthermore, the various components of an epoxy formulation have different wetting behaviour as shown in Fig. 3 [58]. Pure amine hardener has an excellent wetting behaviour (wetting angle of 0° is achievable), hardener free epoxy shows wetting angle up to 25°. The wetting behaviour of hardener/epoxy mixtures is strongly influenced by the hardener components. Increasing amount of acidic surface oxides on the carbon fibres, as obtained by oxidation treatment is a precondition for such excellent wetting.

Dimensional properties, primarily the SSA, and the micropore volume, are normally measured by low temperature sorption of nitrogen, argon, or krypton. Unless the CF has been vacuum degassed at ca. 900 °C, the micropore volume is small, but can increase substantially when treated with some oxidants, like HNO_3 [22]. For most commercial ex-PAN fibers, the SSA ranges from 0.3 to 0.7 m^2/g; for the ex-rayon ones – 1 to 2 m^2/g. Finally, the combined effect of the SSA and the chemical functionality can be assessed by measuring of the surface charge density [30] or surface capacitance [31] of CF immersed in aqueous electrolyte solutions.

On basis of the surface properties, the surface treatments can be divided into

1. those that increase the oxygen content and chemical functionality with little or no increase in the SSA, and
2. those that increase both the SSA and the oxygen content.

The second kind of treatment is typically applied to the high modulus fiber when a relatively large, 3–4 fold, increase in their strength is desired.

5 Flaws

Since the strength of CF has been correlated with its flaw content and severity, the appearance of the flaws has been studied extensively, primarily by electron and optical microscopy, e.g. [32], also references cited therein. The flaws are found both in the interior and at the surface. The interior flaws that are mostly particulate inclusions, round or needle-like voids, and a combination of both, are related to the quality of the precursor fiber [33]. The surface flaws on the other hand are cracks and grooves, apparently incurred during the fiber manufacture, small highly graphitic "microflakes", as well as the above mentioned inclusions and microvoids located at the surface. Surface treatments, capable of reducing the severity of the surface flaws to raise the mean fiber tensile strength by up to 50%, have been reported [34]. The effect of these flaws manifests as a distribution of CF fracture strength that can be described statistically, as has been done for fiberglass [35].

6 The Effect of CF Surface Condition on Composites' Fracture

The interfacial fracture behavior is determined not only by the surface structure of both components, but also by any additional structural features introduced during the interface formation. Although the surface of the epoxy resin binder, after contact with the CF surface, apparently has not been investigated, studies on other polymers, nylon or PE, show that these form epitaxial, highly oriented, strongly adherent coatings on CF surfaces [36]. Such coatings are likely to have mechanical and chemical properties different from those of the bulk resin to change the expected residual stress pattern as well as the adhesive strength. Furthermore, proprietary binder systems are often modified with reactive additives to increase the fracture toughness, to change the T_g, improve the wettability, etc. Also, many CF grades sold have been prewetted with proprietary sizing agents by the manufacturer in order to facilitate the fabrication of preimpregnated goods ("prepregs"). In addition, the interface may contain numerous microvoids (1–3 fiber diameters in size), caused by incomplete wetting as well as large microvoids formed by coalescence of the former. Any of the above phenomena will affect the interfacial adhesion strength and make the contribution of the original CF surface more difficult to assess.

7 Shear

In composites, the effect of the surface treatment manifests itself most dramatically as a greatly enhanced shear strength. About 30–40% of the total increase appears to be due to the removal of the original weak boundary layer on the as-manufactured surface, and the rest is caused by a) the strong bonding of the acidic oxygen-containing groups and the weaker bonding of the basal oxygen-containing groups and atoms with the monomers of the resin, and b) increase surface rugosity to give a larger SSA to bond with that arises from preferential oxidation of the edge crystallites, exposing additional edge and basal planes. For the medium modulus (250 GPa) CF, having a relatively large edge carbon content, 20–30% of total surface, a mainly adhesion-promoting surface treatment may be adequate, whereas for the higher modulus (>400 GPa) CF a treatment increasing the rugosity is usually necessary in order to obtain acceptable shear strength. A larger than 2–3 fold increase in the SSA, however, is likely to result in reduced tensile strength due to increased severity of the etched out surface flaws.

As the rugosity increases, the interfacial stress which is pure shear on a perfectly smooth surface becomes a combined stress of shear and tension, or shear and compression when acting on the edgewise or basally exposed crystallites inclined at different angles with respect to the fiber axis. Consequently, the difference between the shear adhesion and tensile adhesion strength ought to increase with increasing rugosity of the fiber surface. Examination of manufacturer's data indicates that the difference between the composite short beam shear strength and transverse tensile strength is indeed the least for the more rugous high modulus CF, such as Celion GY-70 (30–40 MPa), becoming larger for the less rugous medium modulus type II CF (40–55 MPa) and still larger for the S-glass reinforced epoxy, (50–70 MPa). Although these data, are obtained on composites comprised of different epoxy systems and tested under different conditions, they may nevertheless be indicative of the trend. Consequently, the differentiation between the bond strengths in tension and in shear maintained in some micromechanical models [37] is less meaningful for CF with high rugosity. Among the techniques that have been proposed and used to measure the interlaminar (ILSS) shear strength, the so-called "horizontal shear strength" test, ASTM D 2344, also called "short beam shear" (SBS), is by far the most popular because of the extreme simplicity in specimen preparation and testing.

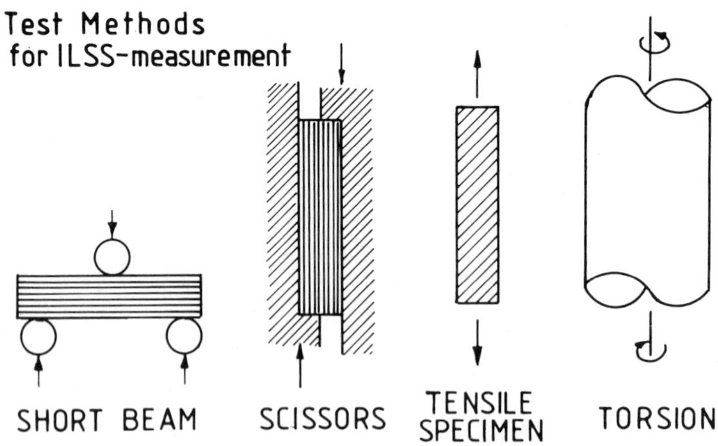

Fig. 4. Test methods for shear strength of oriented CF composites

Unfortunately, it is highly unsatisfactory theoretically [38], and is recommended for quality control purposes only. Nevertheless, its use persists, and shear data on other than short beam shear specimens are rather scarce in the literature. Other techniques, utilizing oriented composite test specimens, Fig. 4, are available, but none has been unequivocally accepted. New methods are being proposed continually, e.g. [39], and may eventually replace the SBS method. Typical SBS strength values range from 55–70 MPa for the high modulus (> 400 GPa) fibers, and 100–125 MPa for the medium modulus (250 GPa) ones, as compared to ca. 15 and 50 MPa, respectively, before the surface treatment. Intermediate values may be obtained, if desired, by manipulating the surface treatment process conditions.

Methods to measure the shear adhesion of single filament interfaces, such as the shear debond method [40], or the fiber pullout [41], Fig. 5, used successfully on fiberglass, have been less successful with the CF. The first method caused premature fiber fracture [42], while the second one gave shear strengths of 5 and 17 MPa for untreated HM and HT fibers, respectively, and 31 and 57 MPa for the treated ones [43] – about half of their respective SBS shear strength values. Taking into account the inverse dependence of the pullout strength on the embedded length might possibly give higher shear values [44]. Another approach utilizes measurement of the fragment lengths of fractured resin-embedded filaments to determine the so-called "critical length" (ineffective length, shear transfer length), L_c, to calculate the maximum shear adhesion stress τ [45]. Since the CF filament strengths show a broad distribution, the fragmented lengths will also reflect this distribution, so that the determination of the L_c and hence, the τ becomes quite complicated. Despite that, two estimates of L_c are available, both for the RAE type I CF (now discontinued). For the treated fiber the L_c was ~0.1–0.14 mm [46], or ~0.14 mm, as compared to ca. 1.2 mm for an untreated one [47]. The calculated shear adhesion strengths were 47 MPa and 6 MPa respectively, again substantially lower than similar SBS values. When comparing the shear data obtained on single filament and composited fila-

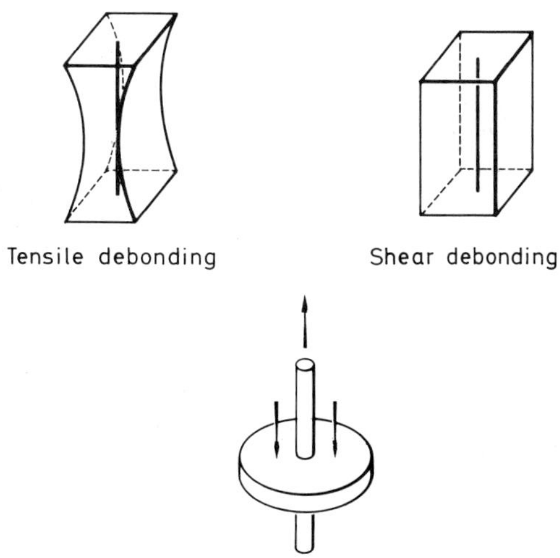

Fig. 5. Test methods for filament-resin adhesion strength

ment bundles, it should be remembered, however, that the axial shear adhesion depends also on the filament misalignment [48], the void content [49] and the chemical nature of the binder. The misalignment is expected to increase the apparent shear, the voids will decrease it, whereas the effect of the binder is often quite specific and may be either beneficial or detrimental.

8 Transverse Tension

The ultimate tensile strength of unidirectional composites increases after a typical (manufacturer's) surface treatment from ca. 15 to 30 MPa (to 0.35–0.40% ultimate strain) for the $>$400 GPa modulus fiber composites and from ca. 20 to 60 (to 0.5–0.6% ult. strain) for the 250 GPa modulus ones. The increase is due to the same causes as the increase in shear, i.e. removal of the weak boundary layer and strengthened bonding between the functional groups and the binder. The tensile adhesion, however, is affected by residual microstresses caused by the thermal shrinkage and the cure of the resin. For high fiber contents, the estimated residual stress is compressive in the direction of adjacent filaments, and tensile in between, [50], imparting a non-uniform adhesive stress circumferentially. Also, microvoids and matrix strain magnification effects can contribute substantially to the transverse failure [51]. However, in view of the low observed fracture strain values that are 6–10 times smaller than the bulk fracture strain of typical epoxy resins, the transverse fracture must start at interface. Unfortunately, detailed studies of the nature of the interfacial tensile adhesion in CF composites are lacking. Attempts to measure the tensile debonding strength of embedded single filaments failed owing to premature fracture [42]. Consequently, the transverse tensile strength data of unidirectional composites is currently the only means to assess the tensile adhesion strength.

9 Axial Tension

Although surface treatments have been reported to increase the fiber tensile strength, ordinarily the effect is small, 10% or less, and has little significance for increasing the composite axial tensile strength, S_a. Without a surface treatment, however, the fracture strength is relatively low, because as the applied stress increases the weak fiber-resin interface debonds and fails prematurely. When the composite shear strength is at or above a certain fraction, ca. 5%, of the S_a, the latter is independent of the shear level. At still higher shears, 8–10% of the maximum S_a, the strength decreases, not only because of possible surface damage by over treatment, but also by the inability of the fibers to fully dissipate their fracture strain energy into the pullout, resin cleavage, or debonding modes [45], resulting in more extensive microfracturing and subsequent failure at a lower applied stress.

10 Compression

Axial compression studies on untreated unidirectional CF composites indicated a need for increased interfacial bonding, since the loaded filaments were observed to undergo transverse tensile debonding simultaneously with filament fracture [52], but the optimal adhesion strength level is not known. Voids, residual stresses, fiber misalignment and the intrinsic fiber structure, however, exert a substantial effect on the compressive strength [53], and may negate attempts to assess the effects of increased adhesion. No experimental studies were found in the literature concerning the effect of interface on the transverse compression.

11 Impact and Fracture Toughness

Charpy impact resistance studies of uniaxial HM-CF specimens showed that high shear (manufacturer's surface treated) specimens have very low impact resistances and a SPS strength of 50–60 MPa (intermediate adhesion) gives the best balance between the overall failure characteristics [54] (Fig. 6). Such behaviour can be explained by crack propagation from fibre to fibre in case of good adhesion, whereas weaker adhesion will cause delamination and thus energy consumption and deviation or stop of the crack propagation [61]. In case of HT carbon fibre reinforcement

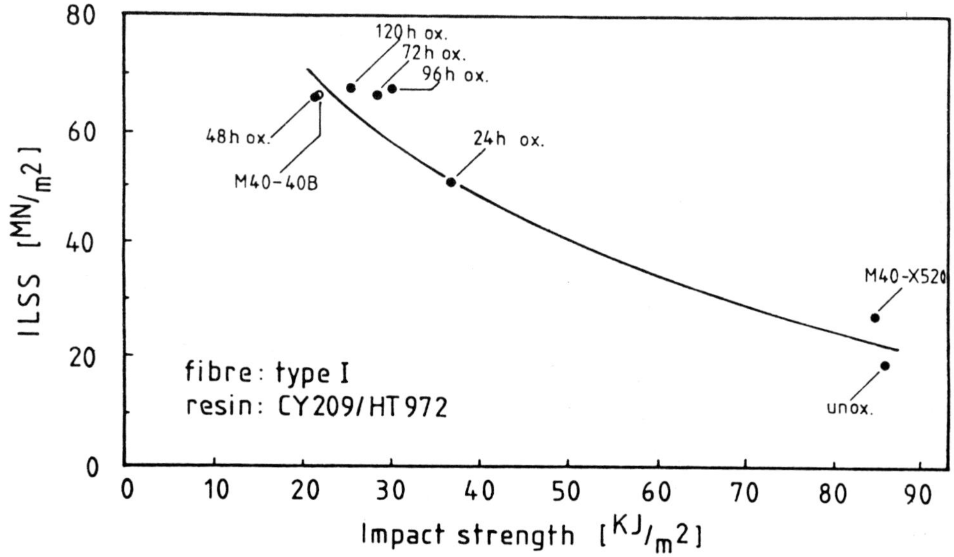

Fig. 6. ILSS of various surface treated HM-C-fibre (PAN based) reinforced epoxy composites as a function of impact strength

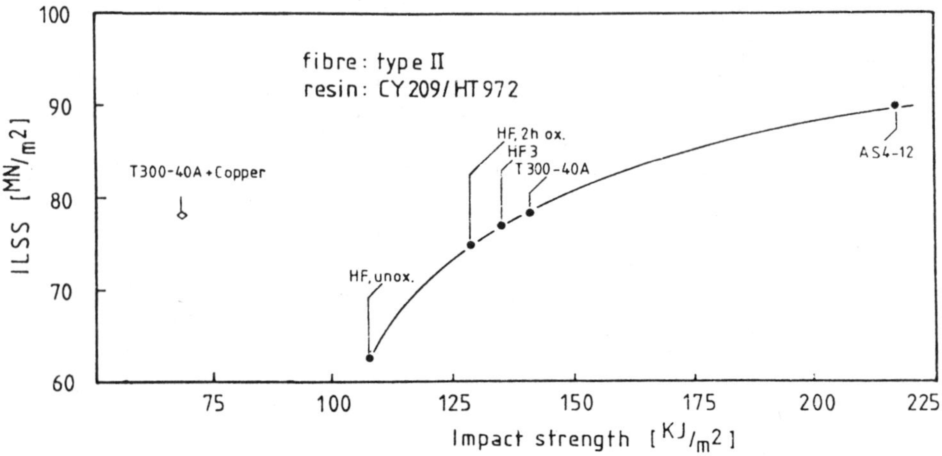

Fig. 7. Correlation between ILSS and impact strength of various HT-C-fibre (PAN based) reinforced epoxy composites

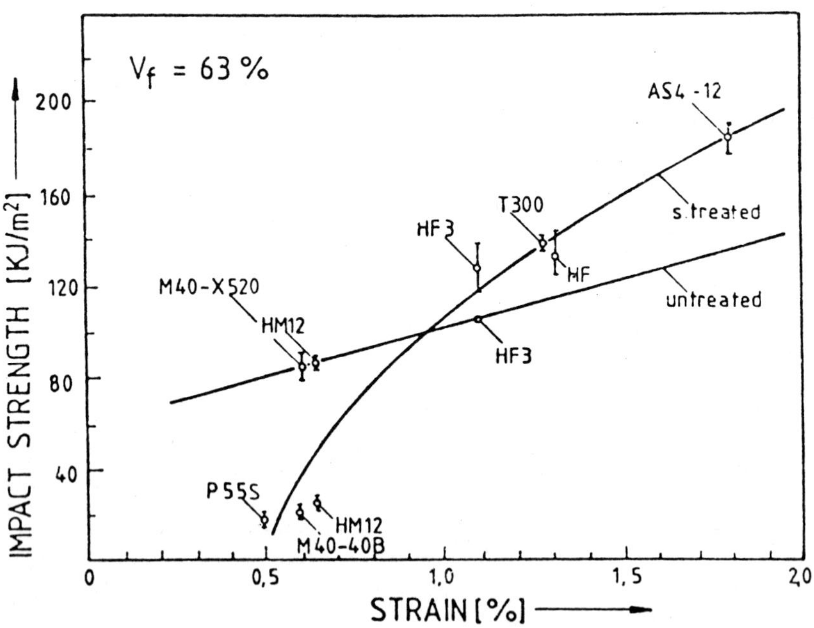

Fig. 8. Impact strength of UD composites increases with increasing strain to failure of the reinforcing fibres

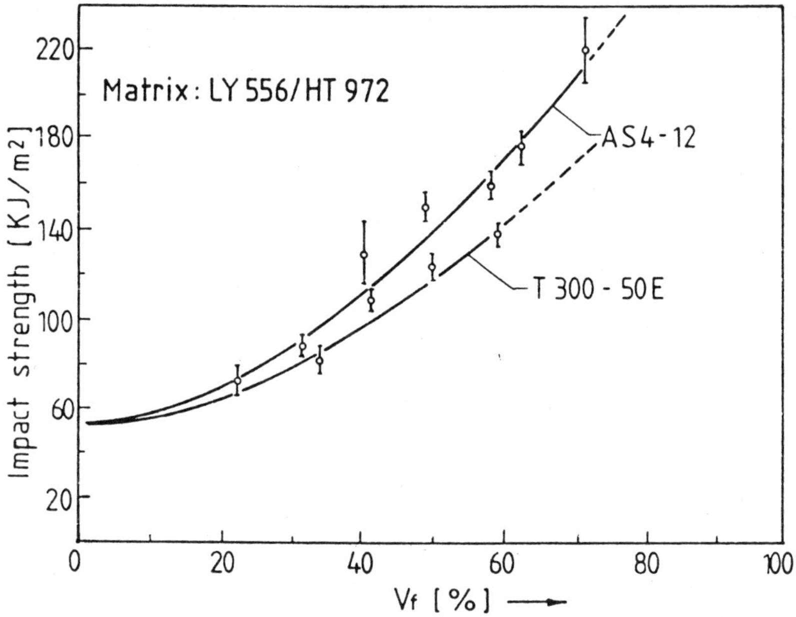

Fig. 9. Increase of impact strength of HT-C-fibre reinforced UD composites as function of volume fraction of carbon fibres

improved adhesion has been found to improve impact strength (Fig. 7) [60]. This controversial behaviour is explained by the strain behaviour of the fibres. In case of strain to failure values higher than 1% (see Fig. 8) increased shear strength will increase impact resistance because of conversion of fracture energy into elastic deformation of the high strain fibres [62, 64]. Also the fibre content was found to be important for the impact toughness of the UD composites (see Fig. 9). With increasing fibre content impact strength is increased in case of high strain fibres [63]. Further studies of fracture toughness of uniaxial composites, made with type I RAE CF (now discontinued) showed that at a 60 vol.% fibre content, the K_{Ic} of surface treated fibre composites was about half of that of untreated ones, ca. 25 vs. 45 MN · $m^{-3/2}$, but that at lower fibre contents the difference was greater [46].

12 Summary and Conclusions

A survey of surface treatments for carbon fibers shows that the predominating ones are those involving surface oxidation. The state-of-the art treatments, applied routinely to commercial CF, provide 2–4 fold increases in the fiber – resin adhesion strength, both in shear and in tension. In composites, this adhesion normally is fully adequate for obtaining satisfactory axial tensile performance. However, it is too high for optimal impact resistance or fracture toughness, and too low for satisfactory transverse tensile strength. Consequently, in order to utilize effectively any further increases in the interfacial adhesion, the CF composites should be toughened first, e.g. by hybridization. Alternately, the transverse tensile strength might be increased by development of tougher and more compatible binder resins. Some promising approaches toward this objective have, in fact, been recently reported [55, 56].

13 References

1. Anon. "Graphite Composites from Space to Auto", Automot. Ind. (1976), *154,* (9), p. 39
2. Clark, D., Wadsworth, N. J., Watt, W.: *Carbon Fibres* (Plastics & Poly. Conf. Suppl. No. 6), The Plastics Institute, London (1974), pp. 44–51
3. Amon, F. H.: US Pat. 2,439,442 (1948), April 13
4. Johnson, J. W.: J. Appl. Poly. Sci., Applied Poly. Symposia (1969), *9,* p. 229–243; Cunningham, A. L., US Pat. 3,816,598 (1974), June 11
5. Molleyre, F., Bastick, M.: 4th London Intern. Conf. on Carbon & Graphite, Sept. 23–27, 1974, Society of Chemical Industry, London (1976), pp. 190–200
6. Kalnin, I. L.: US Pat. 3,723,607 (1973), March 27
7. Goan, J. C.: US Pat. 3,634,220 (1972), Jan. 11; US Pat. 3,776,829 (1973), Dec. 4
8. Wadsworth, N. J., Watt, W.: US Pat. 3,476,703 (1969), Nov. 4
9. Barr, J. B.: US Pat. 3,791,840 (1974), Feb. 12
10. Goan, J. C., Joo, L. A., Sharpe, G. E.: Proc. Ann. Tech. Conf. of RP/Composites Inst., The Society of Plastics Ind., New York (1972), vol. 27, 21-E
11. Ray, J. D., Steingiser, S., Cass, R. A.: US Pat. 3,671,411 (1972), June 20
12. Chapman, D. R., Paterson, W. C.: US Pat. 3,759,805 (1973), Sept. 18
13. Shyne, J. J., Milewski, J. V.: loc. cit. in [10] (1969), vol. 24, 18-D
14. Pinchin, D. J., Woodhams, R. T.: J. Mater. Science (1974), *9,* pp. 300–306
15. Riess, G., Bourdeaux, M., Brie, M., Jouquet, G.: loc. cit. in [2], pp. 52–7
16. Fourdeaux, A., Perret, R., Ruland, W.: *Carbon Fibres* (Plastics & Poly. Conf. Suppl. No. 5), The Plastics Institute, London (1971), pp. 59–67
17. Diefendorf, J. R., Tokarsky, E.: Poly. Eng. Sci. (1975), *15,* (3), pp. 150–9; Lemaistre, C. W., Diefendorf, R. J: SAMPE Quart. (1973), *4,* (4), pp. 1–6
18. Barnet, F. R., Norr, M. K.: loc. cit. in [2], pp. 32–43
19. Johnson, D. J., Crawford, D., Jones, B. F.: J. of Mater. Sci. (1973), *8,* pp. 286–9
20. Wicks, B. J.: ibid. (1971), *6,* pp. 173–5
21. Laine, N. R., Vastola, F. J., Walker, P. L., Jr.: J. Phys. Chem. (1963), *67,* pp. 2030–4
22. Rand, B., Robinson, R.: Carbon (1977), *15,* pp. 257–63
23. Idem, pp. 311–315
24. Nichiokat, Murayama, K., Matsubara, I.: (in Jap.), 1st Symposium on Composite Materials, Tokyo, 1975, Sept. 19–20
25. Ehrburger, P., Herque, J. J., Donnet, J. B.: loc. cit. in [5], pp. 201–8
26. Moller, P. J., Fort, jr., T.: Coll. & Poly. Sci. (1975), *253,* pp. 98–103
27. Fowkes, F. M.: J. Adhesion (1972), *4,* pp. 155–9
28. Kaelble, D. H.: in *Physical Chemistry of Adhesion;* Inter-Science, New York (1971), pp. 139–170
29. Kaelble, D. H., Dynes, P. J., Cirlin, E. H.: J. Adhesion (1974), *6,* pp. 23–48
30. Dietz, R., Peover, M. E.: J. Mater. Sci. (1971), *6,* pp. 1441–6
31. McLean, A. F., Kalnin, I. L.: US Pat. 3,864,626 Feb. 4, 1975.
32. Sharp, J. V., Burnay, S. G., Matthews, J. R., Harper, E. A.: loc. cit. in (2), pp. 25–31
33. Thorne, D. J.: J. Appl. Poly. Sci. (1970), *14,* pp. 103–113
34. Johnson, J. W.: J. Appl. Poly. Sci. – Symposia (1969), No. 9, pp. 229–43

35. DONNET, J. B.: Carbon (1968), *6,* pp. 161–176
36. BAER, E., KOENIG, J. L., LANDO, J. B., LITT, M. H.: loc. cit. in [10] (1971), vol. 26, 20E
37. SKUDRA, A. M., KIRULIS, B. A.: Mekh. Polimerov (1974), *10,* pp. 246–251 (Russ.)
38. BERG, C. A., TIROSH, J., ISRAELI, M.: in *Composite Materials:* Testing & Design, 2nd Conf., ASTM Spec. Tech. Publ. 497 (1972), pp. 206–218
39. CHAMIS, C. C., SINCLAIR, J. H.: Exp. Mech. (1977), *17,* pp. 339–46
40. BROUTMAN, L. J.: loc. cit. in [10] (1970), vol. 25, 13-B
41. ANDREEVSKAYA, G. D., GORBATKINA, Y. A.: ibid., paper, 16-F
42. HAWTHORNE, H. M., TEGHTSOONIAN, E.: J. Adhesion (1974), *6,* pp. 85–105
43. FAVRE, J. P., PERRIN, J.: J. Mater. Sci. (1972), *7,* pp. 1113–8
44. ANDREEVSKAYA, G. D., GORBATKINA, Y. D., IVANOVA-MUMZHEIRA, V. G., EPIFANOVA, S. S., Mekh. Polimerov (1974), *10,* pp. 37–42 (Russ.)
45. KELLY, A.: *Strong Solids,* 2nd ed., Clarendon Press, Oxford, G.B. (1973), p. 172 et ff.
46. BEAUMONT, W. R.: J. Adhesion (1974), *6,* 107–37
47. WADSWORTH, N. J., SPILLING, I.: Brit. J. Appl. Phys. (J. Phys. D.) (1968), sec. 2, vol. 1, pp. 1049–58
48. CLAUS, W. C.: loc. cit. in [10] (1972), vol. 27, 9-C
49. HANCOX, N. L.: J. Mater. Sci. (1977), *12,* pp. 884–92
50. CHAMIS, C. C.: *Composite Materials,* vol. 6, PLUEDDEMANN, E. P., ed., Academic Press, New York (1974) p. 55f.
51. GRESZCZUK, L. B.: *Composite Reliability,* ASTM Spec. Tech. Publ. 580 (1975), pp. 311–25
52. HANCOX, N. L.: J. Mater. Sci. (1975), *10,* pp. 234–42
53. HAWTHORNE, H. M., TEGHTSOONIAN, E.: J. Mater. Sci. (1975), *10,* pp. 41–51
54. BADER, M. G., BAILEY, I. E., BELL, I.: Brit. J. Appl. Phys. (J. Phys. D) (1973), *6,* pp. 572–86
55. STEVENS, G. T., LUPTON, A. W.: J. Mater. Sci. (1976), *11,* pp. 568–70
56. MCGERRY, F. J., ROWE, E. H., RIEW, C. K.: loc. cit. in 10 (1977), vol. 32, 16-C
57. FITZER, E., WEISS, R.: Processing and uses of carbon fibre reinforced plastics (1981), VDI-Verlag, ISBN 3-18-404074-7
58. FITZER, E., HÜTTNER, W., WEISS, R.: Conference on Newer Frontiers of Polymers, Madras (7. 1. 1980)
59. FITZER, E., WEISS, R.: Intern. Conf. on Interface-Interphase in Composite Materials, Liège-B (16.–20. 10. 1983)
60. FITZER, E., JÄGER, H., WEISS, R.: Extended Abstracts, Carbon Conf., San Diego (1983), pp. 471–472
61. HANCOX, N. L.: Composites, March (1971), pp. 41–45
62. FITZER, E., JÄGER, H.: Int. Wehrtechnisches Symposium CFK, Erding-FRG (8.–10. 5. 1984), Z. Werkstofftech. *16,* 215–222 und 232–238 (1985)
63. FITZER, E., JÄGER, H.: ACS-Symposium "Carbon and other High Performance Fibres", Miami Beach (29. 4.–1. 5. 1985), Composites, in press
64. FITZER, E.: Technical Status and Future prospects of carbon fibres and their application in composites with polymer matrix (CFRP's), see page 25 this volume

Chapter 2

Polymer Matrices

Effect of Resin Choice on Prepreg Processing

B. J. Hayes

A wide range of resin chemical types, fibres and processing techniques are used in the multitude of products manufactured under the general description of fibre composites. This paper considers those resins currently suitable for making the highest strength laminate and sandwich composite structures for applications up to a maximum temperature resistance of 130 °C. This represents some 85% of the total market and hence is easily the most important sector. The resins are mostly used with the fibres in a preimpregnated form (prepreg).

The objective is achieved in two stages:

i) The range of matrices available is briefly considered and reduced to those best adapted to the stated purpose.
ii) More detailed discussion is given to the processing aspects of the selected class.

1 Matrix Materials Selection

Resins are generally divided into thermoplastics and thermosets, of which the following could be considered representative members:

- thermoplastics – nylon, polyetheretherketone (PEEK), polybutylene terephthalate, polycarbonate, polyethylene, polysulphone;
- thermosets – epoxy, melamine, phenolic, polyester, polyimide, ureas.

Most thermoplastic resins can be eliminated because of inadequate mechanical performance at high temperatures, as well as in many cases susceptibility to, or even solubility in, commonly used solvents. Those speciality thermoplastics which have adequate temperature resistance, although not necessarily fully solvent resistant, should also be removed because the majority of the currently available autoclaves and presses required for processing prepregs have a maximum temperature limitation of around 200 °C. These speciality thermoplastics require a forming temperature well in excess of 300 °C. This constraint will no doubt be remedied in the future if these materials continue to show promise.

Polyester thermosets are omitted because their higher curing shrinkage leads to greater internal stress generation and hence poorer mechanical properties – especially with carbon and aramid fibres. Ureas, melamines and phenolics generally require rather high forming pressures and, although excellent in many applications, do not give particularly attractive performance with these same high performance fibres. Nevertheless, phenolics are still one of the best choices where low smoke, low flammability and low toxic gas evolution are an additional essential requirement.

As stated at least 85% of the applications for high performance composites are satisfied by a maximum use temperature of approximately 130 °C. Most of the military aircraft requirements are below this level and the civil aircraft, general industrial and sports goods users rarely demand anything higher than 93 °C. This group is well catered for by the epoxy resins which are available in a wide range of formulations to satisfy the majority of applications and which can be processed readily in most existing equipment. However, there will always be special requirements for prolonged performance significantly above 130 °C. It is in this area that the formulated bismaleimides may become of value, with polyimides of the PMR and LARC type taking over when long term performance above 180 °C is required. The majority of prepregs based on these higher temperature resin systems are certainly capable of being processed into complex parts, but demand great care and attention when doing so, as they are not nearly as forgiving or simple in manufacture as the formulated epoxies.

2 Processing Aspects of Epoxy Matrix Resins

In about 80% of all applications the starting material for part production is a prepreg.

a) What are prepregs?
b) Why are they so important? What are their advantages?
c) Are they all the same? What are the differences?

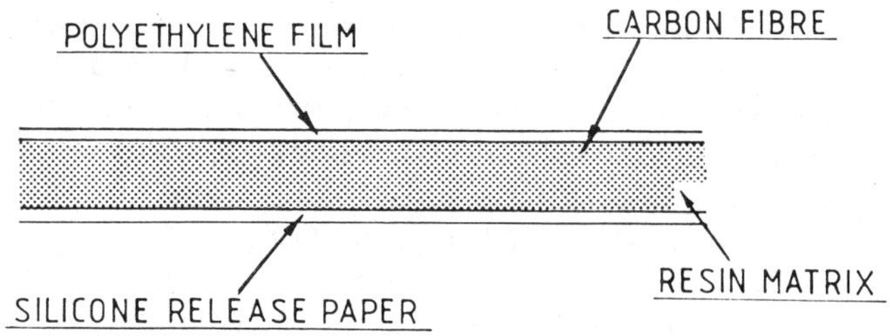

Fig. 1. Schematic representation of a unidirectional prepreg

a) A prepreg is a mixture of unidirectional or woven fibre and a preformulated resin in a particular weight ratio protected against contamination on one or both sides with easily removed separators.

Unidirectional prepregs (see Fig. 1) have a standard width of 300 mm but can be obtained from very narrow tapes (e.g. 5 mm) up to 1000 mm in special cases. The resin content and cured ply thickness can be varied as required but typical figures would be 34% resin and 0.125 mm thickness. They are produced continuously and usually purchased as a 250 metre long reel.

Woven prepregs are manufactured at standard widths of 1 m and 1.2 m. The resin contents and weave styles may be varied as necessary, but typical would be a 5 harness satin weave fabric with 37% resin, moulding to approximately 0.28 mm per ply.

They are normally purchased with one or two interleaving protective layers at a length of about 50 m on a spool. This represents a handleable weight.

b) The importance of prepregs and their advantages are:

i) They are the only practical way of handling true unidirectional fibre arrays
ii) They give precontrolled resin to fibre ratio
iii) They give predetermined total weight
iv) No mixing of resins/hardeners/catalysts is required
v) Tack and drape are designed into the product to ease the lay-up process
vi) Pot life at room temperature will be weeks to months as opposed to a few hours for wet resin processes
vii) Storage life deep frozen is months to years
viii) Viscosity of the resin can be accurately controlled giving rise to easier processing, low porosity laminates and good bonding to honeycomb
ix) A wide range of cure cycles and mechanical properties can be obtained
x) Some prepregs with net resin contents can be easily handled leading to much reduced cost
xi) All handling and toxicity hazards are reduced
xii) Assurance of resin/fibre performance can be given by Q.C. testing before component fabrication

c) Are they all same? No they are not. All prepregs possess some of the advantages listed above, but some possess the majority of them.

Selection of the best epoxy prepreg for the fabrication and use of any given item can only be made satisfactorily after the following features have been considered.
- nature of the part to be produced
- mechanical performance required
- equipment available for fabrication
- cost

The most important factors relating to these are outlined below

Nature of the Part to be Produced	– *Consideration*
Simple unidirectional flat laminate	– handleability? suitable for tape laying?
Multidirectional reinforcement	– tack? fabric? suitable for tape laying?
Multicurvature	– level of tack? fabric? adhesion? porosity? surface appearance?
Honeycomb sandwich	– controlled flow? adhesion? porosity? surface appearance?
Size and complexity	– out life? tack? consistent flow?
Thick or thin section – or both	– cure reaction rate? zero bleed?
Mechanical Performance Required	– *Consideration*
Basic strength and moduli	– adequate to make composite construction worthwhile?
Temperature limits	– are these really known and realistic?
Environmental performance	– What property loss when exposed to use conditions?
Adhesion to honeycomb	– is separate adhesive film necessary?
Toughness	– suitable for likely damage environment?
Fatigue	– will it perform over the full life intended?
Flammability	– will fire hazard be a problem? self-extinguishing?
Equipment Available for Fabrication	– *Consideration*
Autoclave	– versatile but expensive to purchase and operate?
Press – matched tool	– probably less expensive plant but tooling often very expensive!
Ovens – vac bag, expanding rubber	– cheapest method, when feasible!
Tape laying	– save labour costs where suitable!

Cost

Although the price of prepreg is important many other factors must be taken into account as some of them will have a much greater impact on the cost of the finished item. These include:

i) shop floor handleability and shelf life
ii) need for removal of excess resin

iii) type of tooling required
iv) complexity of moulding process
v) reproducibility
vi) scrap and rejection rates

There are a wide range of matrix formulations available to the user to select from but these can be divided basically into two types:

Group 1 Those curing at 170 °C–180 °C
Group 2 Those curing at 120 °C–130 °C

each of these can be subdivided further into low viscosity and controlled high viscosity systems.

Group 1 tend to have the higher temperature and environmental resistance whilst Group 2 offers potentially lower processing costs resulting from the lower curing temperature and the best adhesion to honeycomb.

The majority of aerospace structural and semi structural applications use Group 1 materials although there are many demanding applications which are satisfied by those from Group 2.

2.1 Group 1 170 °C–180 °C Curing Systems

In order to illustrate the different processing characteristics of epoxy matrices within Group 1 comparison will now be made between the widely used typical low and high viscosity systems. For the comparison to be convincing reference will be made to two CIBA-GEIGY products, FIBREDUX 914 and FIBREDUX 922.

FIBREDUX 914 is representative of the controlled high viscosity matrices and also possesses a long out-time at room temperature. It has been in use in Europe for some eight years and has gained wide acceptance with special emphasis on processing ease. The viscosity characteristics are obtained by careful chemical formulation and not by advancement of the chemical reaction by heat during the prepregging operation. Thus the viscosity is accurately controlled and is a permanent feature of the product which does not change rapidly with extended shop floor out-time.

FIBREDUX 922 is typical of the family of very low viscosity, short shelf life products on which a large number of civil and military applications have been based in the U.S.A. These matrices employ as the primary resin CIBA-GEIGY MY 720,

Table 1. Processing properties of FIBREDUX 914 and 922

Property	FIBREDUX 914	FIBREDUX 922
Cure cycle	1 hr at 170 °C Post cure optional	2 hrs at 177 °C No post cure
Out time/tack at 22 °C	>2 months	<3 weeks
Minimum process viscosity	45 Poises	1–2 Poises
Direct bond to honeycomb	Yes	No
Zero bleed potential	Good	Difficult
Max temp./mechanical properties	130–135 °C	140–150 °C
VAC-BAG Processing	Yes	Yes

and as main hardener pp′ diamino-diphenylsulphone. As this composition is quite reactive, even at room temperature, all resin mixing operations, prepregging processes, shipment, storage and shop temperatures have a major effect on its handling and processing characteristics.

The mechanical properties of the two systems are not greatly dissimilar with 914 having greater strain capability and toughness and 922 having the edge in mechanical performance above 130 °C.

Table 1 and subsequent discussion covers the differences in the two systems and in particular those concerned with processing.

2.1.1 Cure Cycle

The ability to cure for one hour in an autoclave and then post cure in ovens separately can be an advantage, depending on autoclave capacity and the need to post cure. Conversely a full cure in two hours could also be considered an advantage. However, these cure cycles are the times taken once the curing temperature has been reached. In many instances complex heat-up schedules are needed for the low

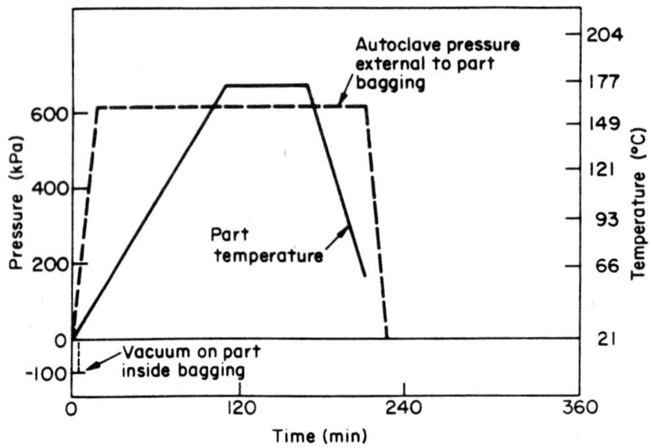

Fig. 2. "Straight-up" simple cure cycle

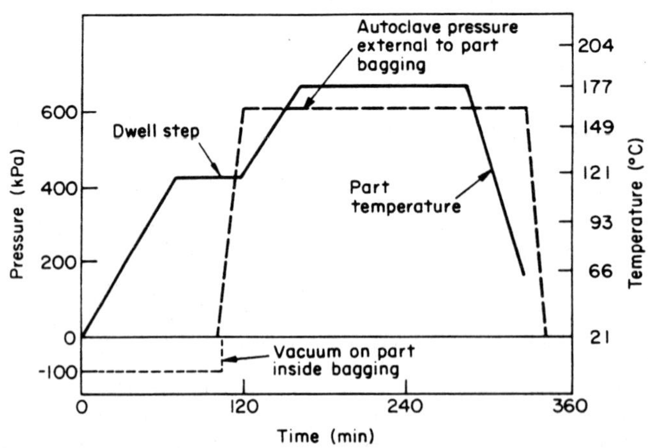

Fig. 3. Dwelled complex cure cycle

viscosity systems, including a dwell step in order for the reaction to advance, thus increasing the viscosity, before applying full pressure. This can be very difficult to control with certainty if an autoclave load consists of different size parts heating up at different rates. This is confused by the reaction occurring at different speeds at all temperatures in the heat-up cycle and great care must be taken to ensure that full pressure is applied before any part of the load is gelled.

914 does not react at any significant rate for processing purposes up to 100 °C and is still slow enough to permit a dwell at 120 °C–130 °C if necessary to ensure consolidation and resin flow from thick section parts. Apart from this possibility 914 has the ideal "straight-up" simple cure cycle.

The total cure time difference in the autoclave can be as much as one third in favour of the simple cure cycle as illustrated in the temperature, pressure, time charts (Fig. 2 and 3).

2.1.2 Out-Time/Tack at 22 °C

The tack of 914 diminishes slowly at 22 °C and a substantial amount remains even after two months. Although no longer tacky the prepreg has been found perfectly usable after six months ambient storage. Additionally the tack is firm but not aggressive allowing the removal of misplaced prepreg without disturbing the previously laid plies.

The tack of completely fresh 922 is good, but rather sticky due to the low cohesive strength of the uncured resin. However, it diminishes fairly rapidly at 22 °C and has effectively disappeared after four weeks. During this time the ability to tack to complex curves and to themselves and to be removed is constantly changing as are the flow properties of all these MY 720/DDS based systems.

It is thus clear that other than for relatively simple layups the advantages for this property lay with 914. There is no need on a production line to return unused material to deep freeze at the end of work time, no defrosting on restart, no resultant condensation of water and no fears of varying flow properties on a large, long lay-up time job.

2.1.3 Minimum Process Viscosity

914 is over a magnitude greater in minimum process viscosity than 922. At first glance this may not appear an obvious advantage – in practice it certainly is (Fig. 4).

Generally the viscosity of 914 is high enough to avoid the need for edge damming a composite part during processing as the resin flow is sufficiently low to avoid resin starvation and consequent voiding at the edges – this saves time and money. 922 definitely needs either edge dams or extreme care during lay-up as otherwise flow can be excessive (Fig. 5).

Additionally, the viscosity of 914 permits full pressure to be applied at the outset of the cure cycle thus permitting the "simple cycle" previously referred to. The actual viscosity of 914 is high enough throughout the cure to ensure that "hydraulic" pressure is applied to entrapped air bubbles to effectively collapse them. With 922 the viscosity is far too low for this effect and results in the need to increase the viscosity by dwelling before full pressure can be applied with the nuisance and attendant problems this can cause.

2.1.4 Direct Bond to Honeycomb

The viscosity of 914 is ideal for good filleting directly onto honeycomb on both the top and bottom faces and its chemical nature permits good adhesion. 922 is far too

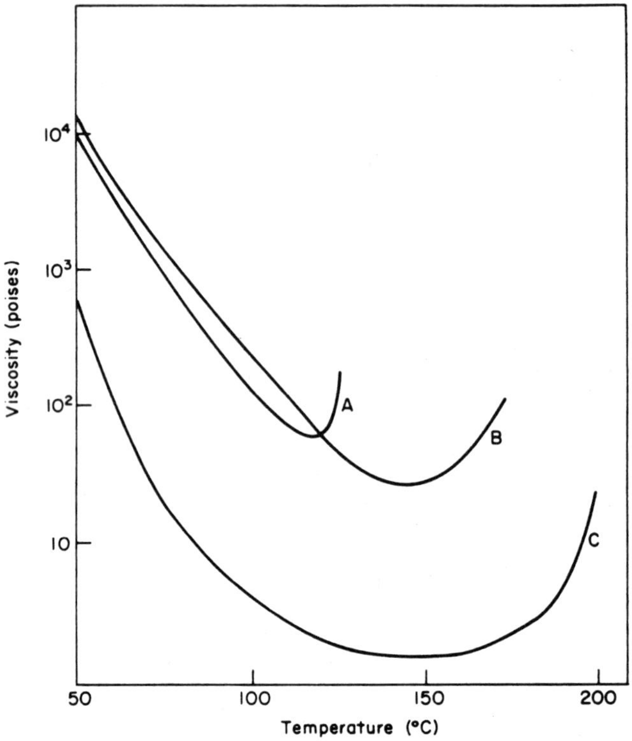

Fig. 4. Matrix resin dynamic viscosity *vs* temperature for: *A – Fibredux 913 controlled high viscosity 120 °C cure system; B – Fibredux 914 controlled high viscosity 170 °C cure system; and C – Fibredux 922, a typical MY 720/DDS system, 175 °C cure*

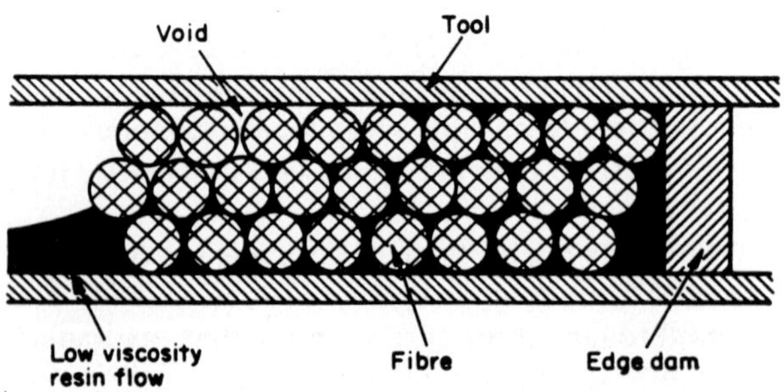

Fig. 5. Schematic of resin and fibre in the tool

low in viscosity for this purpose and will flow from the top face down the cell wall to the bottom (Fig. 6).

Since the 170 °C–180 °C cured resin systems are more brittle than those cured at 120 °C–130 °C it is often preferred to use an adhesive film between the prepreg and the honeycomb. This move is essential with 922 but is optional with 914 whose adhesion in tension and shear is sufficient to "fail" most honeycomb in sandwich structures.

2.1.5 Zero Bleed Potential

Zero bleed or net resin content prepreg is a very important aspect of processing because of its effect on finished part cost.

Prepreg is expensive and it is clearly uneconomic to buy it with the certainty of discarding a proportion of it through intentional resin "bleed out". Additionally and even more expensive is the cost of materials and labour employed to lay-up a suitable controlled resin bleed system – all of it being ultimately waste (Fig. 7).

Two criteria are essential for a good, practical zero bleed prepreg system.

1. The matrix must permit the formation of a good laminate with no excessive resin squeeze out or voiding and this is most readily accomplished by using a controlled flow, high viscosity formulation for the reasons previously discussed.
2. The prepreg itself must be handleable.

The latter is not a big problem per se with woven fabric prepreg but with unidirectional reinforcement at 32–34% resin content by weight it can be a major difficulty. This is because at these low resin contents there is insufficient resin present with some systems to hold the fibres together transversely to permit practical handling. Thus the cohesive strength of the uncured resin is important here. In this case the strength of 914 as a result of its polymer content is many times that of 922 which contains only relatively low molecular weight ingredients. Indeed 914 unidirectional zero bleed prepreg is amply strong enough to be well suited to a variety of manufacturing techniques, including automatic tape laying and gives excellent mechanical performance when so processed. Unfortunately 922 is not as amenable to this approach.

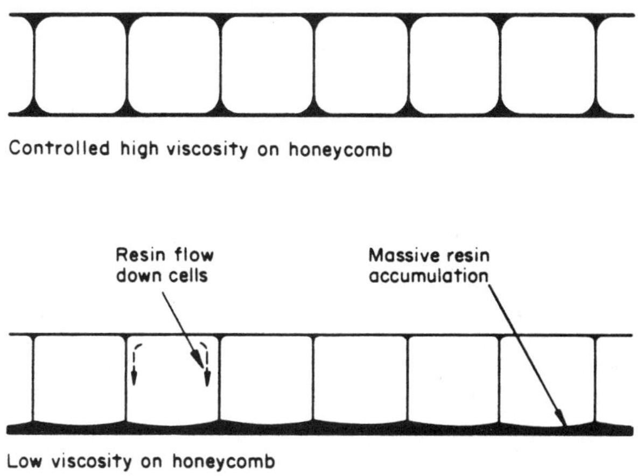

Fig. 6. Fillet-forming capability of different resin systems

2.1.6 *Maximum use Temperature/Mechanical Performance*

It is no use having outstanding processing properties if the composite parts produced will not meet the design requirements and hence no decisions on matrix type can be divorced from mechanical performance.

As the two systems under consideration have such different processing characteristics their chemistry must be significantly different and it might be supposed their mechanical properties also.

Simply stated – the chemistry of 914 gives a slightly lower T_G capability than that of 922 (approx 195 °C for 914, approx 215 °C for 922) but its cured morphology results in a product with 50% higher strain and at least 100% higher fracture energy (Table 2).

The "changeover temperature" for 914 and 922 is around 131 °C. At this temperature the performance of the two systems are about equal in both the resin dominated

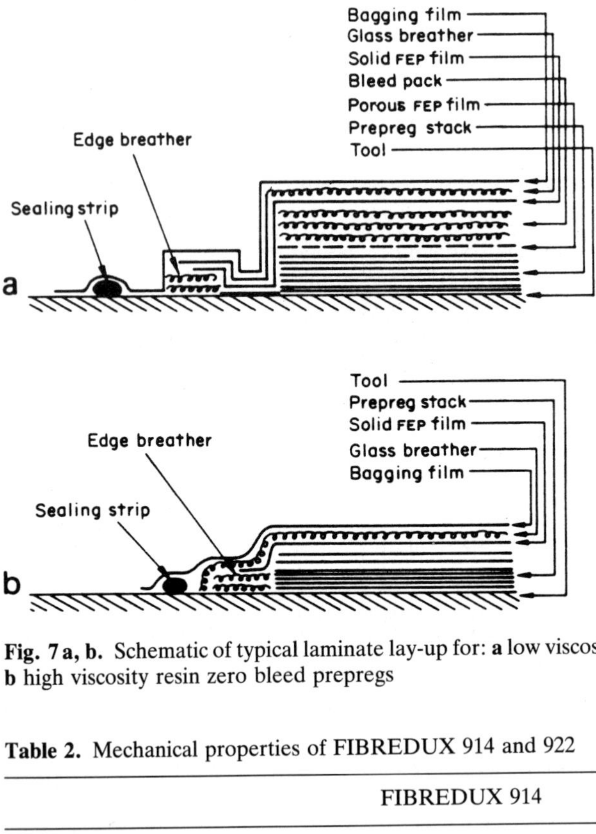

Fig. 7 a, b. Schematic of typical laminate lay-up for: **a** low viscosity resin standard prepregs; and **b** high viscosity resin zero bleed prepregs

Table 2. Mechanical properties of FIBREDUX 914 and 922

	FIBREDUX 914	FIBREDUX 922
Fracture Energy (J/m^2) (Compact Tension)	190	82
Strain to Failure (Cast Specimens in tension)	3%	2%

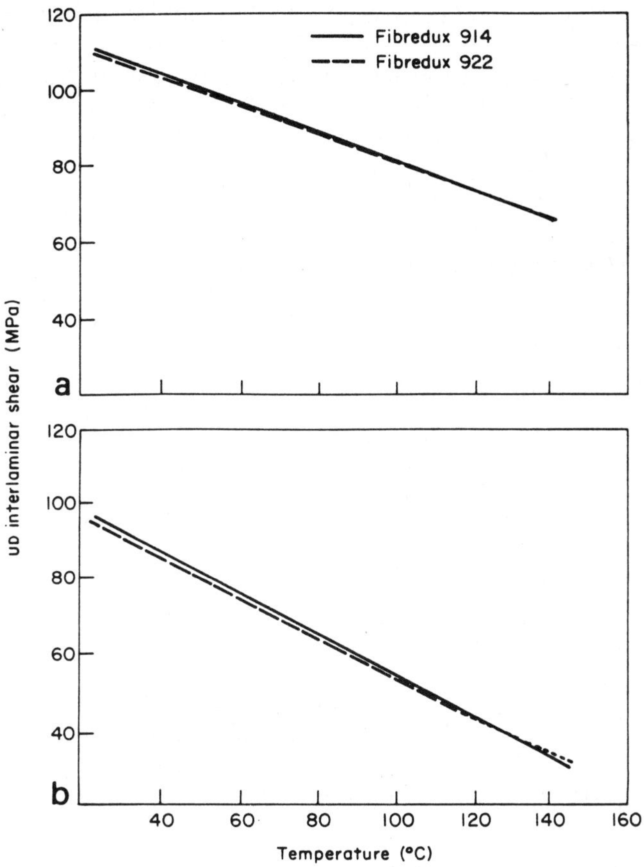

Fig. 8 a, b. UD interlaminar shear *vs* temperature for Fibredux 914 and 922: **a** dry; and **b** after 14 days immersion in water at 71 °C

properties such as short beam shear and resin influenced properties such as compression. Above this temperature, whilst both systems are losing strength, 914 decreases more rapidly (Fig. 8 and 9).

2.1.7 Vac-Bag Processing

It would be ideal if parts could be made under vacuum pressure only in an oven. This could prove economically attractive as well as opening up an almost limitless production capacity as ovens are usually readily available. Recent attempts to establish suitable conditions for processing aerospace type prepregs by this technique have proved promising and indicate that this may well become a practical approach for the production of even quite complex parts in the near future.

The arguments concerning the relative merits of the 914 and 922 type matrices hold good for all handling and lay-up operations but it is likely that all vacuum-only pressure cycles will need a significant dwell period in the curing process to permit the removal of more air. This is probably because there is less pressure to "collapse" remaining air bubbles as some level of vacuum must be maintained throughout the cure cycle – as opposed to autoclave cures where it is normal to "dump" the vacuum at an early stage. Both 914 and 922 have excellent "dwell options" and hence are capable of being processed by this technique – however, once again viscosity considerations favour 914.

One possible limitation of this technique is the likelihood that a maximum fibre volume of around 55% will be achieved with fabric prepregs due to the lower consolidating pressures involved. No such limitation is likely with unidirectional prepreg and thus all comments relating to zero bleed systems still apply.

2.2 Group 2 120 °C–130 °C Curing Systems

This discussion would not be complete without reference to the Group 2 matrix systems which are widely used in aerospace and sports goods applications ranging

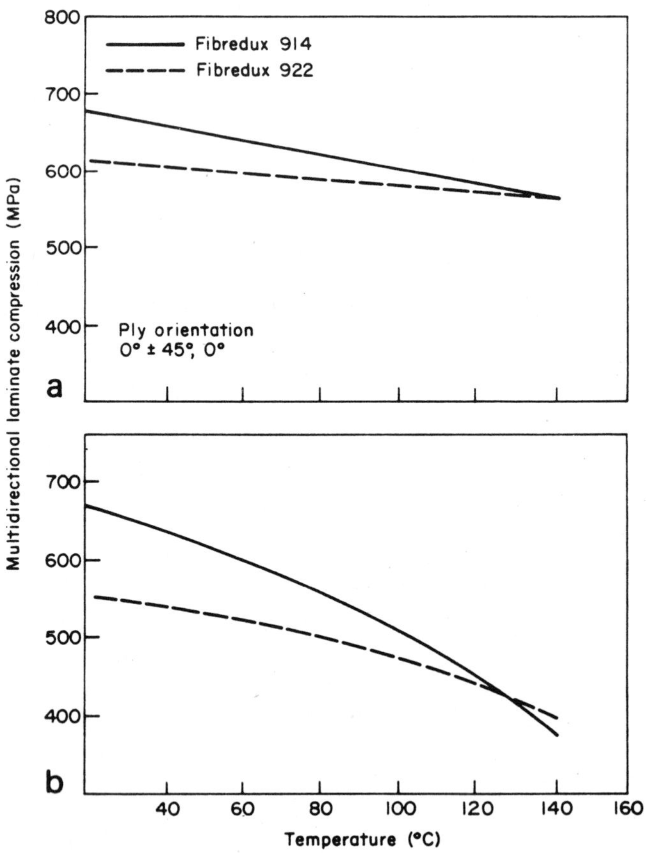

Fig. 9 a, b. Multidirectional laminate compression *vs* temperature for Fibredux 914 and 922: **a** dry; and **b** after 14 days immersion in water at 71 °C

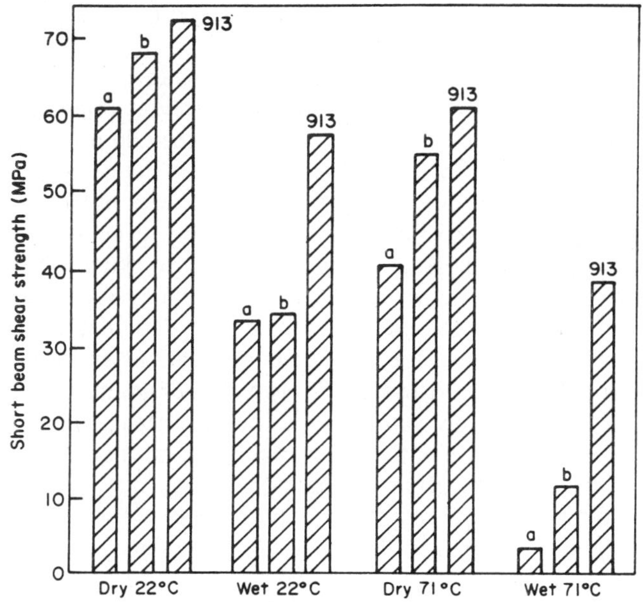

Fig. 10. Short beam shear strength of some Group 2 low temperature curing systems: a) typical traditional glass fabric adhesive prepreg; b) typical current semi-structural acrospace specification resin systems. All systems containing 5 H (G 833) carbon fabric. Wet = after 14 days immersion in water at 71 °C

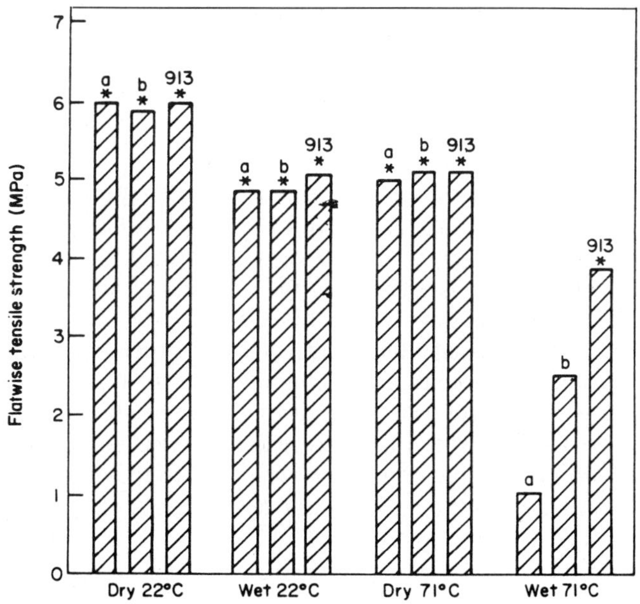

Fig. 11. Nomex honeycomb flatwise tensile tests of some Group 2 low temperature curing systems. Honeycomb density 9 lb/ft^3 (144 kg/m^3), cell size 1/8 in (3.2 mm): a) typical traditional glass fabric adhesive prepreg; b) typical current semi-structural aerospace specification resin-systems. All systems based on 2 plies 5 H (G 833) carbon fabric prepreg per face. * represents 100% honeycomb failure Wet = after 14 days immersion in water at 71 °C

from those which are less demanding to others which are critical, e. g. furnishings, fishing rods, helicopter rotor blades.

Apart from potentially lower processing costs this group attracts applications which need matrices with excellent toughness, impact strength, adhesion to honeycomb and fatigue resistance. Although they do not possess the maximum environmental properties of the higher curing systems, some are eminently suitable for applications requiring operating temperatures of 90 °C–100 °C.

Many of the existing lower temperature curing prepreg systems are old formulations developed for use with glass fabric for skinning honeycomb and as such were not designed to be structural over a very wide temperature/humidity range, although the viscosities were controlled to permit good filleting. There are others which were primarily designed for laminating applications and these tend to exhibit the lower viscosity characteristics of 922 with the resultant disadvantages.

All the previous considerations of processing aspects under Group 1 materials still apply but because of the lower cure temperature constraint it is more difficult to find prepregs in this group which have a long out life, controlled high viscosity and significant retention of properties at higher temperatures. However, one such product is FIBREDUX 913 which has been produced and used in Europe for some eight years. It has an out time of two months at room temperature, is suitable for both laminating and direct bonding to honeycomb and retains a reasonable proportion of its strength under hot/wet conditions. In Fig. 10, 11 it is compared with other existing low cure systems of the following generic type:

a) typical traditional glass fabric adhesive prepreg resin systems
b) typical current semi structural aerospace specification resin systems

3 Resumé

In the earlier comparison of the processing aspects of the two widely used standard products FIBREDUX 914 and FIBREDUX 922, which by mechanical considerations are relatively similar there is little doubt that the long out life, controlled high viscosity system is the most attractive. Such products have big advantages over their, widely used but now “old-fashioned”, low viscosity, short shelf life rivals. The scope for cost saving at almost all stages of processing, coupled with a lower risk of rejected parts makes them economically very attractive and the experience of practical users confirm even more strongly the benefits they confer in production.

That a move towards this type of prepreg is now under way in the U.S.A. appears to substantiate the philosophy which led to these products being actively developed and used in Europe a long time ago in terms of “modern composite history”.

Thermosetting Matrix Systems and Their Influence for Carbon Fibre Reinforcement

H. D. STENZENBERGER

1 Introduction

The outstanding mechanical, thermal and chemical properties of carbon fibres are responsible for their successful application as reinforcement for polymeric matrices. In combination with thermosetting resins (epoxies, polyesters, polyimides, cyanates and phenolics) composites with high mechanical strength and stiffness properties, good corrosion resistance, high thermal stability and good environmental stability can be achieved for the application under static and dynamic load conditions. In the early days the matrix resins were recognized as the binder, whose only function was to fix the fibres in their specific orientation and to give the laminate or component its geometric form. Although thermosetting resins represented, until recently, an almost neglected area of polymer science, engineers and scientists recognized that bulk resin properties and the molecular architecture of neat resins and also the cure conditions to convert the reactive fluid to solid in the thermosetting process, contribute to the physical, mechanical and environmental properties of the composite. It has also been recognized that the fibre-resin interface plays an important role in the performance of a system. In the early days an effort was made to improve the interaction between the resin matrix and the fibre surface and research effort has been concentrated on this aspect. In particular, an attempt was made to generate covalent bonds by chemically modifying the fibre surface and adjusting the resin chemistry to enable resin-fibre surface reactions.

The improvement in the physical properties of carbon fibres, in particular the increase in the strain capability and also the requirements for improved thermal and

Table 1. Matrix resins used for carbon fibre laminates

Thermosetting resins		
Unsaturated Polyesters:	Styrene containing and styrene free systems	
Vinylesters:		
Phenol resins:		
Epoxies:	Diglycidylether BA-types and polyfunctional types	
Cyanate resins:		
Polyimide:	PMR-types (PMR 15, LARC-160 etc.) Acetylene terminated Polyimides (e. g. Thermid 600), Maleinimide (Kerimid, Compimide-types)	
Polystyrylpyridine:	PSP	
Thermoplastic matrix resins		
Polybutylenterephthalate:	PBT	various types
Polysulfone:	PSO	UDEL
Polyethersulfone:	PES	Victrex
Polyetherimide:	PEI	Ultem
Polyparaphenylensulfide:	PPS	Rayton
Polyimide:	PI	(PI 2080, LARC-TPI, $PISO_2$, K-Polymer)
Polyetheretherketone:	PEEK	Victrex (ATP_1, ATP_2)
Modified matrix resins		
Hybrid systems	(Blends from thermosets and thermoplastics)	
Impact resistant systems	(Elastomer modified thermosetting resins)	
Interpenetrating Networks	(IPN's)	

environmental resistance of composites led to increased activity in research for new resin systems. It is the aim of this article to discuss the state of the art resin systems, their physical and chemical properties and also their deficiencies which are the driving force for the development and application of new candidates which also provide better performance.

2 Thermosetting Resins (The State of the Art)

The state of the art matrices utilized in high performance carbon fibre composites are compiled in Table 1. Epoxy resins are the primary matrix materials because of their well-balanced processing and application properties. Many variations are possible by changing the composition and the chemical nature of the monomers, which enables their properties to be tailored to fulfil the requirements of a wide application spectrum. Polyester resins are considered as a potential alternative because it can be demonstrated that differences in the chemical nature as compared to epoxies do not necessarily result in different composite properties. Thermosetting polyimide resins which, like epoxies and unsaturated polyesters, cure without generating volatiles, are more frequently selected, in particular when high temperature performance and hot/wet environmental stability are the main criteria for selection.

Thermosetting matrix resins are based on organic chemicals and are reactive fluids which are converted into solids upon gelation followed by vitrification. The bulk properties of cured resins are closely related to the cure conditions and several other interrelated factors which are as follows:

- the chemical composition and the chemical reactivity of the resin curing agent mixture.

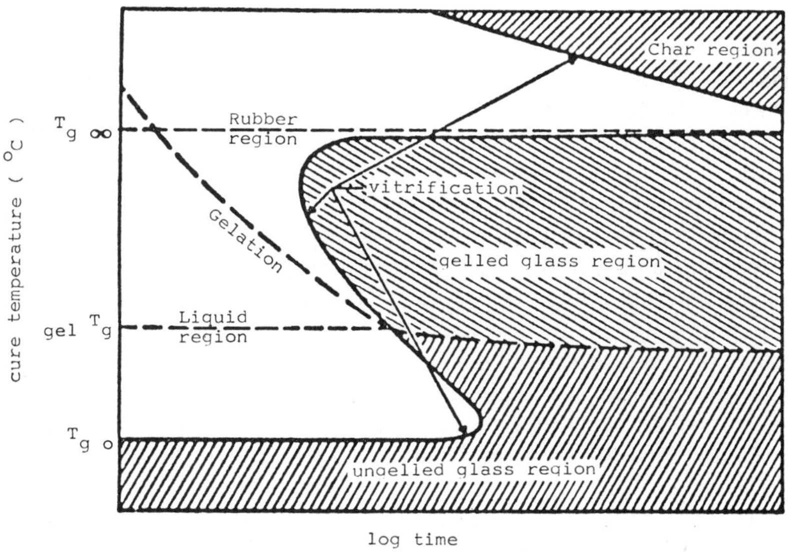

Fig. 1. Time-temperature-transformation (Phase Diagramme)

- the compatibility of all resin constituents in the uncured, B-staged and cured state.
- gelation and thickening rate during the cure reaction.
- phase separation which might occur during or after gelation.
- the vitrification conditions.
- the cure and postcure time.

To describe the cure process and the specific resin states during the cure process and some of the interrelated factors mentioned, a generalized time/temperature/transformation state diagram, Fig. 1, can be used [1]. The diagram describes, in general form, the temperature and time dependent transitions (gelation and vitrification) which are encountered as a consequence of the reactions that convert a reactive fluid to a solid, in the thermosetting process. Consideration of the cure of a liquid thermosetting system at a series of constant temperatures leads to the conclusion that there will be three main types of behaviour, depending on the cure temperature.

- At elevated isothermal temperatures, above the maximum glass transition temperature of the cured system $T_G \infty$, a curing liquid gels and forms an elastomer (rubber) but does not vitrify in the absence of degradation.
- Cure happens at an intermediate temperature (${}_{gel}T_G < T_{cure} < T \infty$). Under these conditions the liquid first gels and then vitrifies after the conversion rate for the selected cure temperature is reached. The ${}_{gel}T_G$ is the isothermal temperature at which gelation and vitrification occur simultaneously.
- Isothermal cure is effected at low temperatures, above the glass transition temperature of the reactants, (${}_{resin}T_G$), but below ${}_{gel}T_G$, in which case the viscous curing liquid can vitrify before reaching the chemical conversion necessary for gelation to occur simply by an increase in molecular weight. After vitrification, further reaction is suppressed due to immobility of the resulting system.

The time to gelation is expected to decrease exponentially with increasing temperature because, according to the gelation theory of Flory [2, Flory 1953], the chemical conversion at gelation for a particular system is constant. The time to vitrification passes through a minimum between ${}_{gel}T_G$ and T_G, [3]. This reflects the competition between the increased rate constants for reaction and the increasing chemical conversion required for vitrification at higher temperatures. The maximum time to vitrify below ${}_{gel}T_G$ arises because of the competition between the temperature and time dependencies of the viscosity of the reacting system. It is apparent that if chemical reactions are quenched by the process of vitrification the glass transition temperature (T_G) of the system after isothermal cure will then equal the isothermal cure temperature T_{cure}. In general, if T_{cure} is less than $T_G \infty$, a reactive material will vitrify and full chemical conversion will be prevented. The material then usually needs to be postcured above $T_G \infty$ for the development of optimum properties. In practice the postcure temperature is increased stepwise so that the T_{cure} is shifted against $T_G \infty$, which is the optimum for a system. For highly cross-linkable resin systems, $T_G \infty$ can be above the limits of thermal stability in which case full chemical conversion of the original network forming reactions would not be attainable. As a general principle, thermosetting systems should be designed so that the cure temperature and the T_G are tailored to meet the processing and application temperature requirements.

2.1 Chemistry of Thermosetting Resins

2.1.1 Phenolic Resins

Phenolic resins have been in technical use for a long time but load bearing structures and high performance composites have only found slight interest. Commercially available phenolic resins are mono-, di- and tri-methylol phenols, in mono-, di- or polymeric form which can be further crosslinked by a condensation process to provide a three dimensionally crosslinked system which carries methylene bridges between the aromatic rings (Fig. 2). Phenolic resins show outstanding temperature resistance and non-flammability properties. As outlined in Fig. 2, the curing reaction may be promoted by heat alone as in traditional phenolic applications, usually at temperatures between 120 and 180 °C, but also may be promoted at ambient temperatures using a strong acid catalyst, [5]. The cure reaction (cross-linking) happens via a condensation reaction, thus generating volatiles which cause problems during the manufacture of void free fibre composites.

However, because of their flammability behaviour, aircraft interior panels, like floors, partition walls, galleys and other interior sandwich laminates, are currently in use, [6].

Phenolic resins are very attractive because cheap and readily available raw materials are used for their synthesis. However, the cured resins are extremely brittle and are therefore of poor structural performance, mainly because of their condensation type cure.

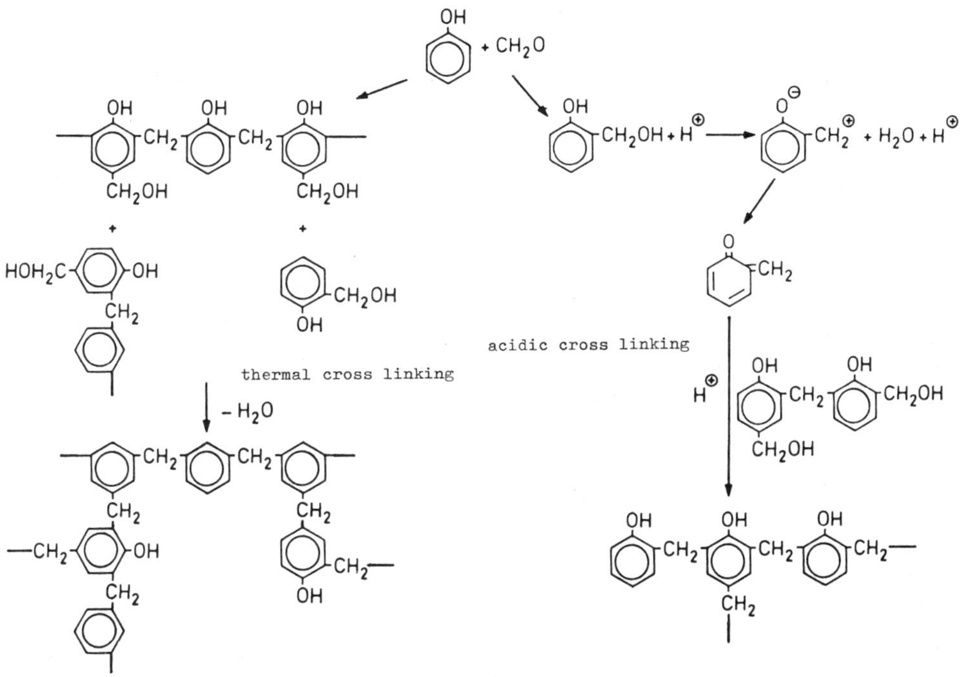

Fig. 2. Chemistry of phenolic resins

2.1.2 *Unsaturated Polyester Resins*

Styrene based unsaturated polyester resins have not been found of interest for carbon fibre laminate applications. Sporting goods for various disciplines such as tennis rackets, golf shafts etc. seem to be successful commercially. Recently, attempts have been made to formulate polyester resins based on diallylphthalate as a reactive diluent, into prepreg systems whose properties are similar to 120 °C curing epoxies [7]. Diallylphthalate based polyesters show the following advantages:

- lower shrinkage on cure, as compared with styrene based polyesters.
- laminate properties equivalent to those of 120 °C curing epoxies.
- the gel and cure behaviour allows storage of prepregs at room temperature.
- good thermal and hygrothermal properties.
- short cure cycles, (15 min. at 170–180 °C).

The chemistry of unsaturated polyester resins is outlined in Fig. 3. They are synthesized from dicarboxylic acids (phthalic, iso-phthalic, terephthalic acid), glycols (propane glycol) and unsaturated acids like maleic acid. Their molecular weight is in the range of 1000–5000 and the backbone chain may be linear or branched. These unsaturated polyesters are dissolved in reactive diluents like styrene, divinyl-

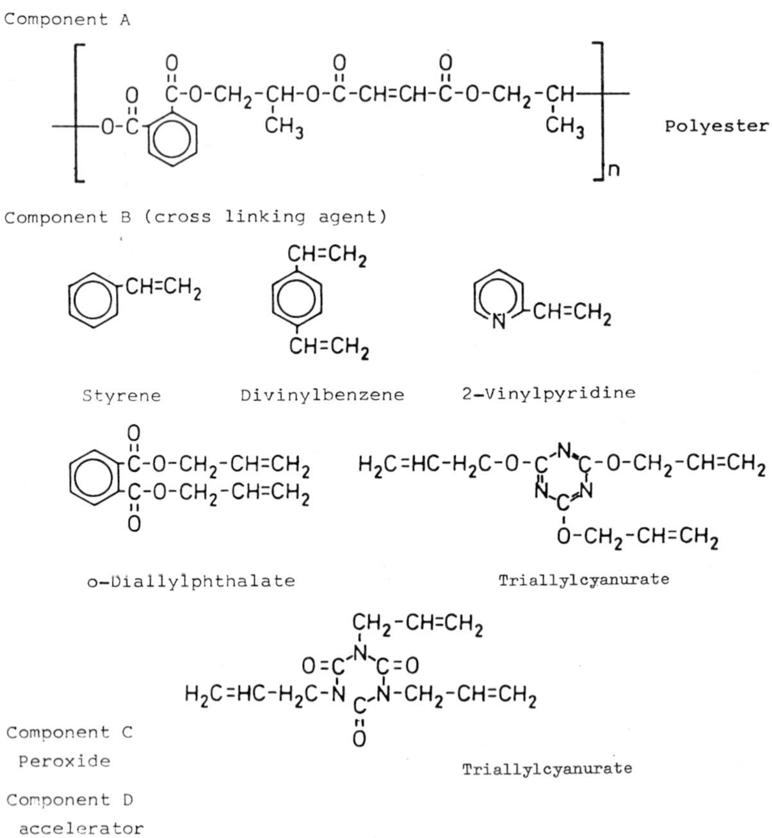

Fig. 3. Chemistry of unsaturated polyester resins

benzene, diallylphthalate and others which are cured in the presence of a peroxide catalyst, [8]. A great variety of products is possible with respect to the backbone chemistry, which allows the physical, thermal and chemical properties of the cured products to be influenced. The cross-link density can be tailored via the amount of unsaturated diacid built into the backbone structure and the processing properties are mainly dependant on the molecular weight of the polyester.

2.1.3 Vinylester Resins

The synthesis of vinylester resins is based on the reaction between acrylic acid and bisglycidyl-bisphenol A epoxy resins or other epoxy resins. The chemical structure of a typical resin is given in Fig. 4. Unlike standard unsaturated polyester resins, vinylesters can be cured without the addition of reactive monomers like styrene, due to their terminal vinyl groups. The terminal vinyl groups are also responsible for the low concentration of residual vinyl groups in the fully cured resin, thus providing an improved thermal oxidative stability over unsaturated polyesters based on fumaric and maleic acid which always contain many unreacted double bonds. The properties of vinyl esters can be varied over a wide range simply by changing or modifying the architecture or the molecular weight of the structural unit between the terminal reactive groups and also by copolymerization with reactive monomers like styrene, divinylbenzene, diallylphthalate etc. Typical physical properties of a vinyl ester resin are given in Table 2. Vinyl ester resins are normally cured in the presence of peroxide type catalysts and accelerators.

2.1.4 Epoxy Resins

Epoxy resins are the primary matrix materials used in carbon fibre composites today. The chemical structures of the most important base resins are given in Fig. 5. In terms of volume, the most important material is 4,4′ bisglycidyl-bisphenol-A, which is available in a wide range of modifications differing in molecular weight (degree of polymerization is between 0 and X). Epoxidized novolaks are important for temperature resistant resin formulations. For applications in the aerospace industry, so-called polyfunctional epoxies like tetraglycidyl-methylene-dianiline (TGMDA) and triglycidyl-p-aminophenol are mainly in use. Cycloaliphatic epoxies play an important role in electrical applications and for outdoor use.

Epoxy resins can be modified in various ways, for example, with low molecular weight monofunctional epoxies to reduce their viscosity thereby improving their processing [9], with toughening agents [10] to improve their impact resistance, or

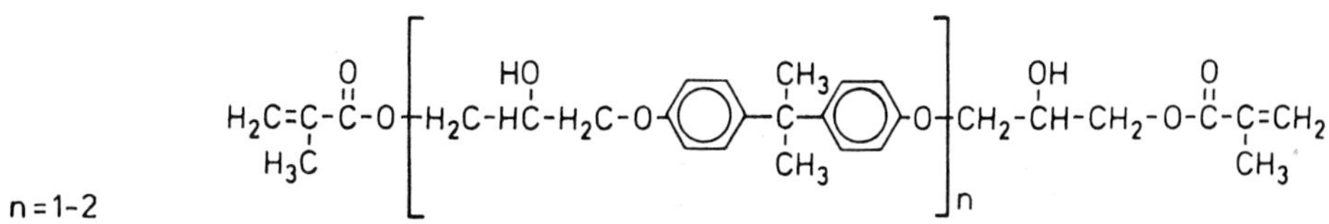

Fig. 4. Chemistry of a vinylester resin

1. Bisglycidyl-bisphenol A-resin

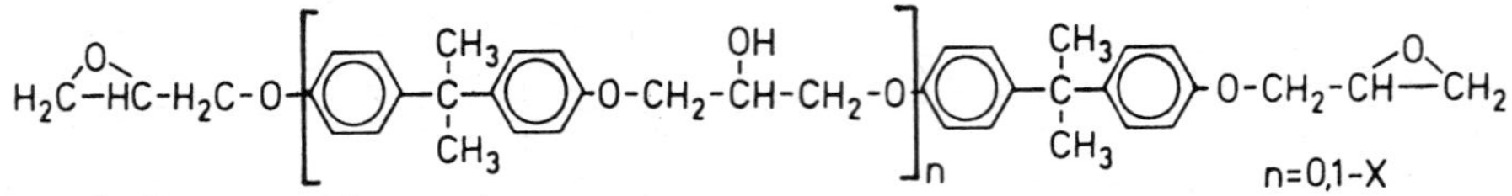

2. Epoxy-novolac-resin

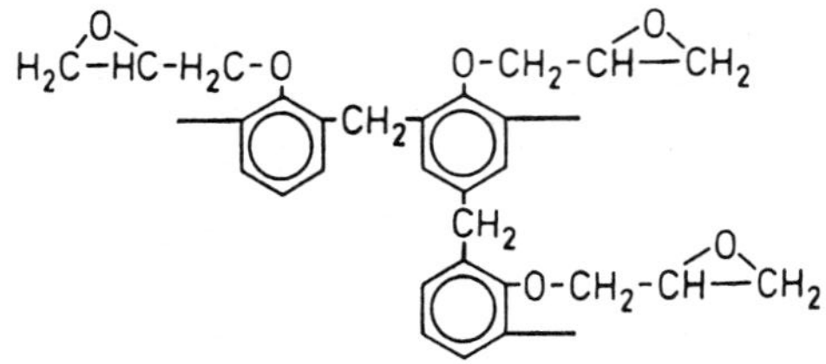

3. Polyfunctional EP-resins

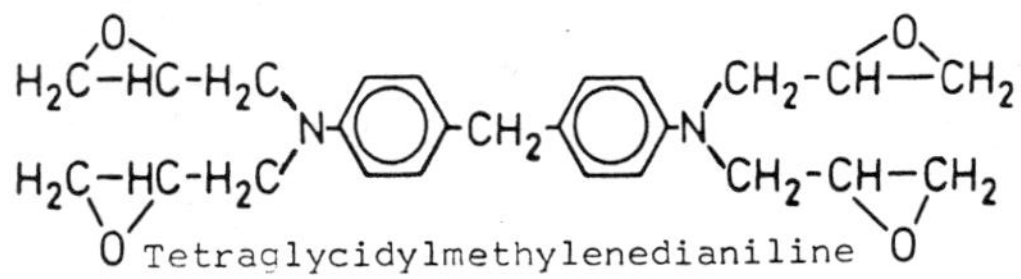

Tetraglycidylmethylenedianiline

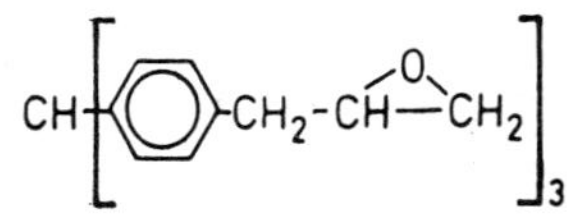

Triglycidyltrihydroxyphenylmethane

4. Cycloaliphatic EP-resin

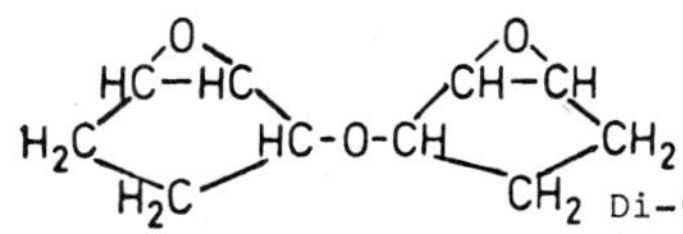

Di-(2,3-epoxy-cyclopentyl) ether

Curing agents:

a) Diethylenetriamine

b) aromatic diamines, e.g.
Diaminodiphenylmethane
Diaminodiphenylsulfone

c) acid anhydrides, e.g.
Methyltetrahydrophthalic anhydride
Methylnadic anhydride
Benzophenone tetracarboxylic dianhydride

d) Dicyandiamide

Catalysts:

Borontrifluoride-complexes

Imidazoles, e.g.
2-Vinylimidazole

Fig. 5. Epoxy resins used to formulate composite matrix resins

Table 2. Mechanical properties of thermosetting resins

Property		Unit	PE a)	PE b)	VE Derakane* 411–45	EP c)	EP d)	EP e)	BMI (Compimide 800)
Density		g/cm³	–	1.31	1.12		1.19	1.24	1.28
Tensile strength	20 °C	N/mm²	20–70		83	78	65	65–70	48–59
	X °C		–		–	–	–	–	–
Strain		%	2–10	2.5	5.0	4.4	3	1.7–2.0	1–2
Flexural strength	20 °C	N/mm²	65–110	97	126	114	105	110–120	85–120
	X °C		–	63 (120)	–	–	77.4 (120)	58–64 (150°)	45–60 (250)
Flexural modulus	20 °C	kN/mm²	3.0–3.6	4.26	3.16	2.9	3.3	3.57	3.95
	X °C			2.52 (120)			2.56 (120)	2.15 (150)	2.64
Fracture energy		J/m²		105		ca. 250	610	158	156
T_g		°C	50–130	– 130	102–150	146	162	>200	287
LOI			–	–	–	0.22	–	–	0.42

a) General properties
b) Technochemie in house development
c) Diglycidyl-bisphenol-A- aromatic amine
d) DGBA/EPN/DICY/Vinylimidazole/7% CTBN-1300 × 8
e) VE 2895/DDS/BF 3500 (Bakelite)

with plasticizers to improve their fracture energy, for instance dibutylphthalate and similar compounds.

The cure of epoxy resins takes place via a polyaddition reaction by use of primary and secondary amines and anhydrides. The chemical structures of anhydride curing agents are given in Fig. 6. Cure can also be effected by employing BF_3-amine complexes. These catalysts promote the ring opening homopolymerization of the oxirane group, thus forming linear polyethers [11–13]. Curing or prepolymerization of epoxies with dihydroxy compounds, as outlined in Fig. 7, is not used frequently for curing laminating resins but is employed for synthesizing solid powder coating resins.

Commercially available prepreg systems, like Fiberite 934 and Hercules 3501, which are based on Ciba-Geigy's MY 720 polyfunctional epoxy resin and 4,4′ diaminodiphenyl-sulfone (DDS) curing agent, contain BF_3-complex catalysts. In these systems the curing agent (DDS) is used lower than stoichiometric because, besides the polyaddition reaction of epoxy and amine, a linear homopolymerization of the epoxy resin takes place consuming oxirane functions. Small modifications and differences in the resin curing agent catalyst concentrations therefore lead to different network structures, network morphologies and resin physical properties, thus influencing the composite long-term behaviour.

Because of the many resin curing agent combinations and the many different curing conditions which may be employed for proper cure, this allows the modification of the following properties:

– heat resistance (glass transition of the resin)
– moisture absorption and performance in hot wet environment
– fracture toughness and impact resistance.

2.1.4.1 Temperature Resistance and Moisture Absorption

The temperature resistance and the moisture absorption of a cured epoxy resin is dependent on
– the chemical composition (monomers) and
– the cure conditions employed to gel and vitrify the reactive fluid.

Both parameters influence and control the cross-link network structure and the microvoid characteristics.

A recent study of the moisture absorption of a wide range of epoxy compositions described in some detail, the influence of the resin/curing agent/catalyst-composition

Anhydride	chem. Structure	Mp/Visc.
Phthalic anhydride	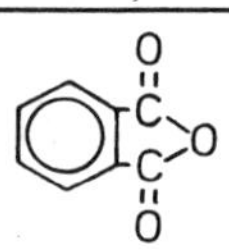	125–130 °C
Maleic anhydride	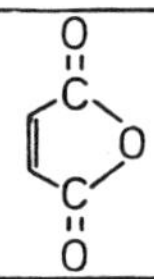	56 °C
Methyltetrahydrophthalic anhydride	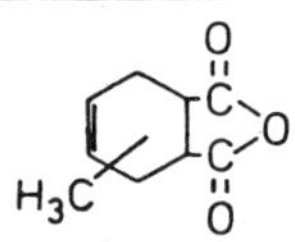	50–100 mPa s
Pyromellitic anhydride	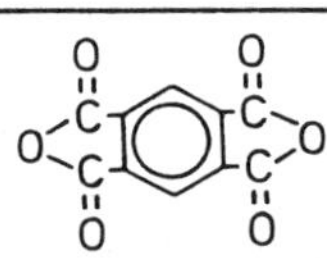	284–286 °C
Nadic anhydride	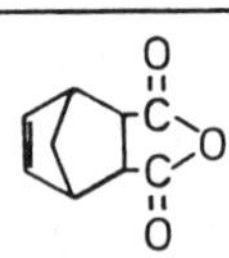	158–162 °C
Methylnadic anhydride	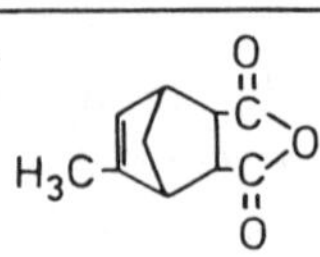	350–450 mPa s

Fig. 6. Diacid anhydride curing agents for epoxy resins

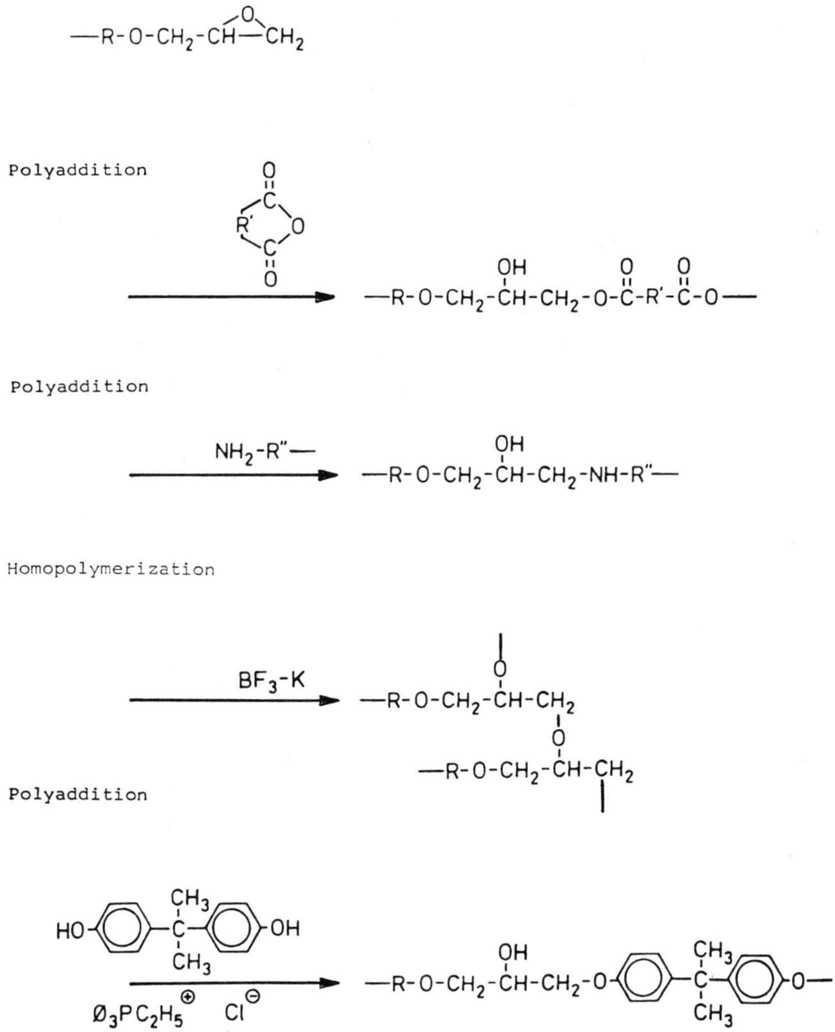

Fig. 7. Chemical reactions involved in epoxy resin cure

on moisture absorption and glass transition temperature [14]. Some of these results are given in Fig. 8, in which the first set of resins is based on 4,4-bisglycidylbisphenol-A and five different aromatic amines and one anhydride as hardeners. The resin shows the lowest moisture absorption when cured with an anhydride (nadic-methyl-anhydride) and the highest when cured with meta-phenylene diamine (mPDA). When cured with other common aromatic amine hardeners, 4,4 diaminodiphenylmethane (MDA) or 3,3′-dichloro-4,4′-diaminodiphenyl-methane (MOCA) showed an improvement over mPDA but were still not as good as the anhydride. The non-stoichiometric formulations (1–3 and 1–4, Table 3) picked up more moisture initially, but the long-term equilibrium studies showed almost no difference between stoichiometric and non-stoichiometric catalyzed versions.

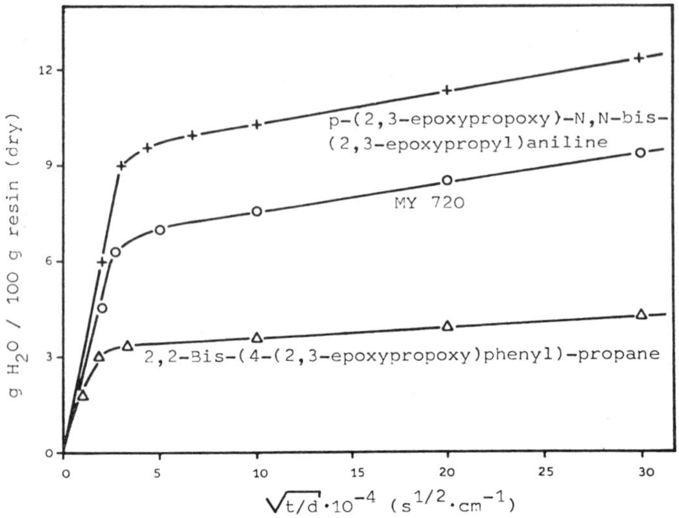

Fig. 8. Water absorption of some commercial resins cured with 1,3-Diaminobenzene (Lit.: P. Johncock, G. F. Tudgey, Brit. Polym. Journal, Vol 15, 14)

Table 3. Moisture absorption of Epoxy Resins (Lit.: B. H. Eckstein, ACS Org. Coatings and Plastics Chemistry, Vol. 38, p. 503 (1978))

No.	Composition	Resin/Cursing-agent/catalyst ratio	HDT °C	% Moisture gain immersion im water at 100 °C		
				24 h	72 h	180 h
1-1	ERL 2774/mPDA	100/14.5	161	1.62	2.41	2.92
1-2	ERL 2774/MDA	100/27	167	1.49	2.08	2.38
1-3	ERL 2774/mPDA/BF 3 MEA	100/3.8/3	172	1.83	2.55	2.81
1-4	ERL 2774/MDA/BF 3 MEA	100/7/3	186	1.84	2.42	2.62
1-5	ERL 2774/MOCA	100/45.8/1.5	122	1.23	1.81	2.05
1-6	ERL 2774/MNA/BDMA	100/80/1	143	1.01	1.21	1.48
2-1	ERRA 4205/MDA	100/55	205	2.95	4.62	5.67
2-2	ERRA 4205/MDA/BF 3-A	100/55/1.5	229	3.52	5.87	7.52
2-3	ERRA 4205/MOCA/BF 3-A	100/85/1.5	158	1.87	2.99	3.90
2-4	ERRA 4205/DDS/BF 3-A	100/58/1.5	204	6.52	8.60	9.11
2-5	ERRA 4205/DDS/BF 3-A	100/23/3	223	7.62	9.67	9.41
3-1	X 810/MDA/BF 3-A	100/50/1.5	232	3.56	4.39	4.76
3-2	X 810/MDA/BF 3-A	100/23/3	–	3.40	3.74	4.00
3-3	X 810/mPDA/BF 3-A	100/25/1.5	194	3.88	5.47	5.79
3-4	X 810/NMA/BDMA	100/46/1	185	2.75	3.83	4.46
3-5	X 810/DDS/BF 3-A	100/58/1.5	230	3.49	4.97	5.29
4-1	ERL 0510/DDS	100/55	245	3.58	5.82	7.18
4-2	ERL 0510/DDS/BF 3-A	100/55/1.5	234	3.89	6.87	8.48
4-3	ERL 0510/DDS	100/22	260	4.04	7.32	9.07
4-4	ERL 0510/DDS BF 3-A	100/22/3	261	3.86	6.41	8.29

X 180 Diglycidylphenyl-glycidylether
ERRA-4205 Bis (2,3-epoxycyclopentyl)ether
ERLA 2774 Diglycidylether des Bisphenol A
ERL 0510 Triglycidyl-p-aminophenol
HDT = Heat distortion temperature (ASTM D-648)

Hence, only the diffusivity appears to be affected, not the equilibrium values. MOCA, an extremely non-reactive curing agent, yields castings with relatively low T_G. The same qualitative correlations are also obtained with Bis(2,3-epoxy-cyclopentyl)-ether as an epoxy resin. The use of diaminodiphenylsulfone (DDS) as a curing agent, either stoichiometric or non-stoichiometric, in the presence of BF_3 catalyst, resulted in high moisture absorption. In conclusion, the following composition/property correlations were observed:

- Methylene-bis-orthochloro-aniline (MOCA) as a curing agent leads to resins with low moisture absorption and low T_G.
- Lower than stoichiometric hardener concentrations provide slightly improved T_G's over the stoichiometric version but no improvement in the equilibrium moisture absorption.
- Diaminodiphenylsulfone as a curing agent provided cured resins with the highest glass transition temperatures but also the highest moisture absorptions.

These correlations also hold for polyfunctional epoxy resins (diglycidylphenylglycidyl-ether and triglycidyl-p-aminophenol). The highest glass transition temperatures are achievable from polyfunctional epoxy resins (triglycidyl-p-aminophenol, tetraglycidyl-methylene-dianiline and trihydroxyphenyl-methane and 4,4-diaminodiphenylsulfone (DDS) as a curing agent. By blending the polyfunctional epoxy resin with diglycidylbisphenol-A the glass transition temperature can be tailored (Fig. 9), [15].

During the last few years, the influence of moisture on the glass transition temperature has been of major concern. Moisture acts as a plasticizer and shifts the glass

Table 4. Moisture absorption of halogenated Epoxy Resins (Lit.: P. JOHNCOCK, G. F. TUDGEY, Brit. Polym. J. *15,* March 1983, 14)

g H_2O/100g dry resin at $\sqrt{t}/d$ $= 5 \times 10^4$ ($s^{1/2}$ cm^{-1})		D × 10^9 (cm^2 s^{-1})	Tg (°C)
5.5	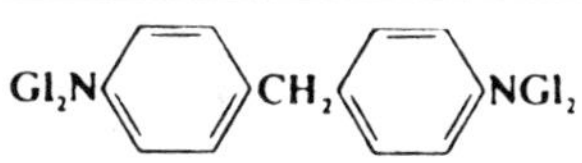	0.6	274
3.3	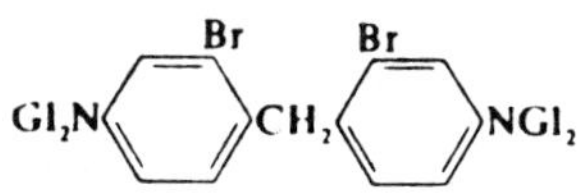	0.5	238
3.7	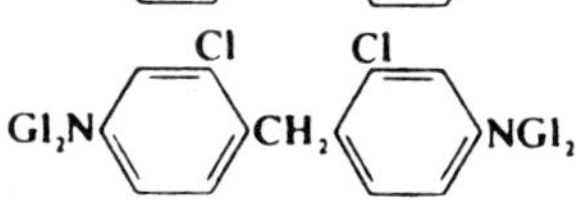	0.6	265
2.9	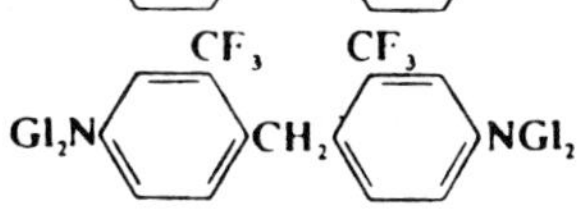	2.0	217
2.8	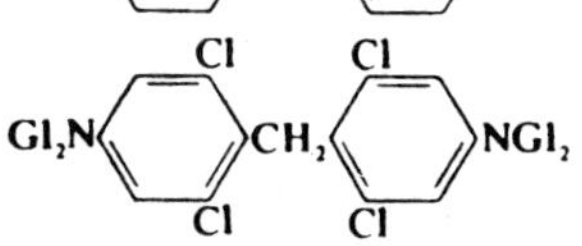	1.1	240

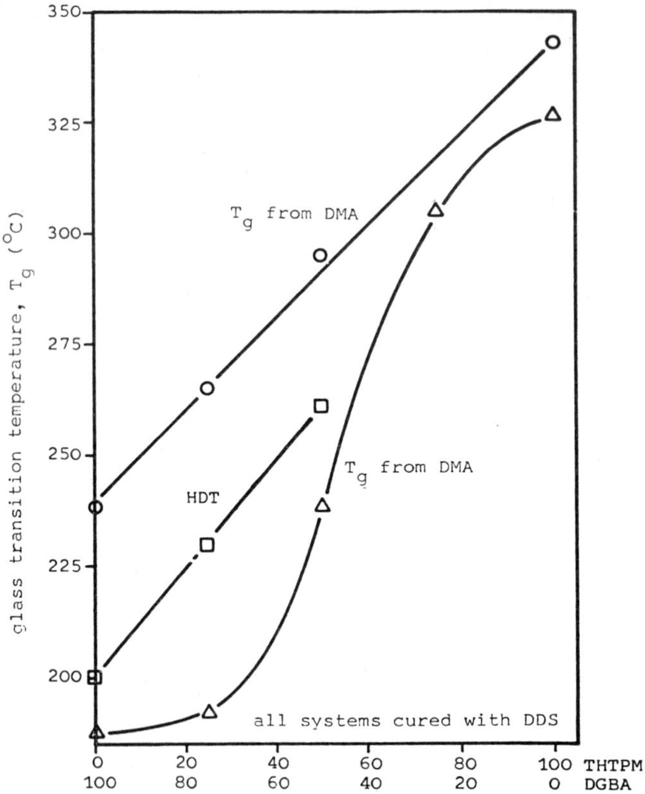

THTPM = Triglycidyl-trihydroxyphenylmethan

DGBA = Diglycidyl-Bisphenol A

Fig. 9. Glass transition temperature as a function of composition (Lit.: K. L. Hawthorne, F. C. Henson, Org. Coat. & Appl. Polym. Sci. Proc., Vol. 46, 1982, 493)

transition temperature towards lower numbers. Chemists have in principle two options for the selection of materials for hot/wet environmental applications:

- Selection of resin curing agent combinations which provide a very high glass transition temperature. These resins show sufficient high temperature capability after moisture saturation, e.g. the wet T_G is above the maximum application temperature.
- Selection of resin systems which show only a very low moisture absorption and therefore the wet T_G is not too different from the dry T_G.

Recently a very interesting publication [16] reported the results obtained for halogenated tetraglycidylmethylene dianiline cured with DDS. The moisture absorption (saturation), as compared with the non-halogenated but structurally identical counterpart, is much lower so that the reduction of the T_G ist not too dramatic, (Table 4).

Besides the chemical composition, the stoichiometry of the reactants and the use of a catalyst, the cure conditions influence the temperature stability (T_G) significantly, as was shown via the TTT diagram. In Fig. 10 the influence of the cure time on the glass transition temperature is given for a standard resin based on bisglycidyl-bisphenol-A cured with DDS [17]. Employing an isothermal cure temperature of 140 °C, the system approaches the maximum glass transition temperature asymtotically. Simultaneously, the density of the system is decreased [18] indicating an increase in the free volume, thus leading to a higher moisture absorption (Fig. 11).

2.1.4.2 Fracture Toughness and Impact Resistance

One of the most severe problems with thermosetting resins in combination with high modulus fibres is their brittleness. Highly cross-linked resins like polyfunctional epoxies and bismaleimides are dense networks of good temperature performance, but low fracture toughness.

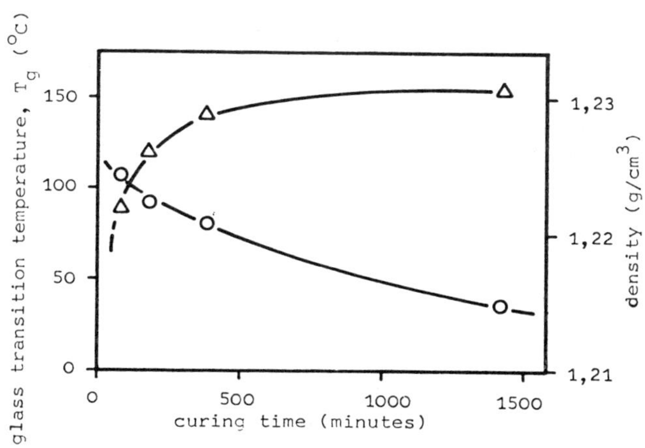

Fig. 10. Glass transition temperature as a function of cure time (Lit.: M. T. AVONHINE, J. K. GILLHAM, ACS, Polymeric Mat. Sci. & Eng., Vol. 49, 576)

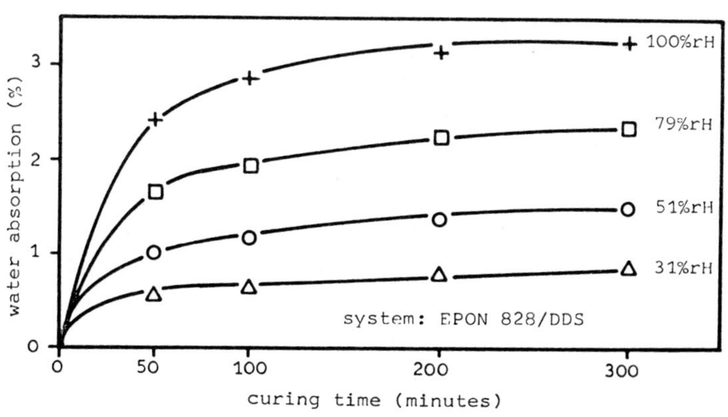

Fig. 11. Water absorption as a function of curing time (Lit.: J. P. AHERNE, J. B. EMS, M. J. DOYLE, J. K. GILLHAM, ACS, Org. Coat. & Plastics Proc., Vol. 46, 1982, 574)

Table 5. Chemical compositions of experimental epoxy resins

Resin code	Resin composition		Comment[a]		
EP 1207	240 g	R 0164	50	T	
	300 g	R 0300	50	T	
	108 g	R VE 2560	8	T	
	6.48 g	1-Vinylimidazole			1%
EP 1208	182 g	R 0164	50	T	
	300 g	R 0300	50	T	
	108 g	R VE 2560	8	T	
	106.8 g	HA 0164/1300 × 8	7	T	
	7.13 g	1-Vinylimidazole			1%
EP 1209	182 g	R 0164	50	T	
	300 g	R 0300	50	T	
	108 g	R VE 2560	8	T	
	106.8 g	HA 0164/1300 × 8	7	T	
	36.0 g	Bisphenyol A	6	T	
	7.13 g	1-Vinylimidazole			1%
EP 1210	240 g	R 0164	50	T	
	36 g	Bisphenol A	6	T	
	42 g	Hycar 1300 × 8	7	T	
	300 g	R 0300	50	T	
	108 g	R VE 2560	8	T	
	7.13 g	1-Vinylimidazole			1%
EP 1801	100 g	VE 2895	100	T VE 2895	
	40 g	DDS	40	T DDS	
	0.65 g	BF_3500	0.65	T	
EP 1806	92 g	VE 2895	92	T VE 2895 + 7.7 T 0164	
	12.7 g	0164/1300 × 13	5	T 1300 × 13	
	5 g	Hycar 1472	5	T 1472	
	40 g	DDS	40	T DDS	
	0.65 g	BF_3500	0,65	T BF_3500	

[a] T = parts

The dilemma is that both high glass transition and high fracture toughness are required for applications in primary aircraft components. It could be shown that the interlaminar fracture energy of carbon fibre laminates is to a major extent dependent on the resin fracture toughness [19].

It is common practice to improve the fracture toughness of epoxy resins by incorporating elastomers. Carboxy terminated acrylonitril butadien copolymers, which can be endcapped with epoxy resin, are blended with the brittle matrix to form a compatible homogenous mixture. During cure the elastomeric compounds undergo a phase separation forming a microvoid structure. The elastomer particle size and concentration contribute to the overall improvement of the resin fracture toughness.

Work performed in our own laboratory with elastomers to improve the fracture energy of 120 and 180 °C curing epoxies showed only a limited translation into the composite. The chemical compositions of the resins investigated are given in Table 5. EP 1207 is a 120 °C curing epoxy resin hardened with dicyandiamide (DICY) in the presence of 1-Vinylimidazole, as a catalyst. EP 1210 is the same basic resin mixture but modified with Hycar 1300 × 8, which was employed as an

EPOXY-adduct. EP 1801 is a tetraglycidyl methylene dianiline resin cured with DDS (4,4′-Diaminodiphenylsulfone) in the presence of BF_3 500 catalyst. EP 1806 is an elastomer modified version of EP 1801. In this formulation two elastomers, Hycar 1300 × 13 (low molecular weight) and Hycar 1472 (high molecular weight) have been employed to optimize neat resin fracture toughness (Tables 6, 7). There is evidence that improvements in neat resin fracture toughness lead to reductions in glass transition temperature.

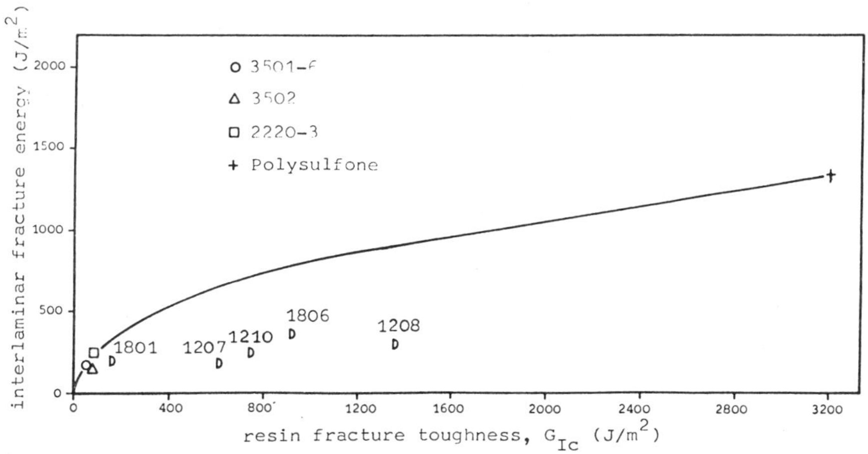

Fig. 12. Fracture toughness-interlaminar fracture energy correlation (Lit.: W. D. Bascom et al., 29th Nat. SAMPE Symposium, 1984, 970)

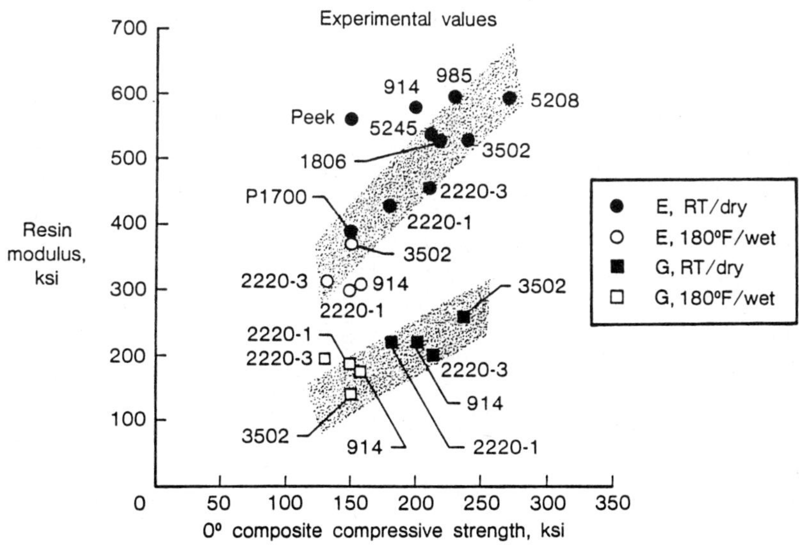

Fig. 13. Composite compressive strength as a function of neat resin modulus

Table 6. Mechanical Properties of 120 °C curing epoxy resins

Resin code	Density	Flexural strength (N/mm^2)			Flexural modulus (KN/mm^2)			Flexural strain (%)			EK mm/mm/°C × 10^6	Tg °C
		20 °C	80 °C	120 °C	20 °C	80 °C	120 °C	RT	80 °C	120 °C		
1201	1.219	126	105	–	3.61	2.89	–	3.92	4.34	–	62.9	ca. 160
1202	1.209	111	90	–	3.14	2.50	–	3.81	4.52		59.9	–
1204	1.202	106	89	–	3.02	2.53	–	3.87	4.33		67.2	–
1203	1.193	64(?)	68	–	2.61	2.15	–	2.47	3.03		62.5	–
1207		105	97	77.4	3.32	2.99	2.56	3.62	4.21	5.0	66.9(1)	162
1208		104	81.5	65	2.81	2.38	1.99	4.45	4.42	6.01	76.5(1)	150
1209	1.181	112	81	39.5	2.86	2.26	1.66	5.19	6.47	9.70	73.9(1)	140
1210	1.187	104	82	61	3.04	2.32	1.95	4.15	4.64	6.92	71.9(1)	147

Resin code	Flexural modulus (KN/mm^2)		K_{IC} (N/m$^{3/2}$)		G_{IC} (J/m^2)	
	RT	70 °C	RT	70 °C	RT	70 °C
EP 1201	3.87	3.24	904	888	212	260
EP 1202	3.37	2.87	1043	1019	323	399
EP 1204	3.14	2.68	1041	566	346	123
EP 1203	2.78	2.32	1114	801	446	282
EP 1207	3.29	2.96	1403	1307	610	606
EP 1208	2.96	2.48	1991	1897	1354	1514
EP 1209	2.82	2.32	1033	1099	397	528
EP 1210	2.88	2.48	1463	962	744	388

Table 7. Mechanical Properties of 180 °C curing epoxy resins

Resin code	Density g/cm³	Flexural strength (N/mm²)		Flexural modulus (KN/mm²)		Flexural strain (%)		EK mm/mm/°C × 10⁶	Tg °C
		20 °C	150 °C	20 °C	150 °C	20 °C	150 °C		
EP 1801	1.238	114	58	3.57	2.15	3.74	3.30	93.7	> 200
EP 1802	1.215	103	50	3.29	1.84	3.34	3.63	78.1	> 200
EP 1803	1.187	–	–	3.24	–	–	–	–	–
EP 1804	1.194	97	54	3.16	1.82	3.35	5.58	–	–
EP 1805	1.224	106	58	3.34	1.86	3.48	4.38	–	–
EP 1806	1.226	87.3	35.9	3.68	1.20	3.29	7.08	79	175
EP 1807	–	85	35	2.53	1.28	3.58	6.45	–	–

Resin code	Flexural modulus (KN/mm²)		K_{IC} ($N/m^{3/2}$)		G_{IC} (J/m^2)	
	RT	120 °C	RT	120 °C	RT	120 °C
EP 1801	3.93	2.2	788	1304	158	720
EP 1802	3.47	2.18	769	757	170	272
EP 1803	3.24	2.00	822	671	209	227
EP 1804	3.01	1.88	794	1808	210	1742
EP 1805	3.27	2.01	1189	1347	501	910
EP 1806	3.09	2.43	1635	1506	920	938
EP 1807	2.54	1.44	1098	1202	483	1006

Table 8. Fracture energy of neat resins and composites

Resin system fibre	Resin properties		Shear strength unidirectional laminate	Interlaminar fracture energy undirectional laminate
	K_{IC} $N/m^{3/2}$	G_{IC} J/m^2	N/mm^2	G_{IC}, J/m^2
EP 1207/Celion 6000	1403	610	88	177.3
EP 1208/Celion 6000	1991	1354	72	295.3
EP 1210/Celion 6000	1463	744	82	240.8
EP 1801/T 300/6000	788	158	111	206
EP 1806/T 300/6000	1635	920	94	360

Of main importance, as said before, is the translation of high laminates to increase interlaminar fracture energy and impact resistance. In a recent publication [21], the correlation between neat resin and laminate fracture toughness was described (Fig. 12) indicating some improvement in laminate fracture toughness, with improved neat resin fracture energy. The results with our own resin modifications are compiled in Table 8 and plotted in Fig. 12, showing lower than expected improvements.

A possible explanation may be that in a composite, due to the restricted space between the fibres, the formation of a large deformation zone is prohibited and therefore low fracture energy is measured.

In neat resins, the crack tip deformation zone can develop without limit. This process is energy consuming. It has to be considered that the elastomer particle size is in the range of the space between the individual filaments.

This mechanism is possibly only relevant for phase separated elastomer modified systems. Obviously, thermoplastic matrices provide much higher laminate fracture toughness [22].

3 The Influence of Matrix Properties on Laminate Properties

At the beginning of this article it was stated that the resin in a composite is only the binder for the fibres. This has to be refuted having discussed the structure/property relations of neat resins in the chapters before.

It has been shown that the chemistry and morphology of the uncured and cured resin and the conditions to cure a reactive resin influence the mechanical and thermal properties of the resin. At least composite properties like

- interlaminar shear strength,
- flexural strength,
- transverse tensile, flexural and compressive strength,
- transverse strain

and their temperature dependency are matrix dominated.

Some other matrix properties and their absolute values influence composite properties. In Fig. 13 the correlation between the elastic modulus of the neat resin and the compressive strength in a 0°-laminate, for some frequently used aerospace

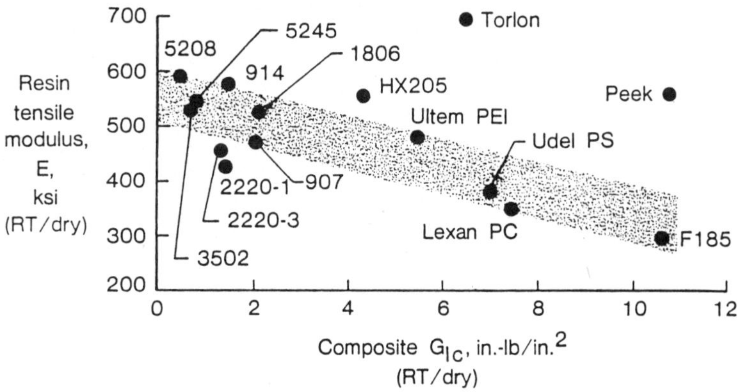

Fig. 14. Composite interlaminar fracture energy as a function of neat resin elastic modulus

resins, is given. Obviously, resins with a high elastic modulus lead to high compressive strength in a 0°-laminate. Modifications of the matrix such as

- modification with elastomers,
- reduction of the cross-link density to improve the strain capability, and environmental effects like
- moisture absorption,
- temperature

lead to a reduction in the neat resin modulus and reduced laminate compressive strength. Combined effects like moisture, temperature and improved toughness (via cross-link density or elastomer modification) can create a dramatic situation. The neat resin tensile modulus is also in good correlation with the composite G_{IC} as can be seen from Fig. 14. The effort put into the modification of matrix resin, as discussed above, to improve the fracture toughness, in most cases leads to significant reduction in neat resin elastic modulus and therefore also to a simultaneous decrease in laminate compressive strength.

4 Final Comment

The ongoing development of new and improved so-called high strain carbon fibres requires new matrix systems with dramatic improved strain capability. The main target for better resin systems is to improve their strain to failure with no loss in neat resin elastic moduli. Researchers all over the world are trying to reach this target, employing the following concepts:

- modification of thermosets with elastomers (reactive or non-reactive elastomers)
- modification of thermosets with thermoplastics (semi-interpenetrating networks)
- synthesis of new thermoplastic and thermosetting backbone structures.
- new cure concepts (cure of low molecular weight fluid into a linear thermoplastic high modulus structure).

New chemical concepts are still required to extend the use of advanced composites.

5 References

1. J. K. Gillham: Polym. Eng. Sci., 19 (10), 676–682 (1979)
2. P. G. Flory: Principals of Polymer Chemistry, Cornell University Pross Ithaca, N.Y., 1953
3. J. K. Gillham: in Developments in Polymer Characterization-3, J. V. Dawkins, ed., Applied Science Publishers, U.K. (1981)
4. A. Knop, W. Scheib: Chemie und Anwendung von Phenolharzen, Springer Verlag 1979
5. K. L. Forsdyke, R. M. Mayer: Use of Phenolic Resins for load bearing Structures, SAMPE, Engineering with Composites Conference, London 1983
6. J. A. Parker: personal communication
7. R. Edelmann, P. E. McMahon: A new Polyester Matrix Resin System for Carbon Fibers, ACS Symposium Series 132, p. 317 (1980)
8. Kunststoffhandbuch Band VIII., p. 248–316, Verlag Carl Hanser, München
9. M. Pilny, J. Mlezira: Kunststoffe 67 (1977), 12
10. A. R. Siebert, R. S. Drake: 27th National SAMPE Symp., May 4–6, 1982, p. 739
11. Y. Tanaka, T. F. Miko: Epoxy Resins, C. A. May und Y. Tanaka, Marcel Dekker, Inc., New York (1973), Kap. 3
12. R. J. Morgan: J. Appl. Polym. Sci., 23, 2711 (1979)
13. N. S. Schneider, J. F. Spronse, G. L. Haghauer, J. K. Gillham: Polym. Eng. and Sci., 19, 304 (1979)
14. B. H. Eckstein: ACS Org. Coatings Plastics Chemistry, Vol. 38, p. 503 (1978)
15. K. L. Hawthorne, F. C. Henson: ACS Org. Coatings, Appl. Polym. Sci. Proceedings, Vol. 46, 1982, p. 493
16. P. Johncock, G. F. Tudgey: The Brit. Polym. Journal Vol 15, March 1983, p. 14
17. J. P. Aherne, J. B. Enns, M. J. Doyle, J. K. Gillham: ACS Organic Coatings and Appl. Polym. Sci., Vol 46, p. 574
18. M. T. Avonhine, J. K. Gillham: ACS, Pol. Mat. Sci. & Eng., Vol 49, 576
19. J. Diamant, R. J. Moulton: 29th Nat. SAMPE Symposium 1984, p. 422
20. D. J. Wilkins, J. R. Eisenmann, R. A. Canin, W. S. Margolis, R. Benson: ASTM STP 775 (1982), Seiten 168–183
21. W. D. Bascom, G. W. Bullman, D. L. Hunston, R. M. Jensen: 29th Nat. SAMPE Symp., 1984, p. 970
22. C. C. Martin May, J. E. O'Connor, A. Y. Lon: 29th Nat. SAMPE Symp., April 3–5, 1984, p. 753

Development of Thermosetting Polyimide Matrix Resins

H. D. STENZENBERGER

1 Introduction

Since the beginning of the space age in the late 1950s, research and development of polymers consisting of aromatic and/or heterocyclic ring structures of high thermal stability have proceeded at an amazing rate. Those familiar with the early literature in this area are still fascinated by the many elegant synthetic routes to high temperature high polymers, examples of which are the polybenzimidazoles, polyoxadiazoles, polybenzothiazoles and polyquinoxalines. In most cases polymer synthesis was successful only because of the ability of the chemist involved to prepare high-purity starting materials. The heterocycle was constructed essentially in one condensation reaction by using the appropriate monomers leading in most cases to intractable low molecular weight materials. A real breakthrough was the introduction of the "precursor polymer" technique for assembling heteroaromatic units into a polymer chain. An intermediate soluble precursor polymer is synthesized in which functional groups pendant to the main chain are arranged so that on thermal or chemical treatment they cyclize to a desired heteroaromatic ring. This technique was employed successfully by E.I. du Pont de Nemours Company chemists for the synthesis of aromatic polyimides (Fig. 1). An aromatic dianhydride is reacted with an aromatic diamine at room temperature in solvents suitable to yield a soluble polyamide, the polyamido-carboxylic acid, containing carboxylic groups ortho, to the amide linkages which propagate the polymer chain. Ring closure is effected either by thermal or chemical treatment. The polyimides thus obtained exhibit unusual chemical, thermal and oxidative stability. Processing to films, foams and

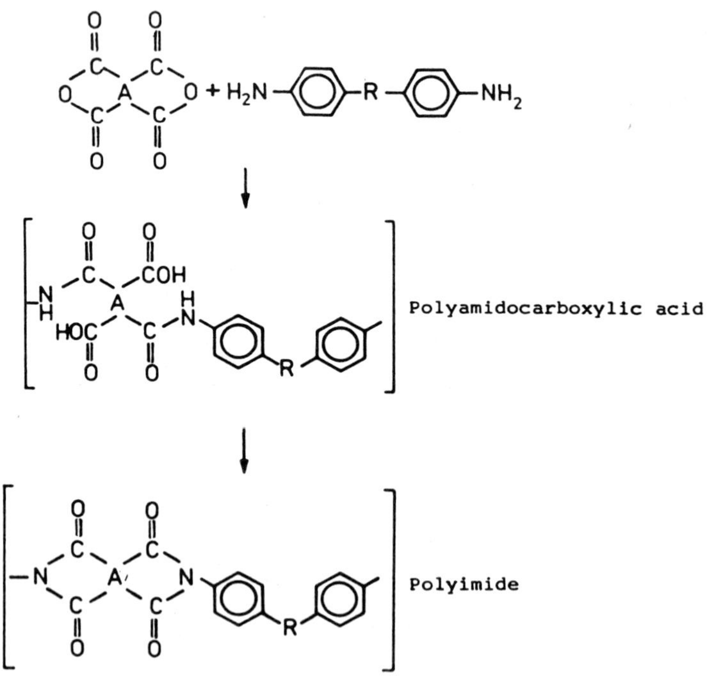

Fig. 1. Synthesis of totally aromatic Polyimides (Lit.: C. E. SROOG et al., J. Polymer Sci., Part A, *3* (1965) 1373)

mouldings includes a 350 °C temperature cure to develop all advantageous properties fully.

Because of their attractive properties, aromatic polyimides have also been tried extensively as matrix resins for fibre composites. Usually the polyamido-carboxylic acid is used in polar solvents like N-methyl-pyrrolidone, dimethylformamide or dimethyl acetic acid amide as an impregnation varnish for fibrous materials. Unfortunately these solvents show an extremely good affinity to the resin and are therefore difficult to remove from the prepreg and the moulded composite. Residual solvent in a laminate acts as a plasticizer and when present in a laminate prevents the development of good high temperature physical properties. In addition, the condensation type cure reaction of the soluble intermediate prevents the formation of void-free laminates. Also, real thermoplastic totally aromatic polyimides were synthesized via the very attractive one step "isocyanate" route (Fig. 2) but did not really overcome the processing problems associated with fibre composite preparation. Because aromatic polyimides failed as matrix resins for advanced composites, a wide variety of thermosetting polyimides was developed with the aim of improving the processing properties.

In general the idea behind all the synthetic routes is the preparation of an imide ring containing backbone that carries terminating polymerizable endgroups providing an imide prepolymer that shows real meltability and therefore advantageous flow properties during composite moulding. Some of the chemical concepts are discussed in this paper.

2 Norbornene Endcapped Polyimides

H. R. LUBOWITZ, working with TRW, discovered polyimide systems that were endcapped with endomethylene tetrahydrophthalic anhydride (ETPA) groups. The synthetic route used again was the classical "amic acid precursor" route. A tetracarboxylic acid dianhydride, an aromatic diamine and ETPA are reacted to form a soluble prepolymer whose molecular weight could easily be adjusted by varying the molar proportions between the reactants. The prepolymer, commercially available (P 13 N – Ciba Geigy) after cyclocondensation, showed a real melt transition at around 270 °C and could therefore within limitations be processed to laminates.

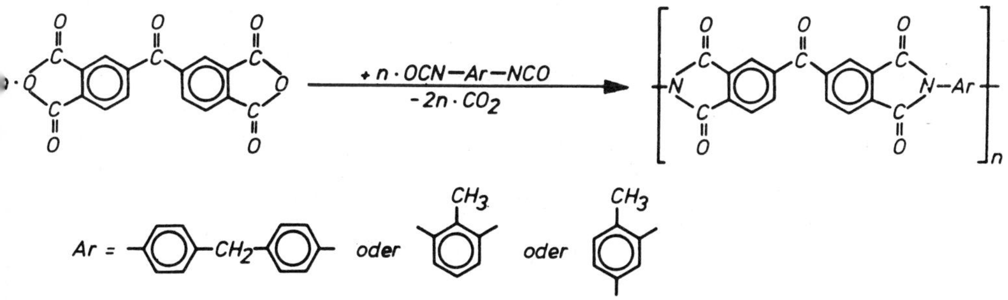

Fig. 2. Synthesis of thermoplastic Polyimides (Lit.: L. M. ALBERINO, W. J. FARRISEY jr., J. S. ROSE, US P 3708458 (1973, to Upjohn))

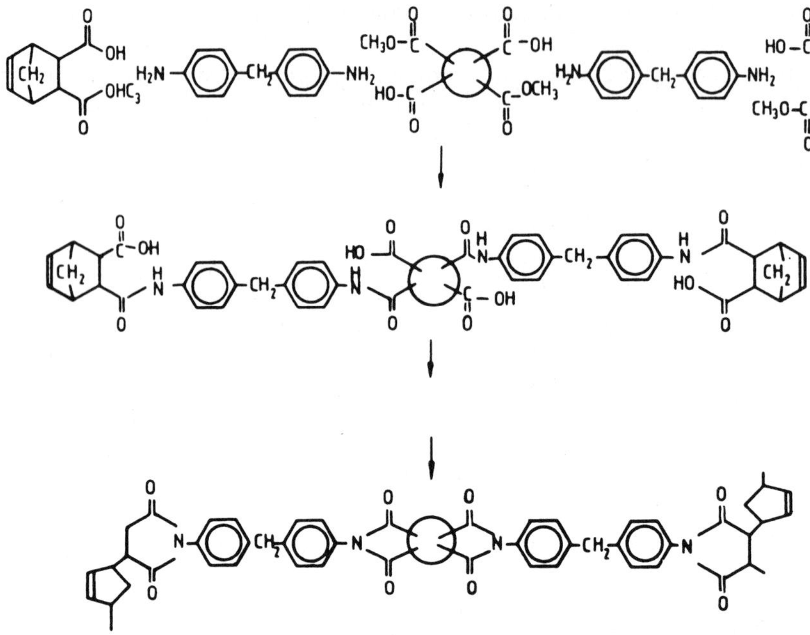

Fig. 3. PMR-Polyimide concept (Lit.: T. T. SERAFINI, P. DELVIGS, W. B. ALSTON, SAMPE Vol. 27, 320)

The norbornene concept was further developed by NASA Lewis Research Center and culminated in the so-called PMR-concept, a novel class of addition type polyimides formed by in situ Polymerization of Monomer Reactants. In this approach a dialkylester of an aromatic tetracarboxylic acid, an aromatic diamine, and a monalkylester of norbornene 2,3-dicarboxylic acid (Fig. 3) are dissolved in a low boiling alkyl alcohol. This solution is then used to impregnate the reinforcing fibres. After removal of the solvent and subsequent heating in the temperature range of 150–220 °C, the monomers undergo in situ cyclodehydration to form a norbornenyl-endcapped low molecular weight imide prepolymer. The cure of the PMRs is generally accepted to proceed via the reverse Diels-Alder reaction of the norbornenyl endgroups to generate maleimide endgroups and cyclopentadiene in the temperature range of 270–310 °C. The highly activated maleimide unsaturation immediately proceeds to copolymerize with cyclopentadiene and unrearranged norbornenyl groups to produce a crosslinked structure without the evolution of void forming volatiles. It is interesting to note that despite the fact that the prepolymer is formed by a classical condensation reaction, cure proceeds in a completely separate step thus providing void-free laminates. The best overall balance of processing characteristics, composite thermomechanical and physical properties, and 300 °C thermooxidative stability was found to be provided by a monomer composition of benzophenonetetracarboxylic acid dianhydride, methylene dianiline and norbornene dianhydride adjusted to an average molecular weight of 1500.

The PMR-concept was developed further by NASA Langley Research Center to provide LARC 160 a melt processable resin, and the replacement of benzophenonetetracarboxylic diester with the diester of 4,4′-(hexafluorisopropylidene) bisphthalic acid leads to improvements in thermal oxidative stability. Both developments could nevertheless not overcome all processing difficulties because moulding of laminates still requires high temperature (285 °C) and high pressure (14 h).

3 Acetylene Terminated Polyimides

The cyclotrimerization of ethynylbenzene leads to 1,3,5 triphenylbenzene, and consequently polycyclotrimerization of bis-ethynylbenzene yields a high temperature resistant polyphenylene resin. Using this chemical approach as a model for an addition type cure, it was only a question of time for the first ethynyl-terminated polyimide to appear in the literature. Figure 4 outlines the synthetic route characterized by simply condensing an aromatic tetracarboxylic acid dianhydride with an aromatic diamine in the presence of an ethynyl-substituted aromatic monoamine providing a fully imidized ethynyl-terminated thermosetting polyimide resin. One resin based on this chemistry became commercially available under the trade mark Thermide 600 (Gulf Oil Chemicals) but has unfortunately to be processed from N-methylpyrrolidone as a solvent to prepregs, leading back to all the problems associated with this high boiling high polar solvent. Nevertheless, the molecular weight of the prepolymer can be tailored to provide a resin that melts at around 190–200 °C and polymerizes when molten immediately as is indicated by its differential scanning calorigramme (Fig. 4).

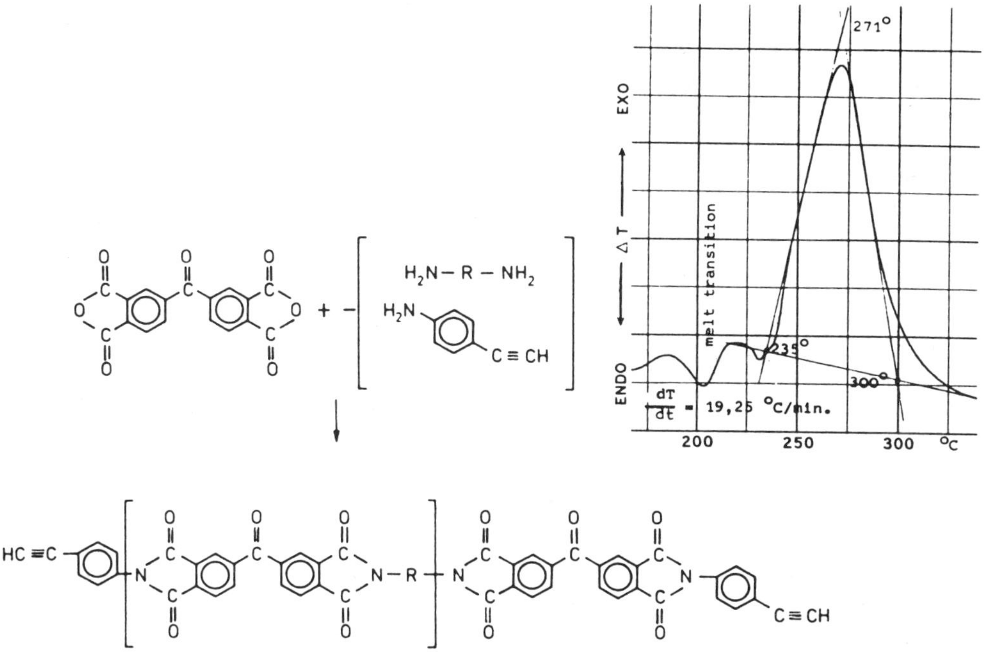

Fig. 4. Thermosetting Polyimides with acetylene endgroups (Lit.: A. L. LANDIS et al., ACS Polymer Preprints *15* (1974) 2, 537, N. A. BILOW, US P 3845018 (to Hughes Aircraft Company)

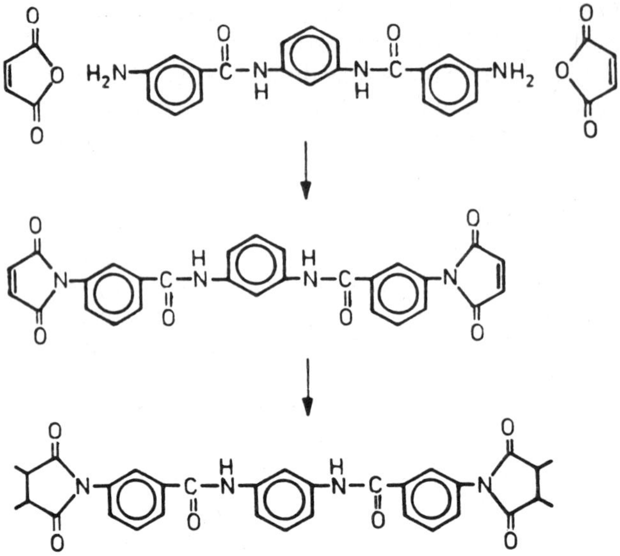

Fig. 5. Synthesis of Bismaleimides (Lit.: US P 2444536 (1948), US P 3018290 (1962), US P 3127414 (1964), Brit. Patent Spec. 1137592 (1968)

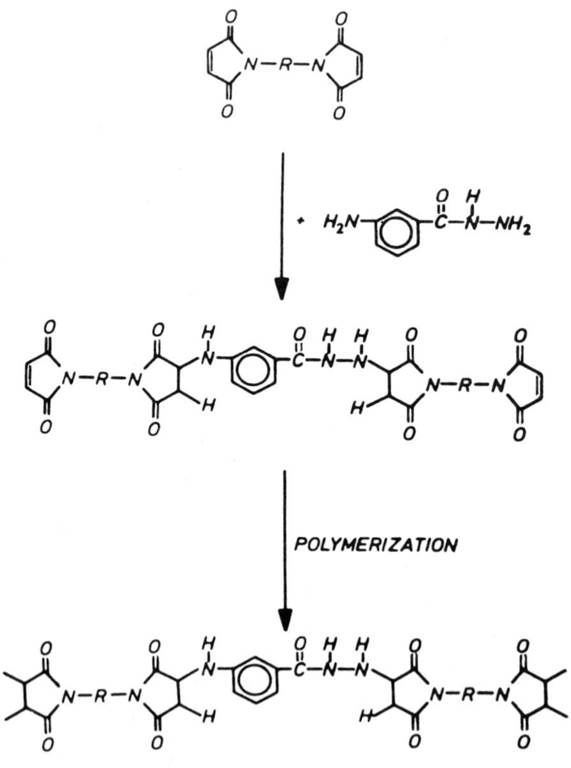

Fig. 6. Polyamino bismaleimide resin concept (Lit.: M. BERGAIN et al., Brit. Patent 1190718 (1968, to Rhone Poulenc))

4 Maleimide Type Resins

Very attractive high temperature thermosets are based on bismaleimides. First Grandschober reported their homo- and copolymerization which can be achieved by simply heating the monomers to temperatures between 150 and 400 °C resulting in the formation of highly crosslinked polymers. The usual synthesis of the monomers starts from maleic acid anhydride and aromatic diamines (Fig. 5) forming a bismaleiamic acid as an intermediate which undergoes cyclodehydration at a temperature of 40–50 °C in the presence of acetic acid anhydride and sodium acetate to form the bismaleimide in high yield. It is easily imaginable that a great variety of bismaleimides are possible by simply changing the structure and molecular weight of the diamine used. In particular resin formulators should pay attention to those bridging units that either provide low melting transitions or improve solubility in low boiling solvents.

The double bond of the maleimide is very reactive and can, for example, undergo a chain extension reaction with primary and secondary amino compounds. This reaction, known as Michael addition, is used to synthesize a well-known commercially available resin, Kerimide 601 of Rhone Poulenc, which is synthesized by reacting 4,4′bismaleimidodiphenylmethane with methylene dianiline either in the melt or in solution to form a poly-amino-bismaleimide (Fig. 6). The material is available as injection and transfer moulding compound and as a laminating resin. Unfortunately, like all the other polyimides discussed so far, the processing solvent is N-methylpyrrolidone but the moulding temperature of 170 °C is fairly low and good high temperature property retention up to 200 °C is achievable when postcured at 250 °C, provided the prepreg used contained only 2% of the residual solvent.

5 The M 751-Poly(amide-maleimide) Resin Concept

The major disadvantage of all bismaleimide resins detected so far is their brittleness. Speaking in terms of mechanical properties, they are high modulus low strength materials accompanied by a very low elongation at break. This is all a consequence of their high crosslink density and the aromatic nature of the starting materials employed for their synthesis. Having been confronted with these problems the authors tried to achieve both a low temperature melt transition and a prepolymer molecular weight high enough to provide a substantially increased elongation at break.

The chemical approach towards these technical goals within a bismaleimide type resin is outlined in Fig. 7. Meta-maleimidobenzoic acid chloride is reacted with 4,4′-diaminodiphenylmethane preferably in methylene chloride as a solvent in the presence of a hydrogen chloride acceptor in such a way that the molar proportions of the difunctional aromatic amine and the acid halide are between 1 : 2 and 1 : 1. The resulting polymaleimide prepolymer consists, depending on the molar proportions of the reactants, of a mixture of the bismaleimide IV and the amino-terminated monomaleimide IV. The polymerization of the resin can be accomplished simply by heating the resin to temperatures between 180–260 °C. The polymerization proceeds controlled in such a way that firstly the free amino groups of the monoimide (III) react via the Michael addition with the double bonds of the bismaleimide IV forming a high molecular weight linear prepolymer which secondly crosslinks by a free radical polyaddition mechanism. The advantage of this chemical approach is that resins with controlled prepolymer molecular weight can be obtained reproducibly by

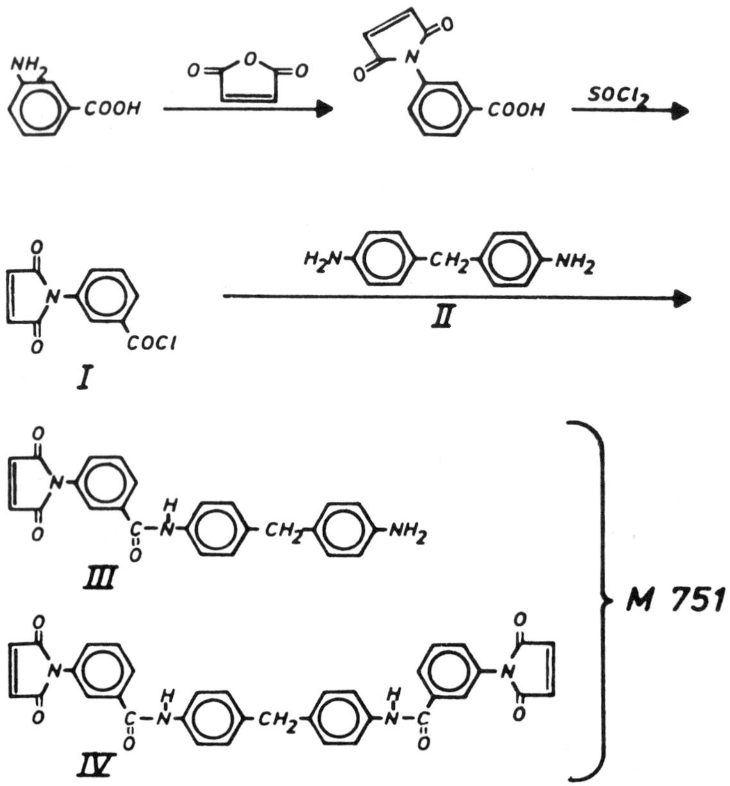

Mechanical properties for resins based on M 751-Bismaleimide concept

Resin No.	FMW	Flex. Strength N/mm^2	Flex. Modulus kN/mm^2	ε %
M 751 - 1	1140	220	4,38	4,81
M 751 - 2	790	180	4,65	3,71
M 751 - 3	670	121	4,28	2,68

FMW = formulated molecular weight

Fig. 7. TECHNOCHEMIE M 751 – Bismaleimide resin concept (Lit.: H. STENZENBERGER, Brit. Patent 1501606 (1976, to Technochemie))

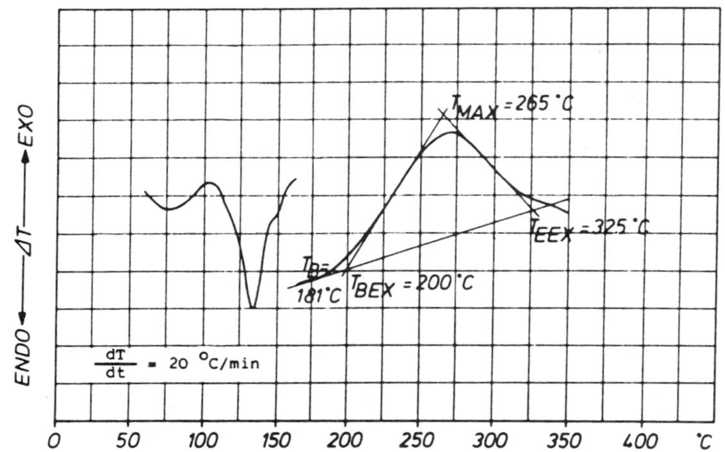

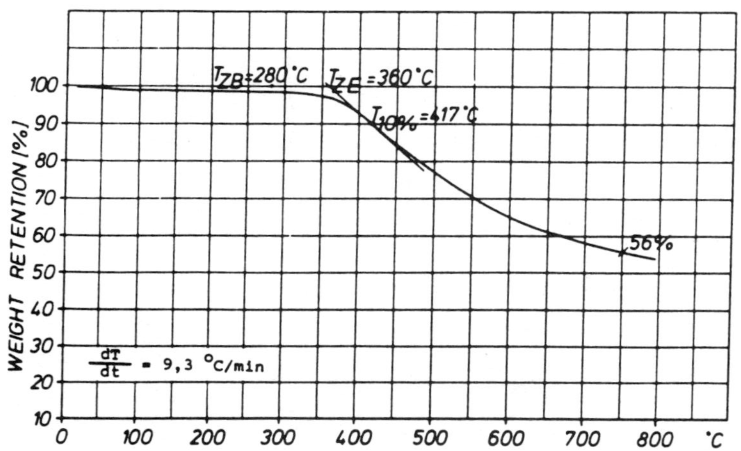

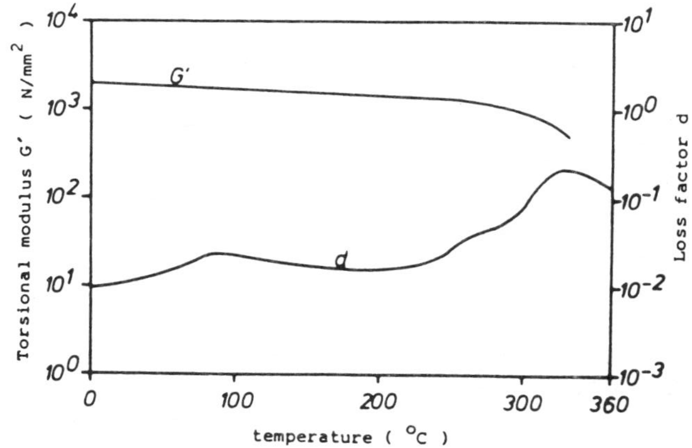

Mechanical Properties of neat Bismaleimide resin

Property	Unit	M 751 RT	M 751 250 °C
Density	g/cm^3	1,32	
Flex. Strength	N/mm^2	210	102
Flex. Modulus	kN/mm^2	5,02	3,17
Flex. Elongation	%	4,8	
Expans. Coeff.	10^6 cm/cm/°C	31	84
Moist. Absorption	%	4,5	
T_g	°C	240	

Fig. 8. Properties of TECHNOCHEMIE M751 resin

simply selecting the molar composition of m-MIC(I) and 4,4'diaminodiphenylmethane. The formulated molecular weight of the commercially available material, TECHNOCHEMIE M751, is 1140, providing, when cured, high mechanical strength properties and good extensibility (flexural elongation = 4.81%). For comparison two resins with lower formulated molecular weights have been prepared and cured under the same conditions and, as was to be expected, they show lower strength properties and lower elongations at break. Interestingly all the three resins show almost the same value for the elastic modulus, leading to the conclusion that the backbone chemistry defines the elastic modulus while the crosslink density influences both the strength and elongation properties.

Figure 8 summarizes the properties of TECHNOCHEMIE M751 resin. A real melt transition is present as can be seen from the differential scanning calorigramme. The torsional modulus curve indicates a T_G for the cured resin of 240 °C after a standard postcure at 250 °C for 15 h. The TGA-thermogramme is typical for cured bismaleimides.

The extensive investigation of the bismaleimide resins from this concept allowed the elucidation of many structure property correlations but because of the starting materials used TECHNOCHEMIE M751 has to be processed from N-methylpyrrolidone to fibre laminates. To overcome this disadvantage, work is in progress with the aim of obtaining modified types which are soluble in low boiling solvents like acetone, methyl-ethyl ketone or methylene chloride.

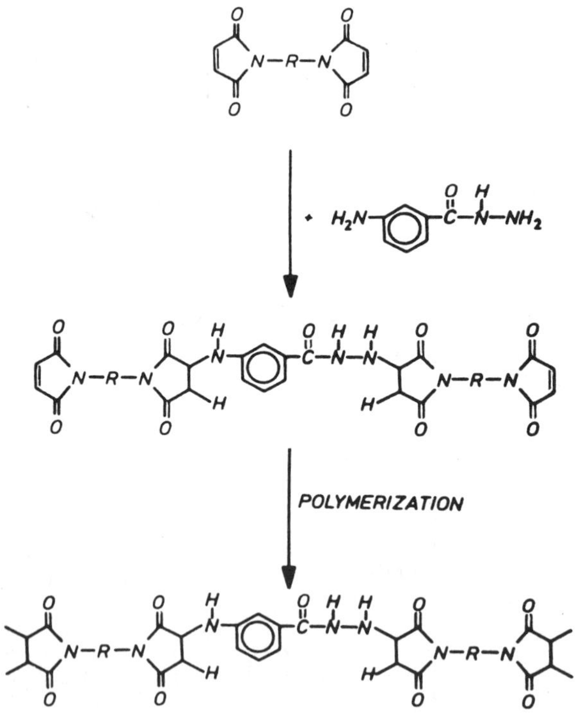

Fig. 9. TECHNOCHEMIE H795 – Bismaleimide resin concept (Lit.: H. STENZENBERGER, US P 4211861 (1980, to Technochemie))

6 TECHNOCHEMIE H 795-Bismaleimide Resin Concept

The aerospace industry requires resin systems that do not emit volatile by-products during cure to allow the production of void-free laminates by using standard 350 °F epoxy-like cure. Tooling, bagging materials, curing procedures and production facilities established for epoxies need to be met by the processing properties of new resin candidates to find acceptance by the industry. Processing of a resin with respect to its use as a matrix in fibre laminates is represented by its prepregging properties and overall moulding characteristics.

Prepregging from solvent is usual for fabrics in order to obtain uniform impregnation of the reinforcement. Low boiling solvents like acetone, methyl-ethyl ketone and methylene chloride are preferred because they can be stripped off easily in circulating air moderate temperatures. Unidirectional tape material can either be produced by a hot melt technique or by a solution dip-coating technology.

Recently TECHNOCHEMIE GMBH-Verfahrenstechnik developed and patented a bismaleimide type resin formulation which shows outstanding prepregging properties. A bismaleimide is prereacted with m-aminobenzoic acid hydrazide either

PROPERTY	TEST METHOD	VALUE	COMMENT
Gel time	DIN 16954	25 - 3 mins	test temperature 170° C
Viscosity	DIN 16954	400 - 2800 mPa.s	complex viscosity
Viscosity build up at 110° C/180 mins	DIN 16954	1200-5750 mPa.s	complex viscosity obtained after aging at 110° C for 3 hours
Residual solvent	-	2 - 6 %	Weight loss at 140° C in vacuum (20 Torr) after 30 minutes
Solubility	-	50 % MEK 50 % Acetone 40 % CH_2Cl_2	clear soluble clear soluble clear soluble
Differential Scanning Calorigramme	DSC	T_B 146 ± 4° C T_{BEX} 179 ± 4° C T_{MAX} 230 ± 6° C T_{EEX} 277 ± 7° C	Heating rate 10° C/min
Polymerization Energy	DSC	265 ± 20 J/g	from DSC-Analysis $\frac{dT}{dt}$ = 10° C/min
Composition	HPLC	comparison with standard	

Abbreviations:

m Pa · s	millipascals · seconds	T_{MAX}	exotherm peak temperature
MEK	Methyl-ethyl-ketone	T_{EEX}	complete reaction temperature
CH_2Cl_2	Methylene chloride	DSC	Differential Scanning Calorimetry
T_B	onset temperature	J/g	Joule/gramm
T_{BEX}	extrapolated onset temperature		

Fig. 10. TECHNOCHEMIE H 795 resin specification

in the molten phase or in solution to provide a prepolymer (see formula III, Fig. 9) which shows solubility in acetone, methyl-ethyl ketone or methylene chloride. The molar proportion of the bismaleimide and the amino acid hydrazide is such that bismaleimide (I), the idealized bismaleimide III and intermediates are present in the reaction mixture providing the solubility characteristics discussed and in addition providing low melt viscosity which is necessary for low pressure autoclave moulding (Resin specification, see Fig. 10).

7 Neat Resin Properties (Fig. 11)

The bismaleimide resin TECHNOCHEMIE H 795 as supplied shows the differential scanning calorigramme given in Fig. 11. The heat of polymerization of 265 ± 20 Joules/g reflects the concentration of reactive maleimide groups, e. g. the state of advancement. The cured resin like all other bismaleimides provides a TGA-thermogramme indicating initial weight loss between 300 and 400 °C and maximum weight loss occurring between 400 and 500 °C. The unmodified resin shows no glass transition below 300 °C as indicated by the torsional modulus curve. The mechanical properties of the unreinforced resin are typical for a highly crosslinked resin. The flexural strength of 100 N/mm^2 at room temperature only decreases to 60 N/mm^2 at 250 °C.

Only a few data about the thermal oxidative stability of bismaleimide resins are available in the open literature. According to our experience with bismaleimides, these resins fill the gap between totally aromatic polyimides and epoxies, which is confirmed by the results obtained for TECHNOCHEMIE H 795 air ageing. Figure 12 summarizes the air ageing result obtained for samples which were cast by use of a steel mould and cured under a pressure of 6 bars at temperatures up to 210 °C and which were postcured at 240 °C for 12 h. The room temperature mechanical properties, flexural strength, flexural modulus were not at all degraded after 500 hours of 200 °C, 220 °C and 250 °C in circulating air. The high temperature flexural modulus of the aged material is increased while simultaneously the high temperature flexural strength is decreased when aged at 250 °C after 500 h. Contrary to that at 200 °C, we can see a simultaneous increase in both the high temperature strength and the high temperature modulus after ageing at 200 °C. These observations indicate that, during air ageing, both postcure and thermal oxidative attack occur simultaneously.

8 Moisture Absorption of Bismaleimide Type Resins

During the last three years extensive attention has been paid to moisture properties of matrix resins and composites. The state of the art high temperature epoxies based on tetraglycidylmethylenedianiline cured with 4,4′-diaminodiphenylsulfone show moisture absorption of up to 6% by weight (neat resin). When used in graphite

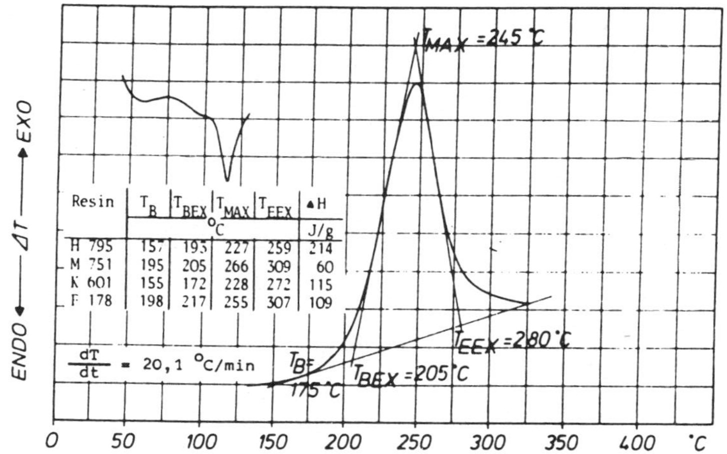

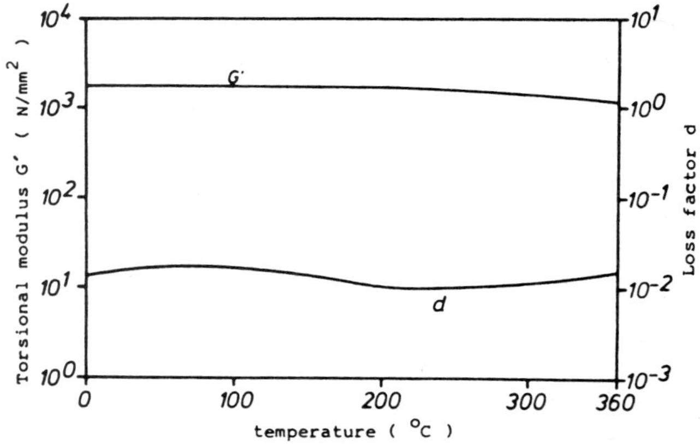

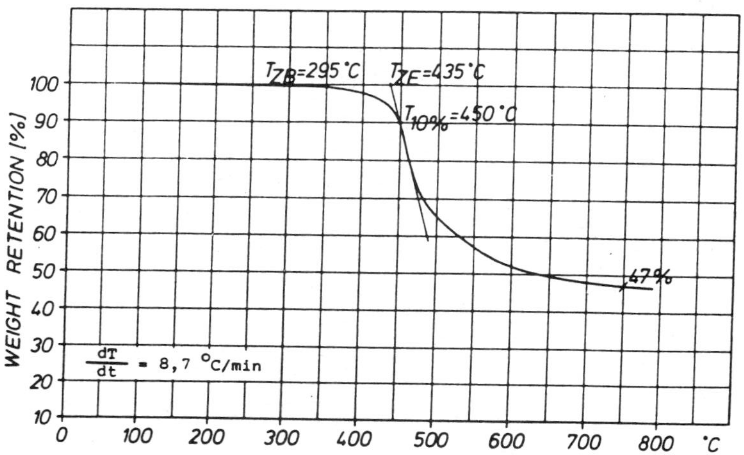

Mechanical Properties of neat Bismaleimide resin

Property	Unit	H 795 RT	H 795 250 °C
Density	g/cm^3	1,32	
Flex. Strength	N/mm^2	90-110	60-65
Flex. Modulus	kN/mm^2	5,3-5,5	3,2-3,4
Flex. Elongation	%	2,2-2,4	2,0-2,2
Expans. Coeff.	10^6 cm/cm/°C	36,6	73,4
Moist. Absorption	%	4,85	

Fig. 11. Properties of TECHNOCHEMIE H 795 resin

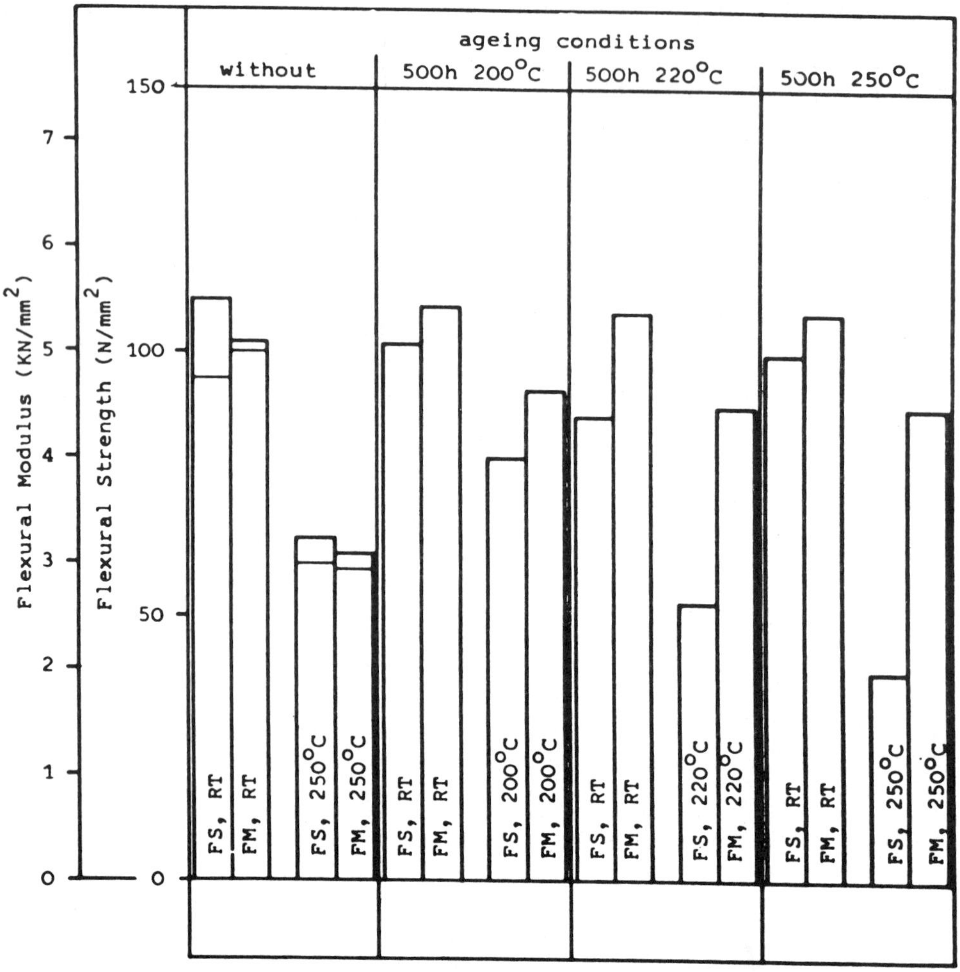

Fig. 12. Ageing of TECHNOCHEMIE H 795 neat resin

composites, this limits their use in so-called "hot wet" applications because the glass transition temperature of the resin is shifted dramatically to lower temperatures. Therefore polyimides are favoured for these applications because of their high glass transition temperature even when loaded with moisture.

Figure 13 summarizes the water absorption of bismaleimides (cured). Samples of TECHNOCHEMIE M 751 and TECHNOCHEMIE H 795 were immersed in distilled water for 1000 h after which time saturation at room temperature was nearly obtained. For comparison neat resin moisture absorption of Kerimide 601 is also presented. All the resins show a moisture absorption of about 4–5% by weight. TECHNOCHEMIE H 795 was also moisturized at 70 °C at 98% relative humidity under which condition moisture absorption takes place much more rapidly but the saturation level is very similar to the cold water immersion. At this point it is of interest to note that bismaleimides absorb moisture much more rapidly than epoxy resins.

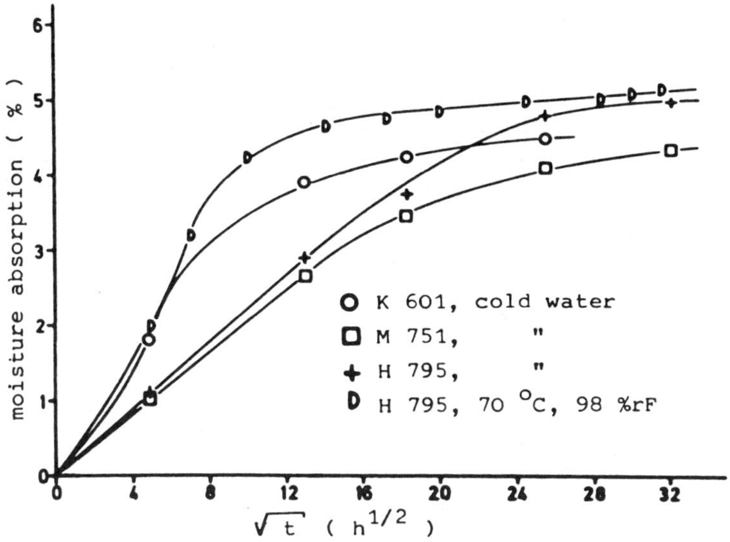

Fig. 13. Moisture absorption of Bismaleimide type resins

9 Processing of TECHNOCHEMIE H 795

9.1 Prepregging of TECHNOCHEMIE H 795 Resin

Prepregging can be performed either from the melt or from solution. TECHNOCHEMIE H 795 resin, which is delivered as a resolidified melt, shows viscosities between 400–2800 mPa.s at 110 °C and the viscosity and the processing temperature of the resin can be reduced further if necessary by blending the resin with "indifferent" solvents like toluene, 2-ethoxy-ethylacetate, diglyme and the like in quantities up to between 10–20% by weight. These solvents show a very low affinity to the resin and can therefore be easily stripped off at low temperatures (50–70 °C).

The bismaleimide resin TECHNOCHEMIE H 795 can also be processed from 70% by weight solutions. This concentration allows the reduction of the impregnation temperature to 50–60 °C but the 70% by weight resin solution has to be kept at a temperature of 50–60 °C during impregnation to prevent the resin from crystallization. The preferred solvent is a 2-ethoxyethyl-acetate (EEA) – methylene chloride (MC) mixture. The EEA-MC ratio should be between 1 : 1 and 3 : 1. Excess EEA prevents skin trapping. Other solvent combinations like EEA and methyl-ethyl ketone (9 : 1, 8 : 2) have also been tried successfully-N-methylpyrrolidone is not recommended because it cannot be dried off satisfactorily.

50–55% by weight solutions are used for prepregging when fibre impregnation is performed by the dip-coating technology and excess resin is allowed to drain back

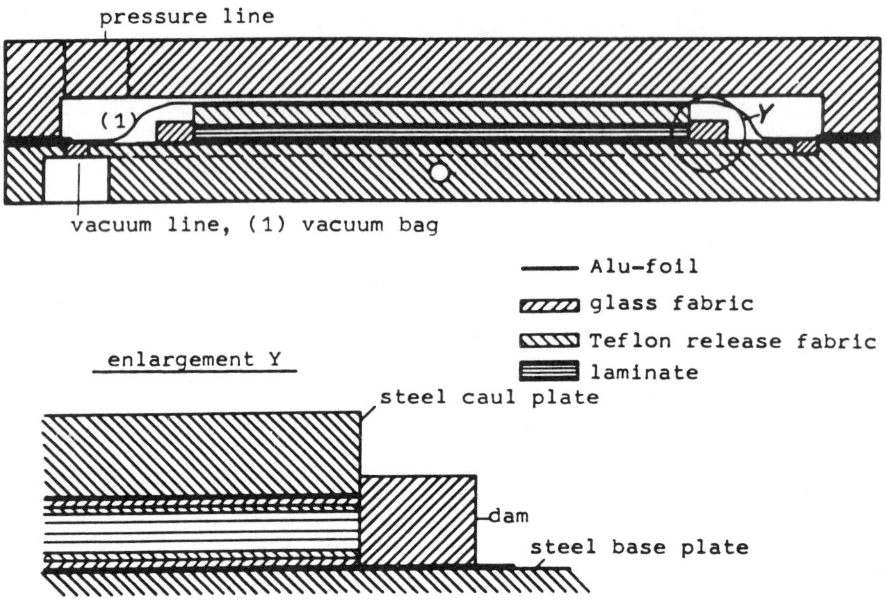

Fig. 14. Processing of TECHNOCHEMIE H 795 resin prepregs

into the resin reservoir. The preferred solvents for 50–55% by weight solutions are 2-ethoxyethylacetate (EEA) – methylene chloride (MC) mixtures. The ratios can vary between 20 : 80 and 80 : 20. Preferably higher EEA concentrations are used to prevent skin trapping.

9.2 Moulding of TECHNOCHEMIE H 795 Resin Laminates

TECHNOCHEMIE H 795 is designed to meet low pressure autoclave moulding procedures. A typical cure cycle is given in Fig. 15 for the basic TECHNOCHEMIE H 795 resin and for an elastomer modified version coded TECHNOCHEMIE H 795 E. The cycle is self-explanatory indicating a maximum cure temperature of 210 °C in the autoclave and a maximum pressure of 4 bars. This cure cycle has been tested successfully within an industrial production-sized autoclave for the manufacture of a complex graphite component. Nevertheless it is possible to reduce the autoclave processing temperature further by using cure accelerators for the bismaleimide resin (H 795 EK, H 795 EKR). When diaza-bicyclo-octane (DABCO) is employed as a catalyst, the autoclave temperature can be reduced to 170–180 °C to provide an "epoxy-like" cure (Fig. 15).

At this point it should be mentioned that bismaleimide cure can be accelerated by many catalysts, examples of which are tertiary amines like trimethylamine, dimethylbenzylamine or diaza-bicyclo-octane, or all types of peroxides. Catalyst type and concentration influence the gel characteristics (e. g. gel times) and the flow properties during moulding. Adversely affected by catalysts is the shelf life of the prepreg, therefore type and concentration of the catalyst employed have to be balanced to obtain the right compromise between shelf life and processing temperature. For TECHNOCHEMIE H 795 we used concentrations up to 0.25% by weight of diaza-bicyclo-octane to obtain the desired reductions of the cure and postcure temperature without losing sufficient shelf life.

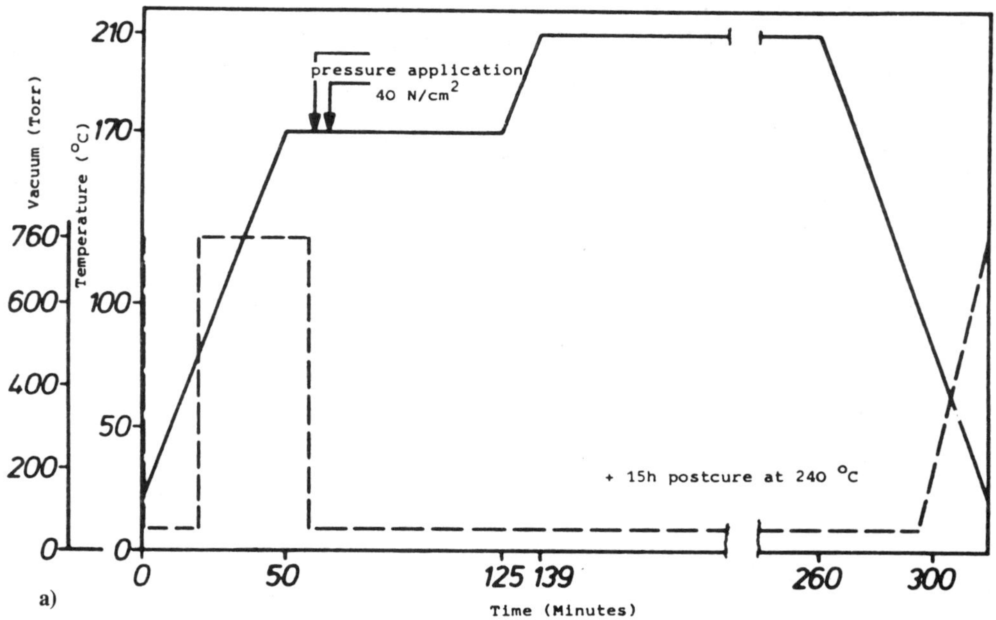

a)

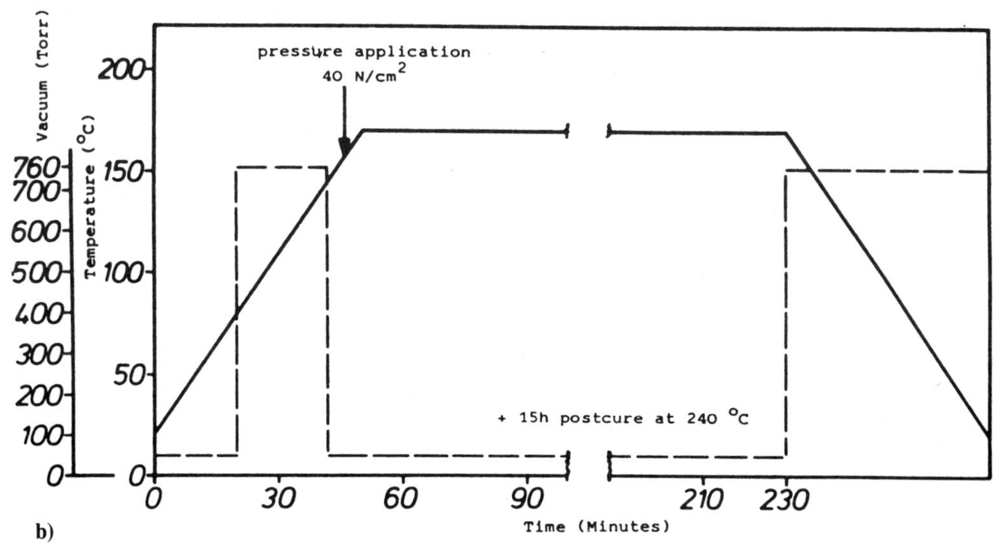

b)

Fig. 15. a Cure cycle for H 795 and H 795 E-laminates, **b** Cure cycle for H 795 EK and H 795 EKR-laminates

9.3 Postcure of TECHNOCHEMIE H 795 Resin Laminates

Processing autoclaves are designed for maximum cure temperatures of 170–210 °C for big parts. These temperatures are not high enough to develop good high temperature mechanical properties in laminates prepared from bismaleimides. Therefore a free standing postcure at high temperatures is necessary.

Practical laminate build-ups consist of symmetric $[0, \pm 45, 90]_s$ lay-ups and we therefore tested the interlaminar shear strength without postcure and after postcure at 210 °C and 250 °C of multidirectional laminates (Fig. 16). Main attention has been paid to the high temperature properties and, as can be seen from Fig. 16, the increase of the postcure temperature increases the high temperature (250 °C) shear strength while simultaneously the room temperature strength is decreased. This reflects the change of the neat resin properties during postcure, in particular the increase of the neat resin elastic modulus.

Similar experiments were performed with the catalyzed TECHNOCHEMIE H 795 resin version (coded TECHNOCHEMIE H 795 EK) and it was found that postcure temperatures of 210 °C are sufficient when the resin is properly catalyzed to develop the high temperature mechanical properties fully.

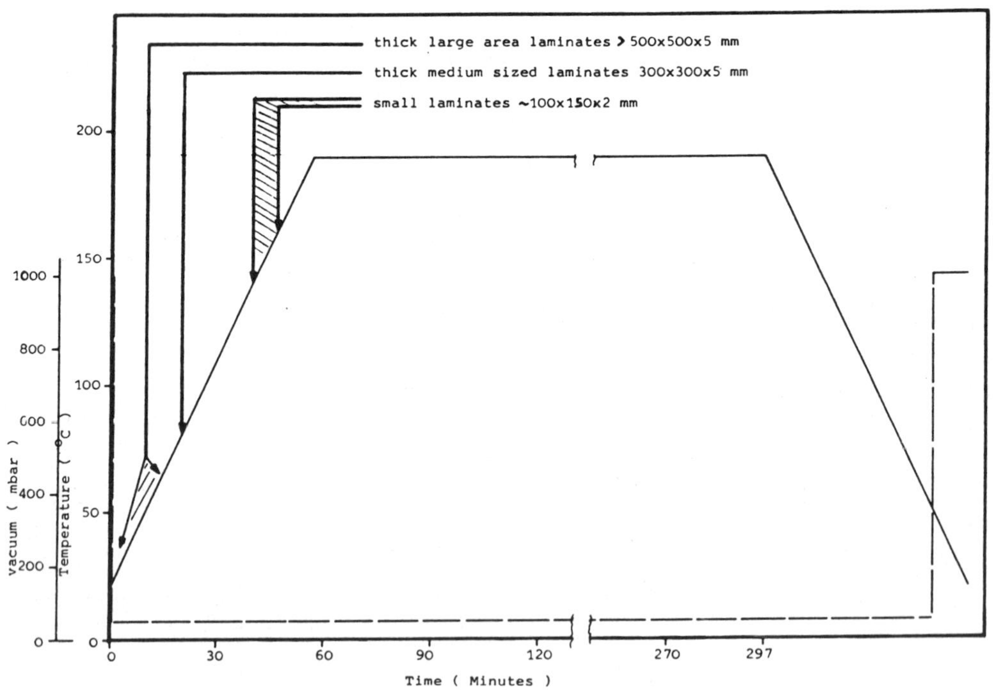

Fig. 15. c Processing of TECHNOCHEMIE H 795 ERK-prepregs

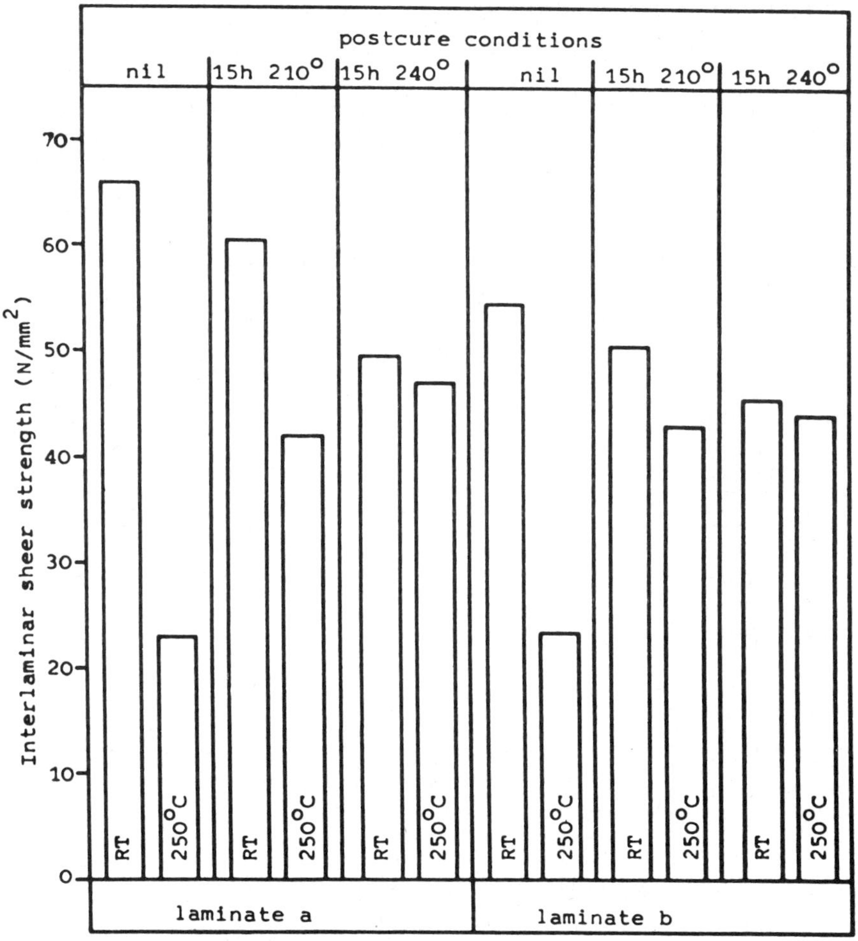

laminate a: 0, + 45, 0, − 45, 0, + 45, 0, − 45, $0)_2$ laminte b: 0, 0, + 45, − 45, 0, 0, − 45, + 45, $0)_2$

Fig. 16. Interlaminar shear strength of multidirectional laminates as a function of postcure conditions (resin: TECHNOCHEMIE H 795, fibre: T 300/3000)

9.4 Laminate Properties

Mechanical properties of laminates prepared from TECHNOCHEMIE H 795 E resin and Celion 6000 (standard epoxy-sized) fibres are given in Fig. 17. Of course the properties at high temperature (250 °C) are of main interest. The matrix dependent properties like the 0°-flexural strength, the 90°-flexural strength and the short beam shear strength are in correlation with the neat resin mechanical properties, e. g. about 60–70% property retention at 250 °C is obvious. The room temperature properties could be expected according to the fibre properties of Celion 6000. The horizontal shear strength of a [0, ± 45] fibre lay-up shows only a very low property

Property	Fibre Direction	Unit	Value 20°C	Value 240°C
Flexural Strength	0	N/mm^2	1850	1240
Flexural Modulus	0	kN/mm^2	117	120
Flexural Strength	90	N/mm^2	53	29
Flexural Modulus	90	kN/mm^2	7,4	5,3
Tensile Strength	0	N/mm^2	1450	1300
Tensile Modulus	0	kN/mm^2	119	128
Compressive Strength	0	N/mm^2	1400	800
Compressive Modulus	0	kN/mm^2	128	120
Horizontal Shear (short beam shear, span to depth ratio 5:1)	0	N/mm^2	96	53
Horizontal Shear	0 ± 45	N/mm^2	48	42

Fig. 17. Properties of TECHNOCHEMIE H 795 E Celion-6000* laminates (* Standard Epoxy-sizing, 1.1–1.3% by weight)

loss as compared with the room temperature value when properly postcured. All multi-directional laminates prepared from TECHNOCHEMIE H 795 E resin are free of microcracks and the void content was found to be negligible.

The mechanical properties of laminates are influenced also by the sizing used on the carbon fibre. It is known that epoxy sizing contributes negatively when used in combination with polyimide matrix resins mainly when these laminates are aged at high temperatures in air. For the bismaleimide resin TECHNOCHEMIE H 795 we tested the standard epoxy-sized fibres and fibres which are sized with a polyimide finish. In this search we employed also the elastomer modified resin TECHNOCHEMIE H 795 E, either catalyzed or non-catalyzed. The results obtained are summarized in Fig. 18. It is obvious that the shear properties of laminates prepared from Pi-sized fibres are higher as compared with laminates made with the standard epoxy-sized fibres. Also the flexural properties perpendicular to the fibre direction are slightly improved when polyimide is used as a fibre sizing demonstrating good compatibility between the polyimide sizing and the poly (bismaleimide) resin TECHNOCHEMIE H 795 or its elastomer modified version TECHNOCHEMIE H 795 E.

Recent research in the area of high temperature epoxy resins indicated negative influence of absorbed moisture on mechanical properties of laminates at high tem-

Property	Fibre Direction	Unit	Test Temp. (°C)	H 795 T 300/EP	H 795 E C 6000/EP	H 795 EK C 6000/EP	H 795 E C 6000/PI	H 795 EKR C 6000/PI
Prepreg Batch No.				x	x	8225	8230	8235
Fibre Content		v/o		61	64	62	67	67
Flexural Strength	0	N/mm^2	23	1646	1850	1761	1992	1957
			250	1378	1240	1217	1575	-
Flexural Modulus	0	KN/mm^2	23	121	117	116	117	121
			250	120	120	114	121	-
Flexural Strength	90	N/mm^2	23	81	53	60	81	62
			250	31	29	37	36	-
Flexural Modulus	90	KN/mm^2	23	8,7	7,4	8,0	8,5	8,2
			250	7,6	5,3	6,3	6,1	-
Horizontal Shear	0	N/mm^2	23	81	96	95	104	109
			250	49	53	53	51	59
Horizontal Shear	0 ± 45	N/mm^2	23	48	48	51	45	54
			250	44	42	43	38	49

Fig. 18. Properties of various TECHNOCHEMIE H 795 resin formulations. H 795 Basic hot melt BMI-resin formulation, E = Elastomer, K = Catalyst, R = Reactive diluent; X average of 3 different Prepreg batches; EP = Epoxy; PI = Polyimide

peratures. Moisture (water) acts as a plasticizer in the laminating resin and degrades the high temperature mechanical properties because of the reduction of the glass transition temperature of the resin. Like many resins, bismaleimides absorb moisture to a high degree (see Fig. 13), but because of their extremely high glass transition temperature (T_G) in the dry state their T_G is still very high when saturated with water. This property makes them prime candidates for application in composites for the so-called hot/wet environment.

We therefore were interested in the moisture behaviour of TECHNOCHEMIE H 795 laminates either prepared from standard epoxy-sized fibres or polyimide (Pi)

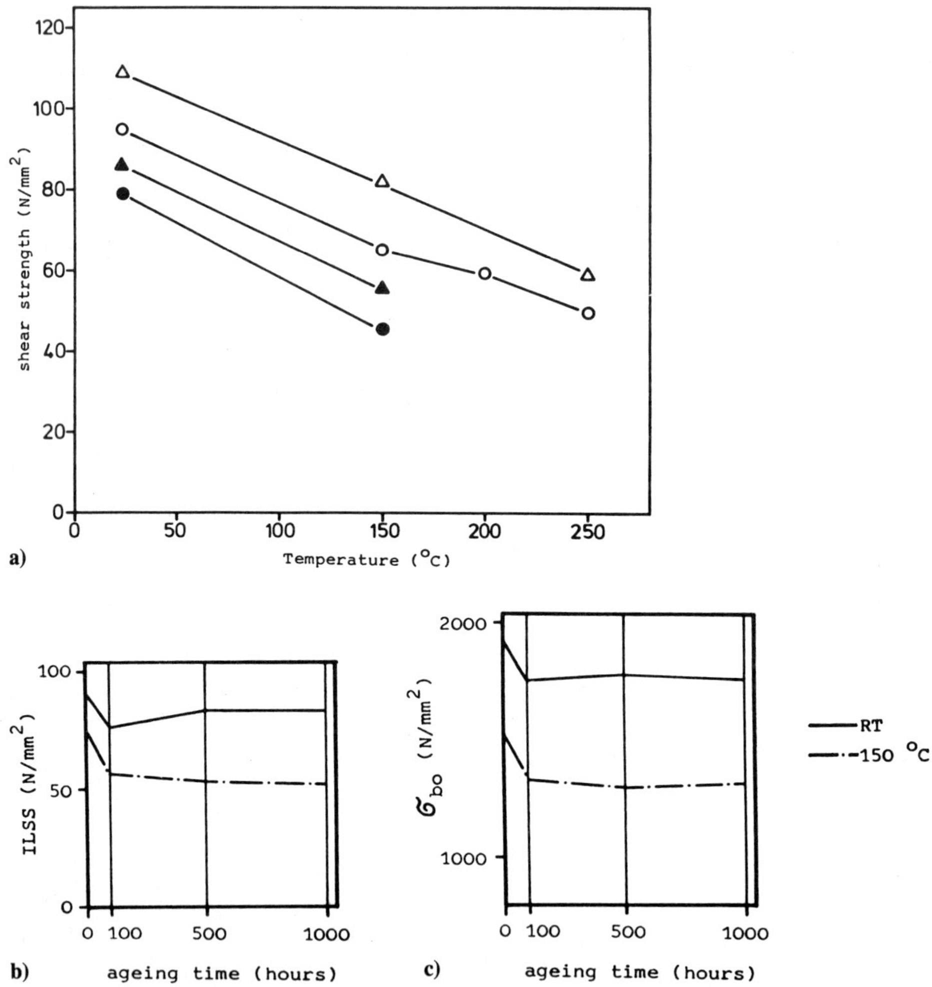

Fig. 19. a Short beam shear properties of TECHNOCHEMIE H 795 resin laminates. △ H 795 EKR Celion 6000/PI, dry laminate; ▲ H 795 EKR Celion 6000/PI, wet laminate (1,71% moisture); ○ H 795 EK Celion 6000/EP, dry laminate; ● H 795 EK Celion 6000/EP, wet laminate (1,61% moisture). Property retention of H 795 ERK/C 6000-laminates, ageing conditions: at 70 °C and 94% relative humidity. **b** interlaminar shear strength, **c** flexural strength

sized fibres. The results are given in Figure 19. Dry laminates of Pi-sized fibres show extremely high shear properties. The shear strength at 150 °C is around 80–85 N/mm^2 for dry unidirectional laminates. The moistured laminate (1.7% moisture absorption) shows a reduced shear strength at both room temperature and 150 °C. At 150 °C the reduction is around 30% as compared with the RT value. It is interesting to note that the epoxy-sized fibres provide shear strength values of around 65–70 N/mm^2 at 150 °C when dry and the shear strength is degraded to around 45–48 when wet. The percentage of property loss is the same as for the polyimide sized fibres which shows that the reduction in shear strength of wet laminates is a matrix dominant property. It is of main importance, according to our observation, that the initial shear values are sufficiently high to provide satisfactory wet shear properties in laminates.

Because of its outstanding high temperature mechanical properties and its good hot/wet properties TECHNOCHEMIE H 795 resin was selected as matrix resin to build a demonstration component, the air brake of the Alpha jet. The component development was performed by Dornier GmbH-Friedrichshafen and the prepregs for this programme were supplied by TECHNOCHEMIE GMBH-Verfahrenstechnik. The demonstration part was tested statically successful at 200 °C and TECHNOCHEMIE H 795 resin will therefore be evaluated further for hardware production.

10 Concluding Remarks

Bismaleimide type resins are prime candidates as matrix resins for graphite fibre laminates for

a) high temperature applications (170–220 °C)
b) applications in hot/wet environment (because of their high T_G in the dry and moistured state).

Because of their easy processability and the chemistry involved, the curing conditions can easily be adjusted by employing curing catalysts, reactive diluents and elastomers. Of main importance is the improvement of the fracture toughness without the reduction of the high temperature properties.

Chapter 3

Fabrication and Design of CFRPs

Fabrication and Design of CFRPs

The Fabrication Processes for High Performance Composites

B. J. Hayes

1 Fabrication Process

The principal methods and usual forms of fibre presentation used for fabricating carbon fibre composites are:

Injection moulding	– short fibre in resin pellets
Pultrusion	– dry fibre/liquid resin or prepreg
Wet winding	– dry fibre/liquid resin
Dry winding	– prepreg
Wet lay up	– fibre as fabric or mat/liquid resin
Resin injection	– fibre as fabric or mat/liquid resin
Vac-Bag-moulding	– usually as prepreg
Press moulding	– prepreg
Expansion moulding	– prepreg
Autoclave moulding	– prepreg.

1.1 Injection Moulding

Of these techniques *injection moulding* is specific to short fibre reinforced resins – particularly thermoplastics – and is usually rather limited in the size of the component that can be produced. Because they do not contain continuous fibre, cannot contain very high levels of reinforcement and yield parts with relatively random fibre orientation – they cannot be considered as primary high performance composites. Nevertheless, these materials are widely used to produce an enormous number of usually relatively small but frequently complex composite parts for a multitude of applications, where the low volume fraction and fibre disposition still gives adequate performance.

1.2 Pultrusion

Pultrusion is a convenient method of manufacturing constant section solid profiles which are largely unidirectional (see Fig. 1). As such it is strictly limited to the relatively small number of applications requiring this type of composite in carbon fibre.

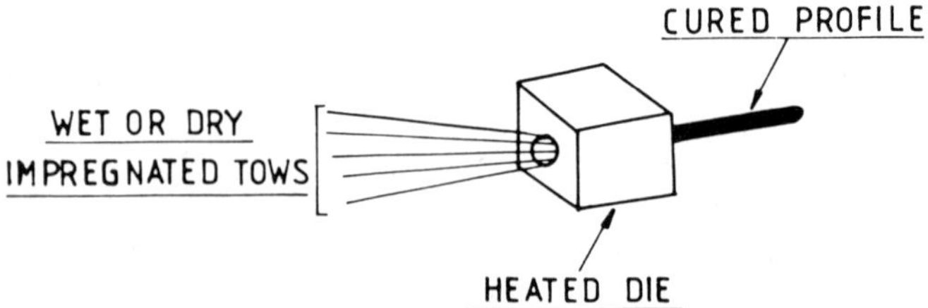

Fig. 1. Pultrusion for manufacturing of unidirectional profiles (schematically)

Fig. 2. Drive shafts made by wet winding technique (power transmission shafts for the LEARFAN 2100) (comp. [1, 2])

1.3 Wet Winding

Wet winding is an important technique but from the CFRP point of view has been mostly limited to pipes as torsion tubes or struts and stiffness reinforcement of pressure vessels and centrifuges. For those intended applications without an axis of symmetry and/or requiring non-geodesic winding or special mandrels, sophisticated computer winding programmes and controls become necessary rendering the whole technique highly specialised. Figure 2 shows as exclusive illustration the long power transmission shafts for the all composite commuter plane LEARFAN 2100, see also [1, 2].

For simple constant section tubes high production rates can be achieved. Tubes 1500 mm × 65 mm with a wall thickness of 3 mm can easily be wound in one minute. This order of speed, however, depends on the use of a roving guide which feeds several resin-impregnated rovings simultaneously and symmetrically to many points on the periphery of the mandrel.

The manufacturing sequence for such tubes, applicable as drive-shafts in automotives, is: –

Resin mixing
Fibre impregnation in dip bath
Winding on mandrel
Wrapping with polyester tape
Gelation
Curing
Mandrel withdrawal
Finishing
Inspection

Resin content is controlled in part by nip-roll pressure to remove the excess after emerging from the impregnation bath and partly by very careful control of tension in the wrapping film to squeeze out any remaining excess.

Finishing operations usually consist of cutting to precise length, machining for end fittings if required, assembly of fittings and varnishing for a good finish. Inspection should consist of measuring dimensions, weight, natural frequency and static and fatigue strength as appropriate.

1.4 Dry Winding

Dry winding and tube wrapping with either prepregged individual fibre tows or narrow tape or even full width prepreg tape is gaining in importance and even represents the main method for the manufacture of fishing rods and golf shafts.

For simple wound shapes as in the previous section, the resin mixing and impregnation stages are omitted, leaving an easier, cleaner operation. Where non geodesic winding is required tacky prepreg can greatly assist in holding the fibre in position. In either case it may prove a production advantage to warm the mandrel during winding. Prepreg tow can also provide tougher resin systems for pultrusion and again omits the use of a liquid resin bath.

1.5 Wet Lay-Up

Wet lay-up techniques are not widely used in carbon fibre fabrication technology. Firstly it is unsuitable for completely unidirectional lay up because of the problems in maintaining fibre orientation – it is thus limited to fabrics. Secondly, wet lay up is not such a controlled process – it is difficult to maintain consistent resin contents, it is difficult to obtain high fibre volume fractions and it is not convenient for the manufacture of honeycomb sandwich structures.

The range of resins that can be used is limited and the cleanliness of the operation, both from workplace and handling points of view, is less than ideal. There have been some developments in aerospace using wet lay-up, but usually this is combined with some form of pressure application during cure. The most likely prospects for carbon fibre wet lay-up in the long term are for repair work or selective stiffening of structures already made in glass, e.g. gliders, racing boats etc.

1.6 Resin Injection

Resin injection is a potentially inexpensive process for the production of a variety of parts but suffers from some restrictions. In principle a low cost mould is required that is capable of withstanding the curing temperature of the resin to be used and which can be opened up to permit the lay up of dry fibre and fabric to the quantity and orientation needed. In addition the mould must have an inlet port for the resin

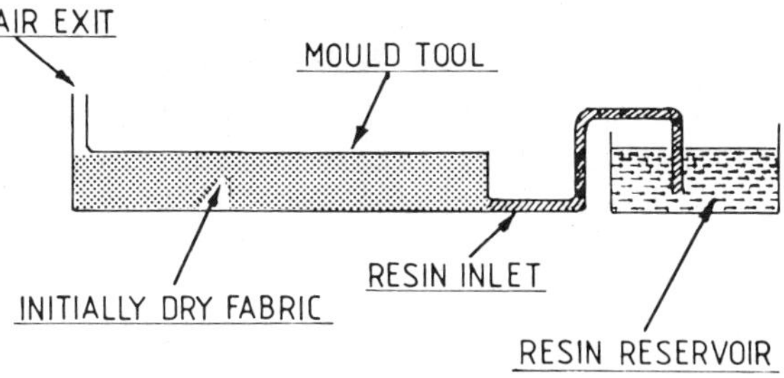

Fig. 3. Resin injection

introduction and an outlet port for the removal of air. When the mould is closed resin is drawn from a reservoir through the fibre, removing the air until the displacement is complete (see Fig. 3). The resin is then cured.

Difficulties may be encountered in placing sufficient fibre in the mould to get a high volume fraction and in finding resin systems with the mechanical properties required after cure but a low enough viscosity to give good impregnation without disturbing the orientation of the fibre.

1.7 Moulding Techniques

1.7.1 VAC-BAG Moulding

VAC-BAG moulding consists essentially of using pressure up to one atmosphere to consolidate the prepreg layers in the manufacture of a part. This is achieved by using a vacuum pump to withdraw the air held between a mould tool, the prepreg and an impervious plastic film which is firmly attached to the mould (see Fig. 4). It is usual practice to heat the prepreg assembly in an oven to cure it, although heated tools may be used as well. This technique is very attractive where expensive autoclaves and presses are not available.

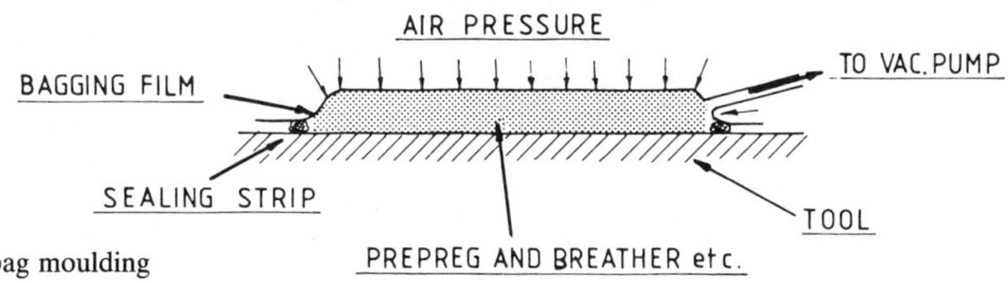

Fig. 4. Vac-bag moulding

1.7.2 Press Moulding

Press moulding is a fairly common procedure and employs matched tools, between which is placed the prepreg needed to make the part. Heat and pressure are applied in the press (see Fig. 5), the resin flows and cures and the part is demoulded usually after cooling. The tooling is often expensive.

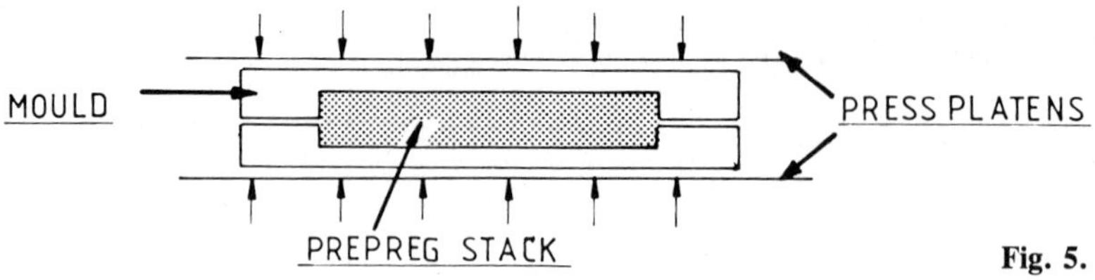

Fig. 5. Press moulding

1.7.3 Expansion Moulding

Expansion moulding is a simple technique requiring some rather sophisticated products and careful part and mould design. The principle is one of producing basically male and female tools from special rubbers – usually silicone – and then laying prepreg between these tools as for press moulding. The whole assembly is then placed in a strong retaining box and placed in an oven at the curing temperature. The rubber tools are so designed that they expand up to the curing temperature when they then exert the correct pressure to consolidate the prepreg and produce a cured part of the correct dimensions (Fig. 6)

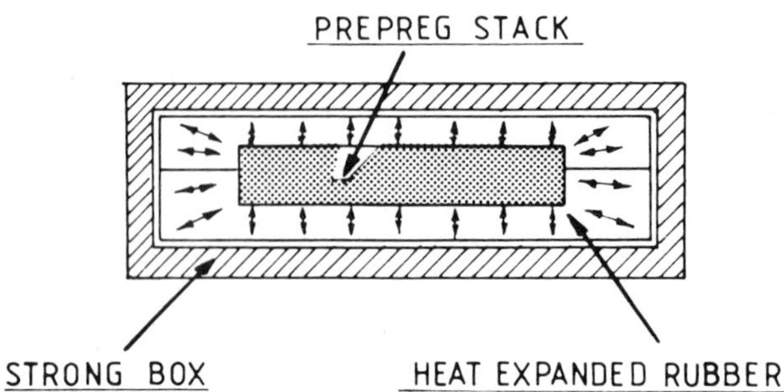

Fig. 6. Expansion moulding

1.7.4 Autoclave Moulding

Autoclave moulding is the principal technique for the production of aerospace carbon fibre composite parts; it operates in a similar manner to vac-bag moulding, except that it has the big advantage of being able to pressurise the part from outside the plastic membrane. This enables removal of air by vacuum inside the bag early in the cure cycle and then application of excess pressure on the outside to ensure good consolidation. The main problem is the high capital cost of the autoclaves in the first place. This important technique is described in more detail by Brunsch [3].

2 References

1. Hayes, B. J.: Experiences in Europe on Industrial and Further Applications of Carbon Fibres, see page 259 this volume
2. Hayes, B. J.: Application of C-Fibre Reinforced Composites in Aerospace, see page 240 this volume
3. Brunsch, K.: Autoclave and Compression Molding, see page 149 this volume

Autoclave, Compression Moulding

K. BRUNSCH

Autoclave and compression moulding are wellknown technologies in use for the fabrication of fibre composite structures. Aside of the general knowledge a lot of details has to be taken care of to ensure troublefree application.

This paper presents detail-information gathered with

- autoclave curing of extremely thin HM CFC (*H*igh *M*odulus *C*arbon-*F*ibre-*C*omposite) face skin of space structures
- autoclave curing of thickness tapered HT (*H*igh *T*ensile CFC skin of an AC taileron
- press curing of thick HT CFC rotorhub components
- and the use of an oven for vacuum bag curing

1 Autoclave-Curing of Extremely Thin HM CFC-Skins

For the OTS (*O*rbital *T*est *S*atellite) and the INTELSAT V satellites CFC solar array structure have been developed and produced. For INTELSAT VI a hybrid CFC-AFC (*A*ramide *F*ibre *C*omposite) solar array is being developed. A common feature of those arrays are the thin face skins.

For example the skin of the INTELSAT V array structure has a nominal thickness of 0.16 mm only (see Fig. 1).

The skins are produced by winding Thornel 75 S on a large mandrel. The uncured laminate is cut off the mandrel and carefully placed on a steel plate.

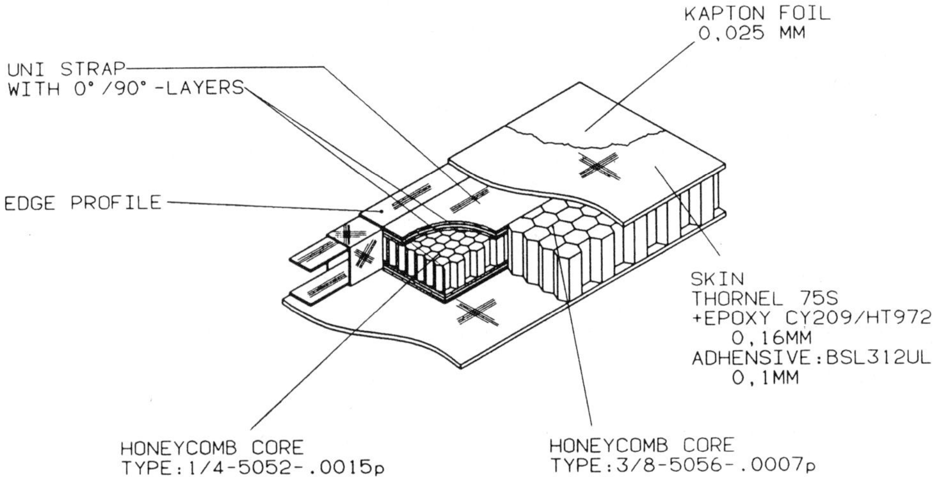

Fig. 1. Intelsat 5 solar array structure

Table 1. Material data

Material	Coefficient of thermal expansion (RT) (K^{-1})	Modulus of elasticity (RT) (kN/mm^2)
Mild steel	11	210
"INVAR"[a]	1.2	200
75 S/CY 209 ‖[b]	-1.2×10^{-6}	298
⊥	40×10^{-6}	3
±45°	-0.5×10^{-6}	11
T 300/CY 209 ‖	0.23×10^{-6}	132
⊥	29×10^{-6}	9
±45°	2.55	16

[a] Composition: Fe 63.1%, Ni 36.2%, Mn 0.4%
[b] Composition: 60 Vol.-%

With curing there is a high risk of damage due to

a) difference of coefficient of thermal expansion of the cured CFC skin and the metal plate and
b) due to memory-effects and coefficient of thermal expansion of auxiliary materials.

The difference of coefficient of thermal expansion becomes effective in the cooling phase. The plate on which the skin is cured contracts and compressive load is introduced to the skin. With the respective material data (Table 1) it can be calculated. Unfortunately the low expansion material "Invar" is expensive and does have long lead time. Therefore a set up using mild steel plate was used for INTELSAT V (see Fig. 2).

The experience was, that the shrinking of the metal tool did contribute to compressive load on the very thin skin and that the skin failed with local instability. The problem was overcome by reducing the curing temperature from 120 °C to 80 °C with vacuum applied until cooling down to 65 °C. To cure the resin completely, a 3 hrs 120 °C post cure was used. Since that no more skins were cured to scrap.

Aside of the metal tool the influence of auxiliary materials like peel-ply and vacuum foil, bleeder lease and so on also must be considered. Production technology of all thermoplastic auxiliary materials includes stretching. At elevated temperature those materials shrink considerably.

Manufacturing trials with hybrid skins, 4 × 4 m wide, 0.14 mm Kevlar and 0.1 mm Carbon fibre fabric composite to evaluate the curing procedure indicated severe problems.

For handling reasons an aluminum plate was used with the intention to let the composite slip.

Therefore the auxiliary materials were not fixed to the plate. The cured skin was scrap because of severe wrinkling.

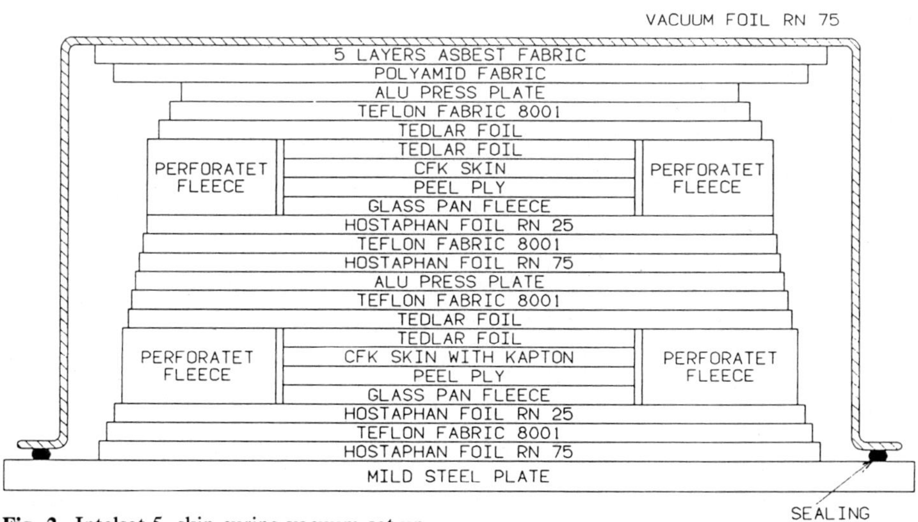

Fig. 2. Intelsat 5, skin curing vacuum set-up

Table 2. Auxiliary material data

Material	Shrinkage 120 °C, 4 hrs	Coefficient of thermal expansion K^{-1}
Peel-ply	1.5%	54×10^{-6}/40%[a]
Tedlar	1.5%	
Pan fleece	1.5%	70×10^{-6}/15%[a]
Perforated foil 4500	3.6%	

[a] percentage of resin picked up/per weight

Thermal behaviour of the materials used was investigated. Both shrinkage of auxiliary material and very high coefficient of thermal expansion (Table 2) of fleece and peel-ply after curing of the picked up resin were found to cause the wrinkling. The problem was overcome by use of a "Kevlar-plate", about 2 mm thick and by fixing everything to this plate.

2 Autoclave-Curing of Tapered Skins of AC Taileron

In the recent years quite a lot of CFC usage for aircrafts concentrated on control surfaces. In the FRG a CFC taileron for the Tornado and horizontal and vertical CFC stabilizer for the Alphajet have been developed. The high concentration of load at the inboard end of the taileron and the service temperature of up to 135 °C made the development a technologywise demanding task.

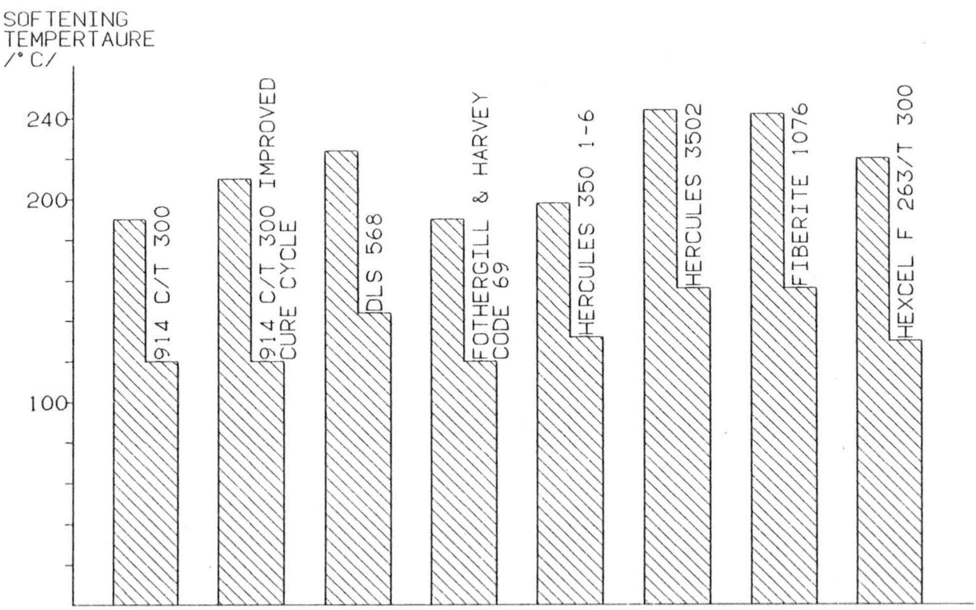

Fig. 3. Softening temperature, dry and after storage at 70 °C 75% rel. humidity/Torrison pendulum test

The curing technology to be used for such components does depend on a few major items: Use of bleeding or zerobleed prepreg and use of a flow controlled or not controlled resin system. For a fast flying aircraft a resin system with a high softening temperature should be used – for ease of production a flow controlled system is desirable. Unfortunately today's ingredients to control flow tend to reduce especially wet softening temperature (Fig. 3).

With a flow controlled system both for bleed and zerobleed an example is presented.

The lay up scheme for the taileron skin is shown in Fig. 4. The curing set up for a bleeding system (6% resin bleed from 40 to 34% by weight) is shown in Fig. 5. It becomes evident, that careful tapering of the bleeder lease layers is to be used to receive uniform resin content. The bleed technology results in low porosity on one hand, but also in high labour cost on the other hand. Therefore, zerobleed technology was investigated, too. The set up for curing the skin with zerobleed material is shown in Fig. 6, the simplification is evident.

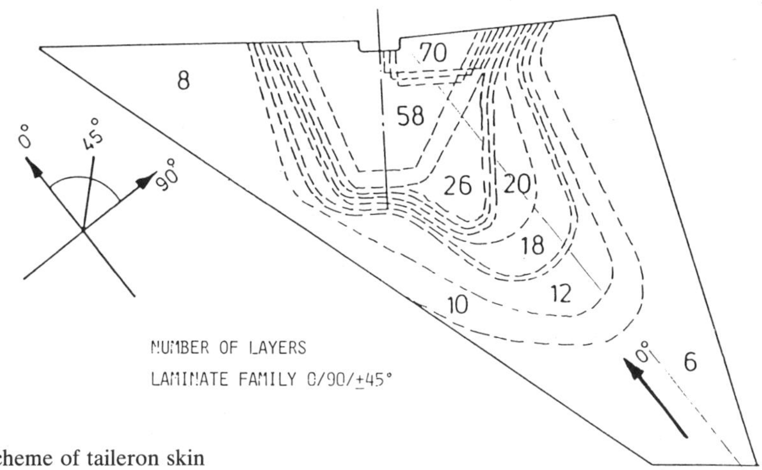

Fig. 4. Lay-up scheme of taileron skin

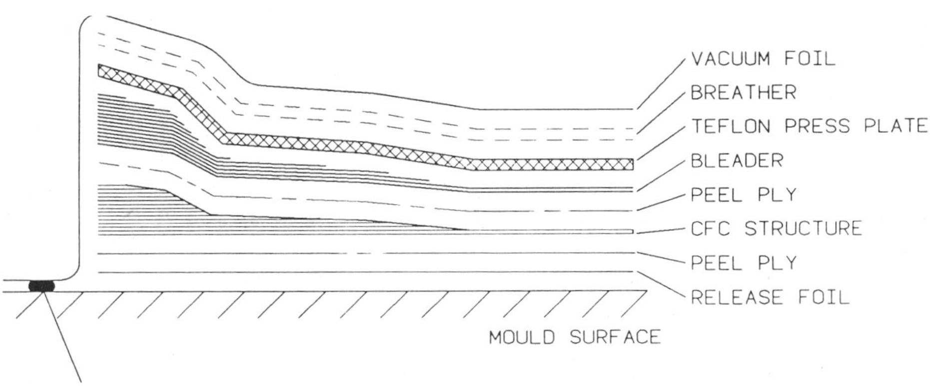

Fig. 5. Curing set-up for bleeding system

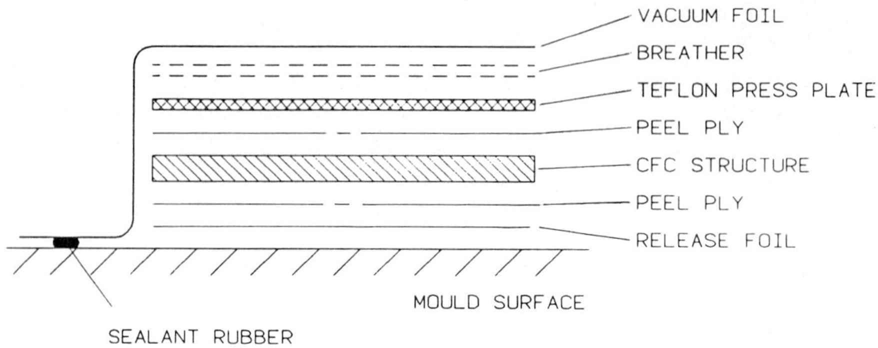

Fig. 6. Vacuum set-up for zero bleed prepreg

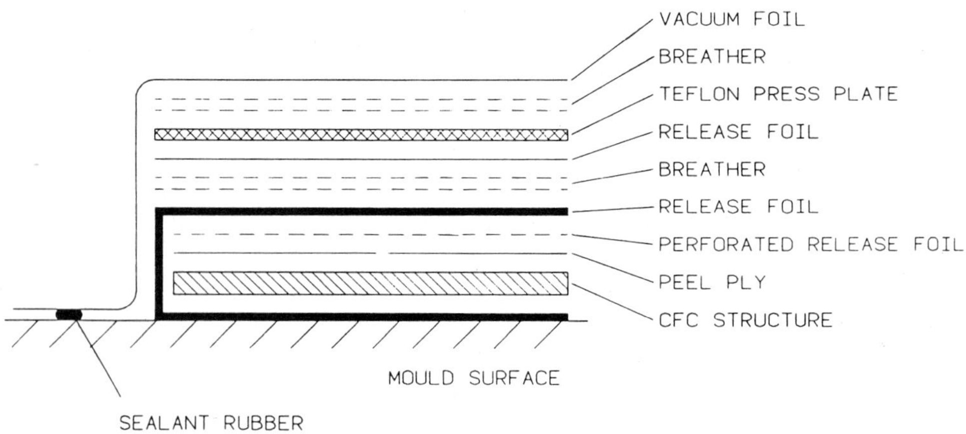

Fig. 7. Vacuum set-up for non flow controlled prepreg

Both examples shown make use of a flow controlled resin system (Ciba 914). Using a non flow controlled system – i.e. low viscosity – a completely different curing set up must be applied. A typical set up for the Fiberite 1076 system is shown in Fig. 7.

With such a system only flat components can be autoclave cured with a rigid mould on one side only. If there is any remarkable vertical extension a second rigid tool surface must withstand the hydraulic pressure of the resin.

Aside of the differences in the technology of vacuum set up, mould concepts are very important for cost effectiveness. For composite components with a thickness about equal or exceeding 2 mm mild steel or electroformed nickel are appropriate materials for the mould. In the recent years the technology of coldforming of mild steel has improved considerably.

The mould shown in Fig. 8 was produced to the close tolerances specified in the drawing. Moderatly double curves shapes also can be made.

In Europe the prices for moulds produced with the technologies mentioned before, are about 3.000.– DM/m^2 for mild steel and 6.000.– DM till 8.000.– DM/m^2 for nickel, respectively, and therefore far below those of NC-milled moulds.

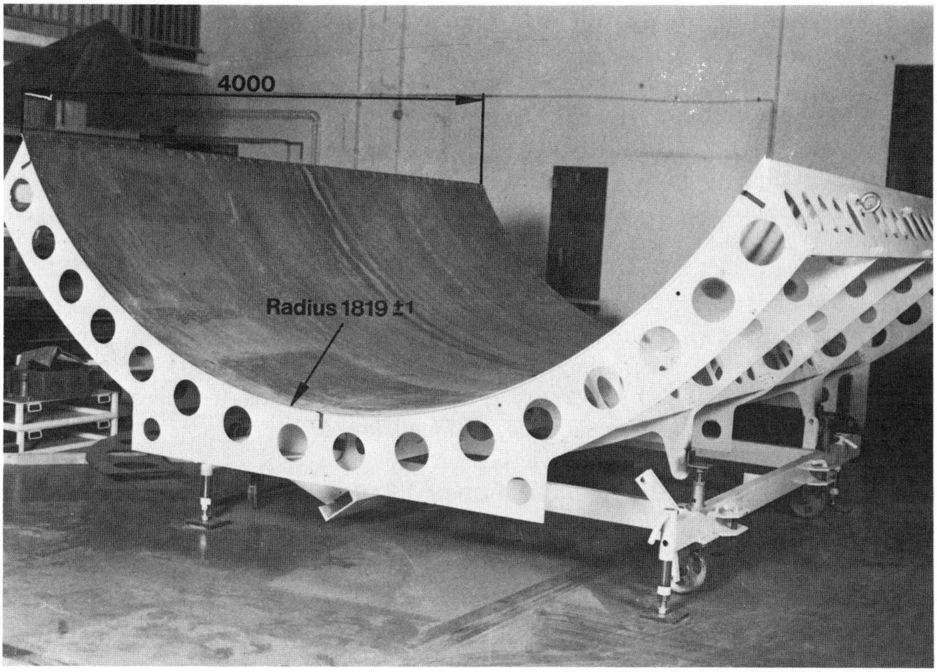

Fig. 8. Sheet metal mould

Fig. 9. Rotor hub-plate

3 Compression Moulding of Thickwall CFC-Components

The fixed wing utilization of CFC normally leads to thin to medium thick wall components. Rather thick components out of composites are used for rotorhubs, for example for the wellknown "Starflex". Whilst this is a glassfibre design most of today's composite hub design use carbon fibres. Experience gained with the development of a carbon *F*ibre composite *E*lastomeric (FEL) bearing rotorsystem is tabled.

The structural components of this hub are two plates, each 55 mm thick, made out of C fabric-prepreg Fibredux 913C-815-40 (Fig. 9). For cost effectiveness machining was to be reduced to the minimum possible, a plate thickness of 55–0.5 mm after curing was requested, also uniformity of fibre volume fraction. To reach this objective the following production sequence was developed:

- Production of 136 oversized single layer cuts
- Presscuring of each 8 cuts to a thickness of 3.2 mm at 85 °C for three hours. This treatment consumes about 22% of the exotherm reaction energy.
- Cutting of the partially cured 8 layer prepreg by means of steel rule die to final dimension − 0.3 mm.
- Laying of the 17 partially cured cuts into the mould (Fig. 10).
- Curing in the closed steel mould – pressure to completely close the mould 5–8 bars, curing cycle see Fig. 11.
- Removal of hot tool from press, disassembly of bolts to prevent the tool from shrinking onto the component.

Fig. 10. Mould of hub-plate

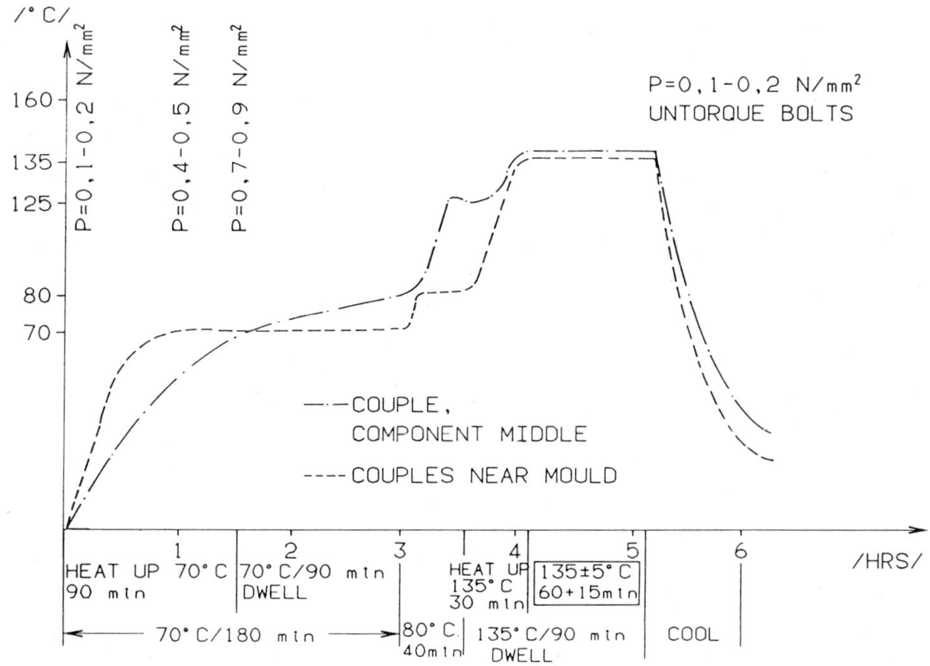

Fig. 11. Curing cycle

Dimensional control:
Since second part the thickness of the plates varies from 54.5 to 54.8 mm, the requirements are met.
The experience to be drawn from this R + D program is:
- Curing of thick prepreg laminates requires special methods to avoid overheating – steel rule dies are very cost effective for medium quantity precision prepreg cuts up to about 5 mm in thickness
- With appropriate tool design conical walls can be avoided
- Rather close thickness and fiber volume tolerances can be met.

4 Advanced Curing Concept

Cost effective utilization of autoclave curing in aerospace industry always creates logistic problems. Small autoclaves are not cost effective and serial quantity production, large ones need to be filled up to be so. Turn around time of individual moulds to produce a given number of parts per day can be manifold of that required if the tools could be used continously. The general tendency, therefore, is to try to make use of curing in an oven with vacuum applied to the laminate.

The MBB-Laupheim facility has installed one of the most advanced production lines for vacuum-oven curing of composite structures. The equipment is used for glass- and aramide phenolic prepregs now but also can be used for carbon fibre components.

The most important features of this productionline are: Computer controlled transportation system, a continuous furnace, four axis adjustable working places for manual prepreg application and a powder-primer spray – and cure equipment.

Production is started by feeding into the computer the drawing number of parts, the number required and the dates when the parts should be completed. The computer then defines the sequence for the moulds to be taken from the store and controlls their transport to the primer application unit. The moulds are coated with powder primer, the powder is cured in approximately 2 min by both infrared radiation and hot air circulation. At the same time the prepreg – precuts necessary for each individual part are assorted into steelwire baskets. These baskets are attached to the coated moulds, which are transported to the intermediate store. From there, again in a sequence defined by the computer, transportation to laminating place becoming free is done. After application of prepreg, peel-ply, release foil and so on a reusable silicon rubber mask is assembled to the mould. The set up is leak proof tested.

During transportation to the intermediate store before the continuous furnace and during storage vacuum is maintained by special vacuum valves which don't allow air to flow in the bag after lines have been cut off. During curing in the continuous furnace movable vacuum lines are connected to the moulds. After curing moulds are automatially transferred to demoulding and mould cleaning stations. Then release is applied and moulds are brought back to the mould store again.

The computer controlled transportation system does provide good traceability in the laminating shop. Calculation of piece work based salaries is eased also.

Although the investment was in excess of 10 Mio DM an average 30% reduction in cost per part was achieved by installation of the equipment described above.

The paper presented gives an overview only. The author will be happy to give detailed verbal information on request.

Design Preparations for Large Space Structures

H. W. BERGMANN

The near future will witness the construction and deployment of large space structures intended to serve as advanced research laboratories, production facilities for high quality materials, communication systems, or the like. The German Aerospace Research Establishment (DFVLR) contributes to this trend by performing design preparations for typical components of space structures. Emphasis is placed on the characterization of fiber-reinforced materials under space conditions, the development of computer programs for optimal design and analysis practices, and the construction of space-type frameworks for test and validation purposes.

1 Introduction

Worldwide, a large number of preparatory studies for large space structures has already been conducted. What form the European engagement might eventually take is currently not well defined. In the expectation, however, that all large space structures will display three-dimensional frameworks as common elements, the DFVLR is directing their efforts towards the response of such frameworks under space conditions. It is stipulated that the material selection will favor carbon-fiber-reinforced resins because of their excellent mechanical and thermal properties, and that the space structure will travel in a geostationary orbit.

The partially completed activities concentrate on three major tasks:

1. the evaluation of several fiber-resin systems before and after realistically simulated exposure to space environments,
2. the development of computer programs for the design and analysis of space structures and
3. the validation of the computer programs against reasonably large test structures exhibiting typical aspects of a space framework.

2 Evaluation of Fiber-Matrix-Systems

Emphasis was placed on the evaluation of the influence of thermal cycling and of electron irradiation on the mechanical properties of the fiber-matrix systems identified in Fig. 1. They included high-strength as well as high-modulus fibers embedded in different epoxy resins with curing temperatures from 120 °C to 190 °C, and one polyimide resin with a curing temperature of 210 °C. The test program was conducted with 195 × 65 mm large laminates, laid up from 0.125 mm thick prepregs

Material Designation	Type of Fiber Type of Resin	Curing Temperature	Symbol
HY-E 1548 A1 B	Celion GY-70 Fiberite 948 A1	120 °C	*
LY 556/HY 917/XB 2692/ T 300	Toray T 300–6000 Unmodified Epoxy	140 °C	□
HY-E 2034 D	Thornel Pitch Fiberite 934	180 °C	△
914 C-TS-5	Toray T 300 Ciba 914	190 °C	○
T3T F 178	UC T 300–3000 Hexcel F 178	210 °C	▽

Fig. 1. List of materials

and autoclave-cured with a volume fiber-fraction of 60%. A stacking order of $[\pm 45]_{2s}$ was chosen to produce matrix-critical test specimens and to facilitate the tension and compression testing. The thermal cycling was simulated in a closed test chamber at a vacuum of 10^{-4} Pa, the heating to 100 °C being applied by infrared radiation, and the cooling to − 160 °C by radiation from a LN_2-filled shroud. One laminate of each material was subjected to 0, 1170, 2295 and 3480 thermal cycles, the latter corresponding to a 10-year service life of a space structure. Prior to cycling the laminates were completely dried and then kept dry by storage in a desiccator. Subsequent to thermal cycling the laminates were dissected into six 110 × 10 mm tension test specimens and six 80 × 10 mm compression test specimens. Tension and compression tests before and after each cycling increment were conducted both at 23 °C and at 100 °C with three specimens per test point [1].

The electron irradiation was simulated in a van-der-Graf accelerometer, at 100 °C and in vacuum. An exposure at 165 μA over a period of 27 h produced ca. 3×10^8 rads, i.e., a multiple of the expected irradiation during a 20-year service life. Laminates from three different materials were irradiated and their mechanical properties before and after irradiation compared.

2.1 *Influence of Thermal Cycling*

Because of the different coefficients of thermal expansion of carbon fibers and polymeric resins, prestresses are induced in the laminates during the curing process. The magnitude of the tensile stresses in the resins depends on the difference between the congealing temperature of the resins and the service temperature of the laminates. Repeated exposures to very low temperatures may lead to accumulating damages particularly in matrix-critical laminates. Figure 2 and 3 summarize, relative to the initial strengths, the residual tension and compression strengths at 23 °C and 100 °C of test specimens cycled 1170, 2295 and 3480 times. The symbols identify the fiber-matrix systems in accordance with Fig. 1. It is apparent that the residual strengths of all epoxy-laminates were only moderately affected, whereas the polyimide-laminate was drastically degraded.

A substantial amount of the strength reduction is due to matrix cracking of the resins occuring already during the first few cycles. Such cracks, as shown in Fig. 4, were quite pronounced in the polyimide-laminate where, apart from the transverse cracks in the 45°-plies, incipient delaminations developed between adjacent plies. Matrix cracks were found also in the plies of the epoxy-laminates; however, as typically displayed in Fig. 5, they were much less severe and no delaminations were detected.

In comparison to the strength reductions, the residual stiffnesses of the laminates tested at 23 °C and 100 °C declined similarly but more severely. Figure 6 displays the decay with increasing numbers of cycles in terms of secant moduli established at constant 0.2% strain values. The symbols correspond to the identification of the fiber-matrix systems in Fig. 1. Figure 7 shows actual stress-strain curves for 914C-specimens at 100 °C. The continuous stiffness decline after increasing numbers of thermal cycles is not easily explainable by matrix cracking alone. Microphotographi-

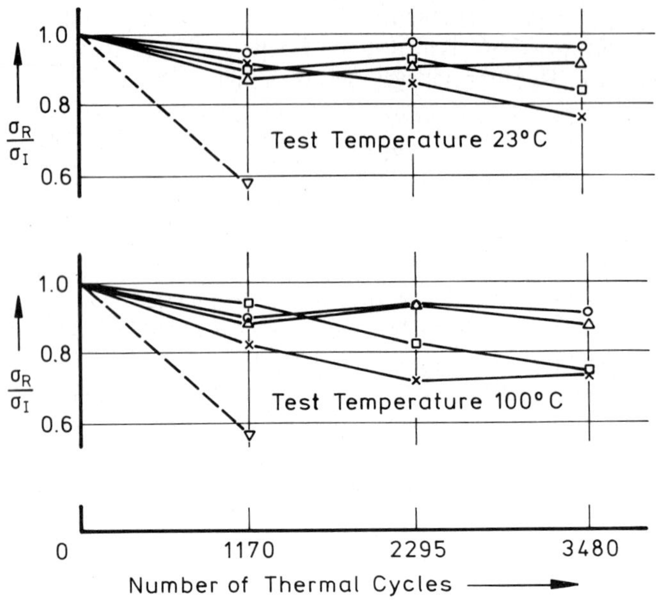

Fig. 2. Residual strength of tension specimens

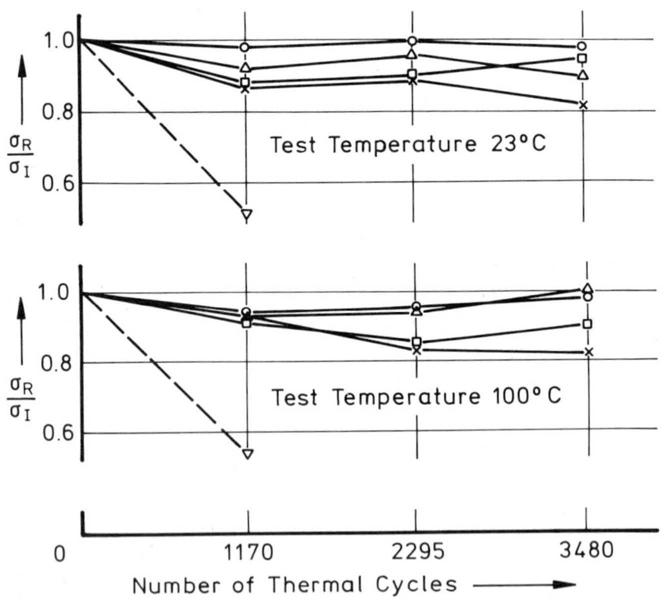

Fig. 3. Residual strength of compression specimens

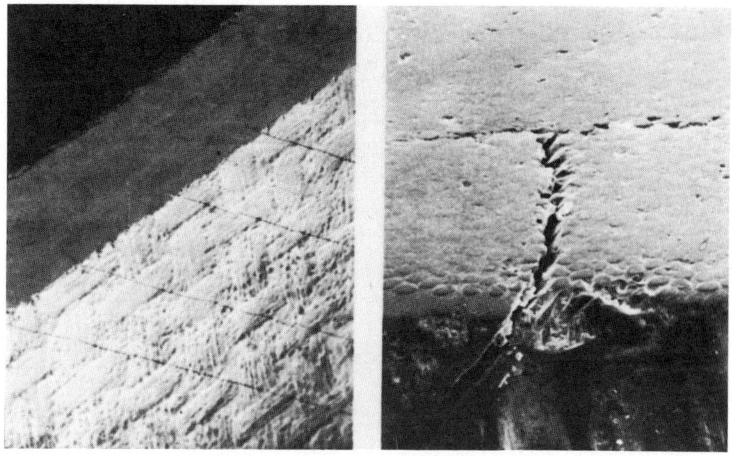

Fig. 4. Matrix cracks and incipient delamination in polyimide-laminate after 1170 cycles

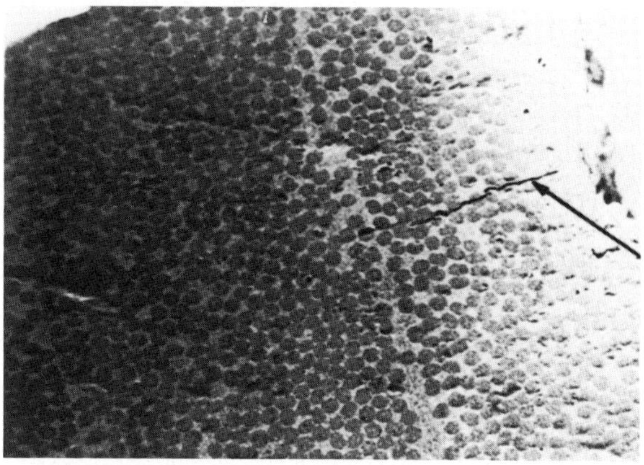

Fig. 5. Matrix crack in epoxy-laminate after 1170 cycles

cal observations of the fracture surfaces of test specimens point to the probability that severe thermal cycling also tends to deteriorate the bond between fiber and matrix. According to Fig. 8 the fiber surfaces prior to cycling were covered with many specks of matrix material, while after cycling the fiber surfaces are relatively smooth, indicating a different quality of adhesion.

The effect of thermal cycling on the coefficients of thermal expansion of the laminates is depicted in Fig. 9. As expected, the reduced stiffnesses because of matrix cracks and bond deterioration lead to a general increase of the coefficients.

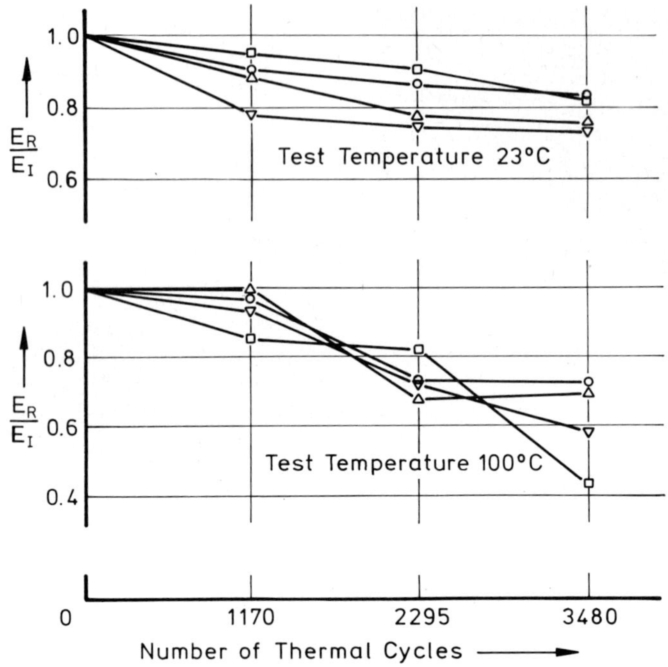

Fig. 6. Stiffness degradation due to thermal cycling

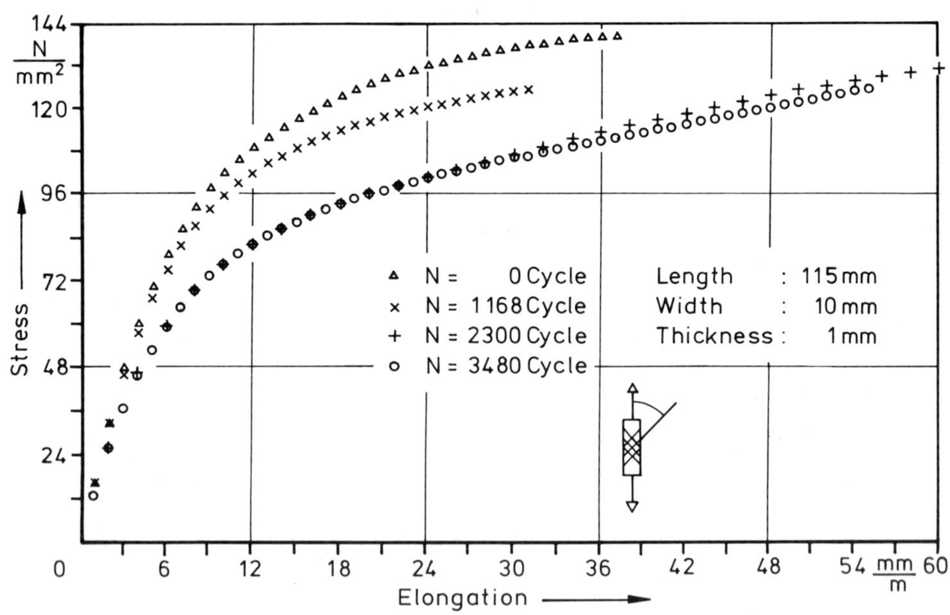

Fig. 7. Stress-strain curves for 914/T300-laminate at 100 °C

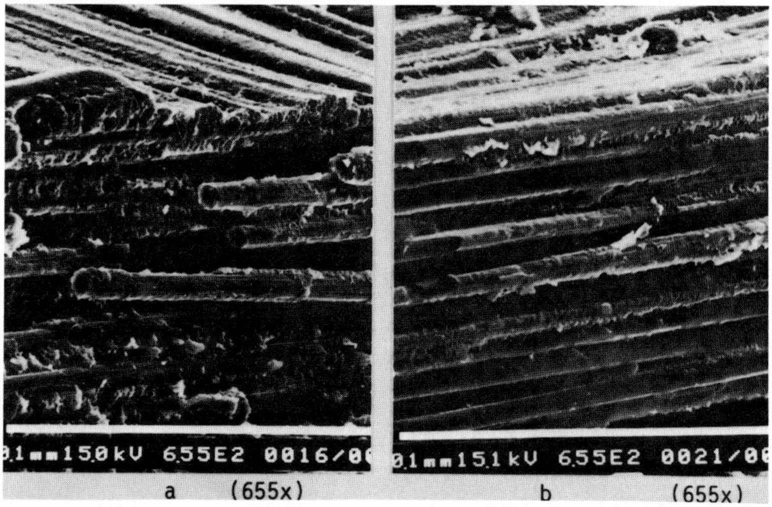

Fig. 8. Fracture Surface of 914/T300-laminate **a)** before and **b)** after thermal cycling

Number of Cycles	Laminate Type (Ref. Fig. 1)			
	○	×	□	▽
0	1.9	0.6	1.5	–0.9
1170	–	–	1.7	–
2295	1.9	0.6	2.0	–0.9
3480	2.0	0.6	2.4	–0.8
	$\times\ 10^{-6}$ m/m °C			

Fig. 9. Coefficients of thermal expansion

2.2 Influence of Electron Irradiation

After irradiation, all of the laminates exhibited a slight reddish tint on their exposed surfaces which was not further investigated.

Figure 10 shows a comparison of the ultimate strengths and elongations at failure of the test specimens before and after irradiation. Whereas only slight variations were noted in the epoxy-specimens, more severe reductions occured in the polyimide-specimen. Non-destructive and microphotographical investigations gave no evidence of cracks or other degradations.

3 Development of Computer Programs

Large space structures are subject to design constraints which include mechanically and thermally induced loads as well as prescribed displacement limitations at a

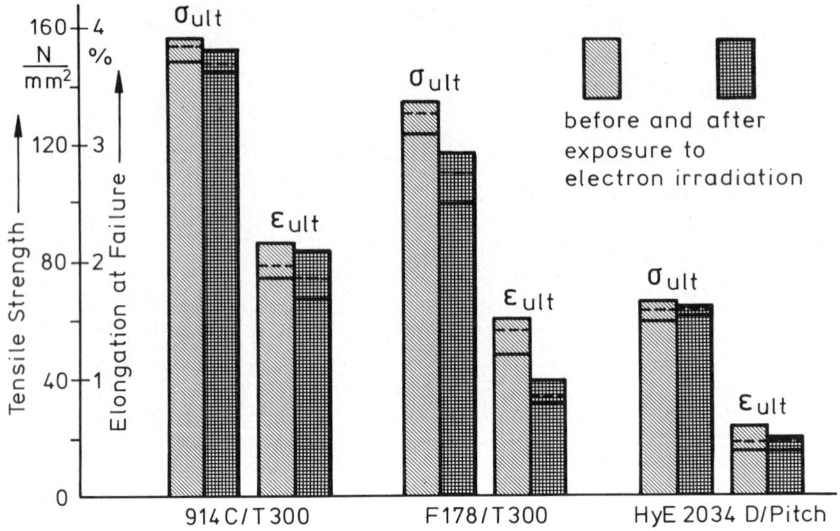

Fig. 10. Degradation due to electron irradiation

minimum of structural weight. The qualification of such structures by ground tests may become physically impossible. The generation and validation of improved analytical tools for the forecast of their structural response is, therefore, a mandatory task. The related efforts of the DFVLR center on the weight optimization of three-dimensional frame works, on the general approach to finite-element modelling of large space structures, and on the non-linear response of large structural systems.

3.1 Weight Optimization of Tubular Components

The three-dimensional frameworks envisioned in this study consist of straight thin-walled tubes interconnected at their nodal points by rigid joints. Of issue here is the optimization of the buckling strength of laminated tubes with the angular directions and the stacking order of the individual plies as design variables. It is stipulated that the plies are arranged pairwise under angles of $\pm \alpha_i$ so that the laminates are balanced but not necessarily symmetrical. Some of the peculiarities of the optimization problem are illustrated in Fig. 11 which displays, in the form of contour lines, the buckling loads of a cylinder consisting of two pairs of plies of thickness t = 0.25 mm and with angles $\pm \alpha_1$ and $\pm \alpha_2$ varying between 0° and 90°. A generalized solution developed by the DFVLR Institute for Structural Mechanics is predicated on a classical buckling theory of ideally shaped cylinders and on appropriate mathematical search routines [2]. Figure 12 demonstrates the potential of the optimization process by the substantial difference between the buckling strengths of cylinders with optimized and pessimized stacking arrangements and wall thicknesses ranging from one to eight pairs of plies. It has been proven that the program is valid

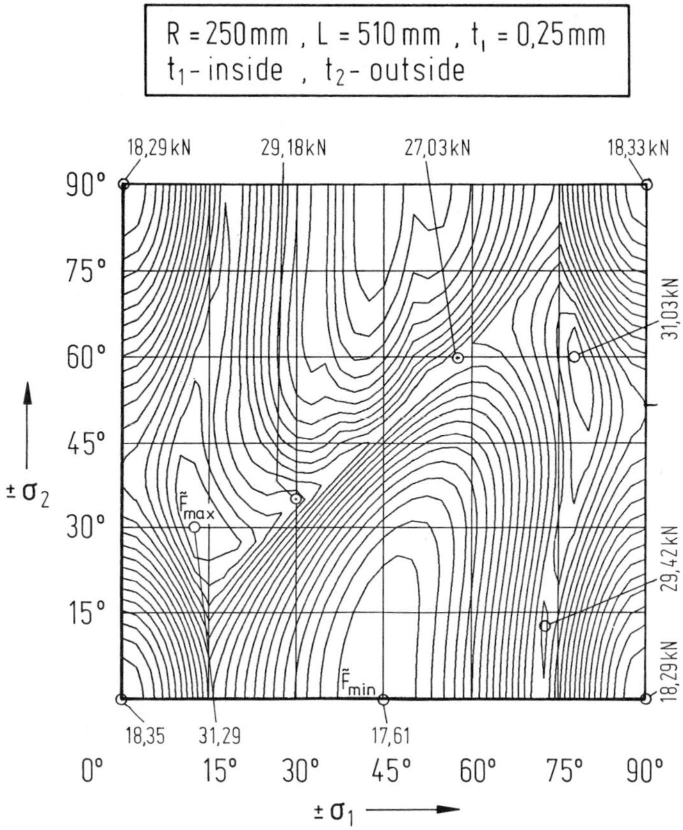

Fig. 11. Buckling strength of cylinder with ply-orientations $\pm \alpha_1$ and $\pm \alpha_2$

for a wide range of cylinder lengths and radii, and that a restriction of the variables $\pm \alpha_i$ to discrete values 0°, 45° and 90° does not significantly affect the maximum buckling strength if the wall thickness exceeds four pairs of plies.

3.2 Structural Modelling Techniques

Taking for granted the numerical realiability of available computer codes, the accuracy of structural analyses depends on the quality of the physical or mathematical models and on the accompanying input data. Depending on the complexity of the design, it may be irresponsible to perform an analysis of the complete structure solely on the basis of material test results, especially, when the analysis cannot be substantiated by a full-scale test. The problem is accentuated in the case of laminated structures with their inherent uncertainties in regard to prestresses induced by curing or environmental effects. The credibility of the structural analysis would gain

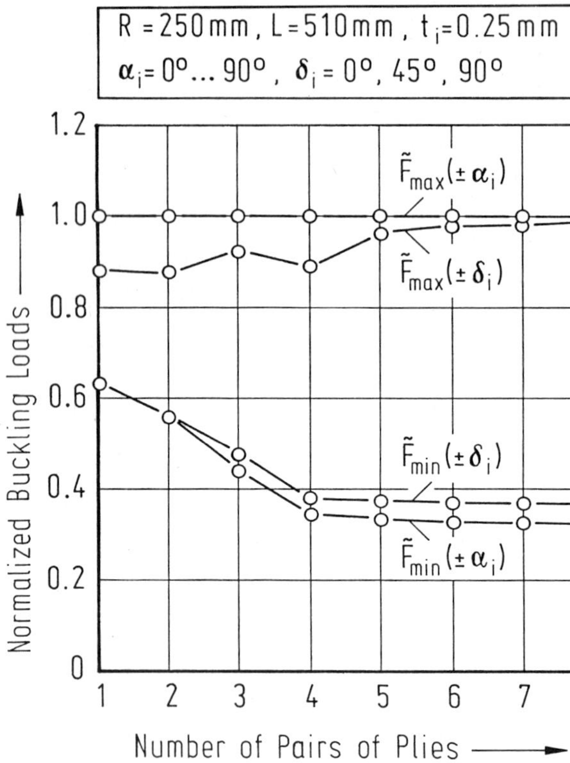

Fig. 12. Potential of buckling coad optimization

by preceding correlations of analytical and experimental results of smaller components. In three-dimensional frameworks with recurring geometry such components could be representative struts or joints. On a higher level, the predicted responses of substructures comprising several components may be substantiated by verification tests and correlation. Figure 13 depicts the various options in schematic form.

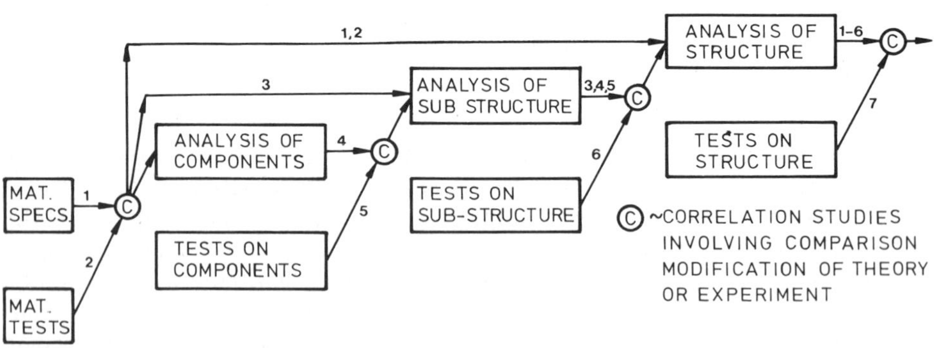

Fig. 13. Alternate paths for design validation

The DFVLR Institute for Structural Mechanics has partially succeeded in identifying the critical parameters for static and environmental analysis of laminated structures and in determining the minimum dimensions of components or substructures for accurate predictions of the static behavior and the eigenmodes of three-dimensional frameworks [3]. The nature of the test structure is described in the following section.

With respect to the dynamic qualification of large space structures a similar modelling concept is being pursued by the DFVLR Institute for Aeroelasticity [4]. Model synthesis techniques are in an advanced state of development which, on the basis of ground-tested eigenmodes, eigenfrequencies, amplitudes and damping properties of appropriately selected substructures, allow the prediction of the dynamic response of the complete structure. As only a truncated set of mode shapes can be accomodated, the convergence of this procedure depends on the descriptiveness of the selected mode shapes and, especially, on the realistic assessment of the coupling conditions during the substructure tests and in the coupling analysis. The fully computerized program has been successfully applied to several objects of the European Space Agency.

The special problem of structural non-linearities also requires a combined experimental/analytical approach. The currently used test techniques, however, lead to difficulties if an accumulation of non-linearities is introduced by, for example, active damping devices. A new approach to this problem is being pursued in which the non-linear elements are replaced, physically, by artificial linear elements representing the actual stiffness energy. The non-linear aspects are investigated in separate tests and then reintroduced into the mathematical model.

4 Test Structures for Validation of Analytical Techniques

A convincing proof of the reliability of analytical procedures can be provided only by comparison to the response of test structures. In the case of large space structures the establishment of this proof is encumbered by the very size and cost of a realistically designed structure. Considering, however, that major aspects of a test program can be satisfied with a test configuration displaying, if not all details, but the important features of a large space structure, a reasonable compromise is feasible. On this premise, the first phase of the DFVLR test program utilizes a two-dimensional ladder-type structure, consisting of straight tubular elements interconnected at their junctions by joints with three outlets. The assembled structure, shown in Fig. 14, is 10,35 m long and 1,10 m wide. The fiber-wound tubes consist of three plies with angular directions of 0° and ± 40°. With an initial length of 2,10 m, a diameter of 90 mm and a wall thickness of 0,4 mm, the weight of a typical tube is 380 ± 5 g. The equally fiber-wound joints, consisting of three axially and two circumferentially directed plies, weigh 101 ± 2 g and are depicted in Fig. 15. After the assembly of all elements by adhesive bonding, the total structural weight was determined as 5531 g.

The recently started test program envisions a thorough investigation of the test structure in regard to the deformation patterns under various loading conditions.

Fig. 14. Ladder-type test structure

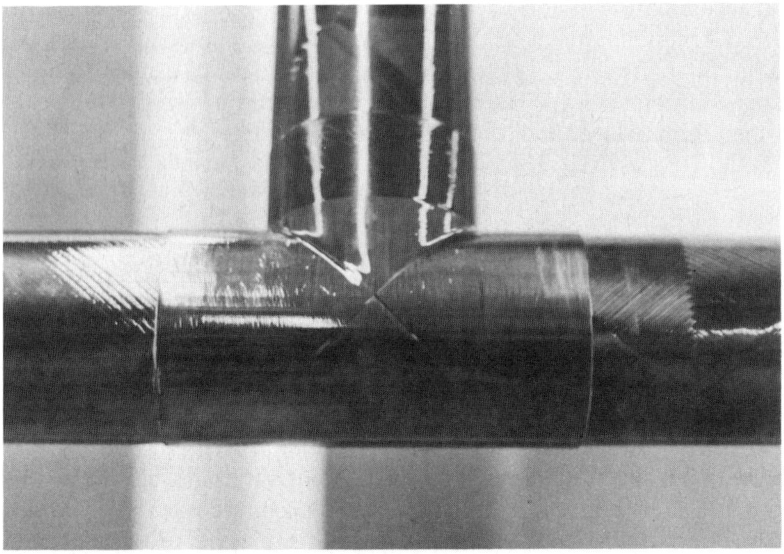

Fig. 15. Fiberwound joint with three outlets

The influence of nonlinear effects, the response to temperature and moisture, the dynamic response including material damping, and others. Following the mechanical tests, it is intended to utilize the test structure for the validation of active damping techniques by means of sensing and actuating devices.

While the ladder-type structure may be well suited for the current test requirements, its lack of stiffness is unrealistic and precludes its usefulness for more advanced test programs. Anticipating the need for those and encouraged by the rela-

Fig. 16. Fiberwound joint with four outlets

Fig. 17. Tool for fiberwound joint with multiple outlets

tively simple poduction of fiber-wound T-joints, the construction of more complicated joints is now in progress. Figure 16, for example, shows a completed joint with four outlets allowing the assembly of shear-stiff plate structures, and Fig. 17 the tool for a joint under development with a rather ambitious configuration.

5 Summary

The evaluation of several fiber-matrix systems for space structures operating between + 100 °C and − 160 °C indicates that the strength and stiffness reductions in epoxy systems curing at low temperature are more pronounced than those in high-temperature-curing epoxy systems. The extent of degradation in the 914- and 934-laminates seem tolerable, while the losses in the polyimide system were so severe that the applicability of polyimides must be questioned in general. The effects of electron irradiation are tolerable in all tested system excepting again the polyimide system. The significance of matrix cracks and interface deterioration needs further investigation especially in orbits with higher numbers of thermal cycling.

The issue of qualification of large structures which defy ground testing because of size or because of operation in a non-simulatable zero-g environment is still wide open. The ongoing DFVLR test program and its philosophy of approach may be insufficient to effect the objective but are certainly steps in the right direction.

With regard to the construction of more advanced test structures, the intended development of three-dimensional fiberwound joints warrants serious consideration of their use as replacement for machined metallic joints in actual space frameworks. Apart from their low cost and low weight aspects, they would avoid the problem of delaminating bond lines in the carbon fiber-metal interfaces during thermal cycling.

6 References

1. Hartung, W., Bergmann, H. W.: Influence of Thermal Cycling on the Behavior of CFRP-Material for Space Structures, DFVLR-IB 131-84/02, 1984
2. Zimmermann, R.: Optimierung axial gedrückter CFK-Zylinder, DFVLR-IB 131-84/43, 1984
3. Nair, P. S. et al.: Theoretical and Experimental Studies on Large Space Structures, DFVLR-IB 131-84/09, 1984
4. Breitbach, E.: Dynamic Qualification of Spacecraft on the Basis of Measured Modal Characteristics, ESA ST 121/1976, 1976

Chapter 4

Testing and Fracture Behaviour of CFRPs

CFRP Testing and Properties Optimization

D. F. Adams

The availability of suitable test specimen configurations and test methods has long plagued the composite materials community. A brief discussion is given of the properties which need be measured, and methods currently available to do so. The advantages and disadvantages of these various techniques are presented, in terms of accuracy, ease of use, and cost. It is suggested that much remains to be done in developing full properties determinations for even static properties of the simplest of composites, the unidirectional ply. Currently, insufficient information is frequently obtained, and inadequately documented.

The costly and time consuming task of material characterization can be reduced somewhat by combining micromechanics analyses with laboratory testing, to reduce the size of the test matrix. Available analyses are described, and Brasil's potential role in this type of work is discussed.

1 Introduction

The proper testing of composite materials has been a topic of continuing emphasis for more than 25 years. Early work was with glass fiber reinforcements; and the problem was not altered significantly when the high modulus fiber such as boron and graphite were introduced in the mid-1960's. The principal problems continue to be the ability to introduce the load into the highly anisotropic composite material in a uniform manner, and to accurately measure the small strains developed. Both problems do become somewhat more difficult to overcome when testing carbon fiber reinforced polymers (CFRP), because of the relatively low shear strengths of these composites relative to their axial tensile strengths, and the fact that a higher stiffness for a given tensile strength translates into an even smaller axial strain to be measured.

An additional complication when testing CFRP is associated with the negative coefficient of thermal expansion of the carbon fiber itself. When these fibers are combined with a polymer matrix material of limited stiffness but very high thermal expansion, a composite of extremely low thermal expansion can result. While this is often a very attractive attribute of CFRP, e.g., in its use in dimensionally stable structures, the adequate measurement of these very small strains creates special problems.

Since most structural polymers absorb considerable amounts of moisture (5 to 10 percent by weight) and swell in the process, while carbon fibers do not, it is also necessary to characterize the moisture expansion properties of CFRP. Unlike thermal expansion, which occurs rapidly with temperature change and quickly stabilizes, moisture expansion is associated with the very slow diffusion of water molecules through the polymer. Even in a relatively thin composite, e.g., 1 mm thick, this process can take many weeks. Since very accurate measurements must be made over long periods of time, special data acquisition problems are encountered.

A variety of other unique measurement problems also exist when testing CFRP, associated with their low thermal and electrical conductivities (which are also anisotropic), high damping properties, low impact resistance, low delamination resistance, etc.

Thus, it is not difficult to understand why CFRP testing continues to be a topic of major interest within the composite materials community.

2 Test Methods

Test methods covers a broad range of technologies. Particular emphasis will therefore be concentrated here on test methods primarily measuring static mechanical properties. In terms of most properties of composite materials, these properties are most difficult to measure for unidirectionally-reinforced materials, since the degree of anisotropy is highest in this configuration. Unidirectional composite, or lamina or ply, properties are of primary importance, of course, since the unidirectional ply is the building block of most actual composite structures. Thus, test methods developments are usually focused on unidirectional composites.

To characterize a unidirectional composite, it is desired to measure the strength and stiffness both parallel and perpendicular to the direction of fiber reinforcement, i.e., both axial and transverse properties. Since different failure mechanisms govern, both tensile and compressive properties must be measured. Shear properties parallel to the fibers must also be determined. Shear properties perpendicular to the fibers, i.e., through the thickness of the unidirectional ply, have not historically been of primary interest, and are not usually easy to measure. However, with increasing concern for free edge (delamination) effects associated with induced interlaminar stresses (both normal and shear), these stresses can no longer be automatically neglected.

In summary, primary static structural properties of the unidirectional ply include:

σ_1^{yt}, σ_1^{yc} – axial yield strengths in tension and compression
σ_1^{ut}, σ_1^{uc} – axial ultimate strengths in tension and compression
σ_2^{yt}, σ_2^{yc} – transverse yield strengths
σ_2^{ut}, σ_2^{uc} – transverse ultimate strengths
τ_{12}^{y}, τ_{12}^{u} – longitudinal shear strengths, yield and ultimate
E_{11}^{t}, E_{11}^{c} – axial modulus, tension and compression
E_{22}^{t}, E_{22}^{c} – transverse modulus, tension and compression
ν_{12}^{t}, ν_{12}^{c} – major Poisson's ratio, tension and compression
G_{12} – longitudinal shear modulus
α_{11}, α_{22} – coefficients of thermal expansion, axial and transverse
β_{11}, β_{22} – coefficients of moisture expansion, axial and transverse

It is recognized that α and β are physical rather than mechanical properties, but they are typically also required for complete analysis. Often the compressive stiffness properties E_{11}^{c} and E_{22}^{c}, and the major Poisson's ratio in compression ν_{12}^{c}, are not measured, being assumed equal to the corresponding tensile properties.

The properties listed above are typically adequate for basic design purposes, where linearly elastic material response is assumed. However, it is often desirable to perform a nonlinear analysis, in which case complete stress-strain curves must be documented. Unfortunately, these data are often not collected. If temperature and moisture absorption are also design parameters, then all of the above properties must be generated over a range of temperatures and moisture contents. This could result in a prohibitively large test matrix (prohibitive in terms of cost, and calendar time required). Thus, often only room temperature, dry and elevated temperature, wet conditions are selected, "elevated temperature" implying the maximum use temperature expected in the application, and "wet" corresponding to the quasi-equilibrium moisture content to be achieved. (A value of approximately one percent by weight in CFRP in a typical environment is representative.) If practical, tests at all nine combinations of three temperatures and three moisture contents should be performed, as a minimum.

Although not specifically addressed in this brief paper, many other properties besides those static properties listed and discussed here must often be evaluated in completing a specific design. These include fatigue, impact, creep, relaxation, multi-axial loading, and high rate loading, to name just a few. The proper determination of these material characteristics is often more difficult than static properties evaluations, and cannot be minimized, although outside the scope of the present discussion.

Having defined the static properties to be measured, test methods for measuring them can now be discussed.

2.1 Tensile Properties (σ_1^{yt}, σ_1^{ut}, σ_2^{yt}, σ_2^{ut}, E_{11}^t, E_{22}^t, ν_{12}^t)

The most suitable tensile test specimen has evolved into a relatively simple form over the years. Various dog-bone shapes have given away to a simple, straight-sided specimen of constant width and thickness. The key is in the tab design, and the adhesive used to bond the tabs to the CFRP specimen. Metal tabs (steel or aluminum) have been replaced by glass fabric/epoxy tabs (printed circuit board material often being used, as being readily available and inexpensive). The tabs are generously tapered on one end, to provide a smooth transition of the applied load into the CFRP specimen. An adhesive which has a very high shear strength, and which forms a thick bond line, is desirable. The requirement of high shear strength may require an elevated temperature cure under some clamping pressure. Of course axial tensile testing is much more difficult than transverse tensile testing, because of the much higher forces to be transferred from tabs to specimen.

Either strain gages or extensometers are usually used to measure the strains. A strain measurement in the direction of loading is required to measure E, and also in a direction normal to the loading if ν is to be determined also. In a transverse tensile test, the strain normal to the loading direction is often not measured (needed to measure the minor Poisson's ratio ν_{21}^t), since its value is typically small and hence difficult to measure accurately. The reciprocal relation [1], i.e.,

$$\frac{\nu_{21}}{E_{22}} = \frac{\nu_{12}}{E_{11}} \tag{1}$$

states that if ν_{12} and the moduli E_{11} and E_{22} are measured, ν_{21} can be calculated. The relation also indicates that ν_{21} will be smaller than ν_{12}, in the ratio E_{22}/E_{11}. Hence, for a highly anisotropic material such as CFRP, where E_{22}/E_{11} is typically on the order of 0.1, ν_{21} will be on the order of 0.1 ν_{12}. While ν_{21} is not listed as a property to be normally measured, if it is, Eq. (1) can be used as a check of the mutual consistency of the data set.

Commercially available extensometers (usually strain gage instrumented, but occasionally Linearly Variable Differential Transformer (LVDT) instrumented) are expensive (currently about US \$ 1000 each), but are reusable and easily installed. Strain gages cost on the order of US \$ 5 per single-element gage, but are not reusable, and require more labor to install. Problems can also be encountered when strain gages must be used at high temperatures (they tend to reinforce the softened matrix surface of the specimen, giving low strain readings) and moist environments (in addition to the problem of a softened matrix, the moisture may weaken the adhesive bond, causing the gage to become partially, or even completely, debonded). Hence the present author prefers to use extensometers whenever possible.

2.2 *Compressive Properties* (σ_1^{yc}, σ_1^{uc}, σ_2^{yc}, σ_2^{uc}, E_{11}^c, E_{22}^c, ν_{12}^c)

Although not commonly recognized as such, compressive testing might very well be the most difficult to perform properly. The problem is that compressive strength is not a uniquely defined property. The mode of compressive failure can smoothly transition from gross buckling, to local instability, to fiber microbuckling, to shear failure, to end crushing (brooming). Unfortunately, any one of these failure modes may be meaningful in an actual application, and no one measurement is a fundamental material property. That is, each depends on the "structure" of the composite. Since there may be as much as an order of magnitude difference in measured "strength" from one extreme to the other, which failure mode a particular test method induces is obviously important. Often the test method is selected which is believed to best represent the failure mode of the structure it is to be used in, obviously a difficult task.

As a compromise, the Celanese [2] or ITTRI [3] test methods have been somewhat standardized for use. By using a common test method, the failure mode tends to be uniform from one material to the next (although not necessarily), and hence at best relative comparisons can be made. However, the use of this type of compression test is by no means universal today, and with good reason, as stated above.

Strains in compression are measured using the same devices previously described for tensile testing, with the same limitations. Although perhaps poor practice, compressive strains are often not even measured. The values of E and ν in compression are simply assumed to be equal to those in tension. Of course, this also means that complete stress-strain curves will not be available.

2.3 *Shear Properties* (τ_{12}^y, τ_{12}^u, G_{12})

The determination of the shear properties of composite materials has perhaps received more attention than any other over the years [4]. The problem is a combination of having many different test methods to choose from, and the fact that many of these tests are either difficult to use, or provide only limited data. For example, thin-walled tubes are relatively expensive to fabricate, and somewhat delicate to handle. Solid rods are an excellent compromise, being easily fabricated and very rugged, but do not produce a uniform state of shear stress. Strength and modulus can both be measured, and a complete shear stress-shear strain curve can be generated, if a strain gage or rotometer is used.

In contrast, many of the commonly used shear tests permit the measurement of either strength or modulus, but not both. For example, the short beam shear or interlaminar shear test, and the notched tensile test, only provide strength, while the plate twist test only provides modulus. Others, such as the 2- and 3-rail shear tests, or the picture frame test, while claiming to provide both strength and stiffness, are suspect in strength determination because of induced stress concentrations at the attachment points.

The Iosipescu shear test has emerged during the past few years as a very promising shear test for composites [4, 5]. The specimen is compact, easily fabricated and easily tested, providing both strength and modulus, for any one of the three possible shear loadings, i.e., the in-plane and two interlaminar shear components. Much study of this test method is currently underway, both analytical and experimental.

2.4 Coefficients of Thermal and Moisture Expansion (α_{11}, α_{22}, β_{11}, β_{22})

Until the past 6–8 years, most applications of CFRP were in ambient temperature environments. Also, moisture absorption effects, while known to be a potential problem, were largely ignored. Thus, there had been little concern with measuring mechanical properties over a range of hygrothermal conditions, and even less concern with measuring the coefficients of thermal and moisture expansion of the various composites. This has now changed. Elevated temperatures and moist environments are now an integral part of the service conditions of composite hardware, and must be accounted for in design and stress analysis. This is particularly true as performance demands increase and the margins of safety decrease. For example, even the curing-induced thermal residual stress can no longer be ignored, as once was the case. To predict these stresses, and to account for them in the design process, the coefficients of thermal expansion, both parallel to the fiber reinforcement, α_{11}, and perpendicular to the reinforcement, α_{22}, must be known. The same can be said for the coefficients of moisture expansion, β_{11}, and β_{22}, when humidity is present. Since β has the units of strain per unit moisture absorption (usually expressed as weight percent moisture relative to the dry composite), it is also necessary to know the moisture absorption rate (moisture diffusivity) and moisture saturation level of the composite.

As previously noted, the axial coefficient of thermal expansion α_{11} of CFRP is difficult to determine since the values are typically very small (on the order of 10^{-7}/°C, or less). Standard quartz tube dilatometers are not sufficiently sensitive, and some type of interferometer is usually required. Thus, the cost of determining α_{11} goes up. On the other hand, α_{22} is usually readily measured using a dilatometer, since values may be in the 10^{-5}/°C range. One distinct advantage of thermal expansion testing is that the specimen can be brought to thermal equilibrium relatively quickly. Thus, testing is rapid.

In contrast, moisture equilibrium can be a very slow process; it may require one to two months for even a relatively thin (1 mm) specimen to reach equilibrium. Thus, testing is slow, and potentially expensive. Also, with limited equipment of this specialized nature available to most laboratories, the volume of data generated to date is still very limited [6]. Although the magnitudes of β_{11} are small relative to β_{22}, the problem of measuring β_{11} is not quite as severe as when measuring α_{11}. Nevertheless, a standard quartz tube dilatometer (equipped, of course, for humidity control) may be marginally adequate.

Often it is feasible to measure the α and β of the neat polymer matrix itself (more easily achieved since these values are typically large), and then use a micromechanics analysis to predict values of α_{11} and β_{11} (see for example, References [6, 7]).

3 Data Reduction Techniques

The days of reading dials and recording data on clipboards have now passed for the most part. Chart recorders have become the primary mode of data acquisition. Data reduction must still be done manually in this mode, however. Material modulus is the estimated best-fit slope of the stress-strain plot (usually a force-strain plot, suitably converted), while yield and ultimate strengths are points picked off the chart. Having only a single hard copy analog plot of the raw data, the tendency is to digitize the data manually into these few points of information. Thus, much potentially useful information is lost to future users.

The current trend is to develop computerized data acquisition systems. That is, the raw data are stored in digital form on magnetic tape or disk, along with information as to specimen identification, specimen dimensions, test conditions, etc. While the raw data can be immediately reduced to usable forms, e.g., moduli, Poisson's ratios, strengths, and stress-strain curves, and displayed on a CRT, in essentially real time as the test progresses, this is not the sole, or even the primary, benefit of computerization. The most important advantage is that the data are permanently on file, available for immediate recall at any time. Thus, if at some future time (days, weeks, or even years later), it is desirable to restudy the data, or reduce them in another manner, or compare them with more recently acquired data, this can be done almost instantly. No information is lost during the lapse of time. This obviously increases the value of testing.

Such computerized data acquisition and reduction systems are currently expensive, however, and demand a considerable number of user hours to get familiar with them and their capabilities. Thus, a testing laboratory must be willing and able to expend considerable funds to get started (perhaps on the order of US$ 100,000), and designate at least one dedicated user. Presently, such systems are somewhat unique; in another decade they will probably be commonplace, and considerably less expensive.

4 Property Prediction Techniques

As emphasized in previous sections, the complete, or even reasonably complete, static material properties characterization of a unidirectional CFRP material is a major undertaking, both in terms of calendar time and cost. While excessive cost is always a strong consideration, calendar time may often be also. If several weeks are required to fabricate and test a laminate, little can be done to speed the process. If a month or two are required to moisture precondition a specimen, the time scale is set. Even when time is not a governing factor, the fabrication, instrumentation and testing of hundreds of specimens, with sufficient numbers of replicates to ensure a reliable data base, is a very labor intensive process.

During the past 20 years, work has proceeded steadily toward the development of so-called micromechanics analyses for predicting unidirectional composite properties based upon known constituent material (fiber and matrix) properties and the composite geometry (fiber volume fraction, fiber shape and packing array, interface bond performance, and thermal and moisture history). While early models were crude, and often very empirical in nature (e.g., rule-of-mixtures, Halpin-Tsai equations, etc.), subsequent models have become very rigorous in their representation. In particular, the finite element method of analysis, which has come into common use during this same time period, and the growth of large scale digital computers, have combined to revolutionize micromechanics analyses.

It is now possible to model arbitrary constituent material stress-strain response, i.e., both time-independent and time-dependent material nonlinearities, crack initiation and propagation, interface debonding, temperature- and moisture-dependent material properties, and anisotropic material response, using either a 2D or 3D analysis [8–13]. Limited analytical/experimental correlations available to date [6, 8, 9, 14] indicate that bulk properties (E, ν, G, α, β) can be predicted very accurately using current methods. Work is now proceeding toward the improvement of methods to predict point properties (strengths), and good progress is being made.

The realistic goal of micromechanics analyses is not to replace composites testing, but rather to supplement it. That is, the analyses would be used to predict those properties not measured experimentally as part of a limited test program, using correlations with measured values as verification of the predictive ability. In this manner, complete data tables could be made available at a fraction of the cost of a fully experimental program, since a computer simulation is much less expensive than an actual experiment.

An important secondary advantage of a demonstrated micromechanics analysis capability is in optimizing existing composites, and developing new systems. A hypothesized combination of existing or proposed fiber and matrix can be analytically "tested" to establish its potential. Only the most promising systems then need be subjected to the more costly and time consuming laboratory testing. This capability currently exists; it may be in widespread use in another decade.

5 Brasil's Potential Role in CFRP Testing

With the possible exception of computerized data acquisition, which does require sophisticated equipment and extensive resources, much of the needed development of composites test methods and properties optimization primarily involves careful thought, attention to detail, and considerable personal effort. Thus, any composite group is a potential contributor to the technology.

Certainly Centro Tecnico Aerospacial (CTA) in San Jose dos Campos is an excellent example of an organization in Brasil which has the facilities to do this type of work. Perhaps the weak element at present is the lack of background knowledge of what has already been done elsewhere. Thus, selected personnel must be provided with current literature, and have the opportunity to visit other laboratories, and talk with other investigators. International conferences such as the present one are an excellent source of such contacts.

6 References

1. Jones, R. M.: *Mechanics of Composite Materials,* McGraw-Hill, New York (1975), p. 38
2. "Standard Test Method for Compressive Properties of Oriented Fiber Composites", ASTM-STD-D3410-75, 1979 Annual Book of ASTM Standards, Part 36, American Society for Testing and Materials (1979)
3. Hofer, K. E. Jr., Rao, P. N.: "A New Static Compression Fixture for Advanced Composite Materials", *Journal of Testing and Evaluation,* Vol. 5, No. 4 (July 1977), pp. 278–283
4. Walrath, D. E., Adams, D. F.: "The Iosipescu Shear Test as Applied to Composite Materials", *Experimental Mechanics,* Vol. 23, No. 1 (March 1983), pp. 105–110
5. Adams, D. F., Walrath, D. E.: "Iosipescu Shear Properties of SMC Composite Materials", Proceedings of the Sixth Conference on Composite Materials: Testing and Design, ASTM STP 787, Phoenix, Arizona, May 1981, pp. 19–33
6. Cairns, D. S., Adams, D. F.: "Moisture and Thermal Expansion Properties of Unidirectional Composite Materials and the Epoxy Matrix", *Journal of Reinforced Plastics and Composites* (October 1983)
7. Adams, D. F., Monib, M. M.: "Moisture Expansion and Thermal Expansion Coefficients of a Polymer-Matrix Composite Material", Proceedings of the Fourth Conference on Fibrous Composites in Structural Design, San Diego, California (November 1978)
8. Adams, D. F., Crane, D. A.: "Combined Loading Micromechanical Analysis of a Unidirectional Composite", *Composites,* accepted for publication
9. Schaffer, B. G., Adams, D. F.: "Nonlinear Viscoelastic Analysis of a Unidirectional Composite Material", *Journal of Applied Mechanics*, Vol. 48, No. 4 (December 1981), pp. 859–865
10. Adams, D. F., Murphy, D. P.: "Analysis of Crack Propagation as an Energy Absorption Mechanism in Metal Matrix Composites", Report UWME-DR-101-102-1, Department of Mechanical Engineering, University of Wyoming (February 1981)
11. Mahishi, J. M., Adams, D. F.: "Micromechanical Predictions of Crack Initiation, Propagation and Crack Growth Resistance in Boron/Aluminum Composites", *Journal of Composite Materials,* Vol. 16, No. 6 (November 1982), pp. 457–469
12. Monib, M. M., Adams, D. F.: "Three-Dimensional Elastoplastic Finite Element Analysis of Laminated Composites", Report UWME-DR-001-102-1, Department of Mechanical Engineering, University of Wyoming (November 1980)
13. Mahishi, J. M., Adams, D. F.: "Energy Release Rate During Delamination Crack Growth in Notched Composite Laminates", Proceedings of the ASTM Symposium on Delamination and Debonding of Materials, Pittsburgh, Pennsylvania, November 1983
14. Adams, D. F., Schaffer, B. G.: "Analytical/Experimental Correlations of Stiffness Properties of Unidirectional Composites", *Composites Technology Review,* Vol. 4, No. 2 (Summer 1982), pp. 45–48

Mechanisms of Fracture in Fiber-Reinforced Laminates

H. W. Bergmann

Structures composed of fiber-reinforced materials may exhibit various kinds of defects incurred in the manufacturing process or during service. The ability to forecast the effects of such damages on the safe operation of aerospace structures presupposes the development of principles of damage mechanics, analogous to fracture mechanics of metals but considerably more complex. The identification of potential failure modes and the judgment of their contribution to the progression of observed damages require extended empirical and analytical investigations. Until the various mechanisms of damage progression are understood and predictable, the disposition of an afflicted structure must be supported by costly and time-consuming tests. The ultimate goal of all such efforts ought to be the establishment of reliable and relatively simple accept/reject/repair criteria for the support of series productions and for maintenance requirements. The related activities of the DFVLR Institute for Structural Mechanics are guided by this goal. The present paper does not purport to offer simple solutions; rather, it aims to convey the complexity of the issue of damage mechanics and to describe several of the more important mechanisms of fracture.

Key Words

Fiber-reinforced laminates, damage mechanics, fracture mechanisms, failure modes, failure analysis, matrix cracks, delaminations, environmental effects.

1 Introduction

The concept of fiber-reinforced materials is not new and certainly not an invention of this century. However, only in recent years has the development of special fibers, in combination with appropriate matrix materials, led to promising applications in the aerospace industry. Of dominant interest are carbon-fiber reinforced epoxy resins which, apart from their high specific strength and stiffness properties, are also fatigue-tolerant and corrosion-resistant.

Inhibiting a wider range of applications are the small elongations to failure of epoxy resins, their susceptibility to environmental effects, and an as yet insufficient comprehension of the significance of damages occurring during the production process of structural parts, or in the course of their service. The disposition of such flaws as matrix cracks, delaminations, misdrilled holes, etc., by analytical means is as yet impossible for a lack of understanding of the mechanisms of damage progression. Consequently, the accept/reject decisions of damaged parts are normally made by time-consuming testing and often after extensive repairs. The knowledge thus gained is seldom transferrable to other cases as it provides no insight into the interpendence of many parameters affecting the response of damaged parts [9].

Anticipating a rapid increase of composite structures in the near future, the issue of damage mechanics was introduced as a major research program at the Institute for Structural Mechanics of the German Aerospace Research Establishment (DFVLR) [2]. It must be fully recognized that the complexity of damage mechanics exceeds that of fracture mechanics by far, partially because of the anisotropy and heterogeneity of the composite materials, and partially because of the much larger variety of possible kinds of damage.

2 Test Program

Considering the large variety of parameters to be encountered, it stood to reason to commence the experimental investigations with flat laminates subject to unidirectional loading. The dominantly used materials are 0.125 mm thick 914C/T300 autoclave-cured tape prepregs; all subsequent statements relate to that material unless otherwise noted. For the purpose of comparing their performance, test specimens are prepared without defects and with typical defects such as delaminations, holes and notches. The majority of the test specimens for tension and compression tests are from 10 mm to 40 mm wide and from 235 mm to 380 mm long, respectively. The substantial width of the larger specimens assures that the response of the centrally introduced defects is unaffected by the specimen boundaries. The stacking orders of the laminates are chosen such that they represent rib or spar chords, webs and skin panels of typical aerospace structures. Under compression loading the test specimens must be laterally supported by anti-buckling guides. In cyclic tests the specimen temperature may rise significantly due to internal and external friction and in dependence on the test frequency. In order to control the temperature effect the anti-buckling guides contain electrically activated cooling devices [3]. The majority of the specimens are exposed to realistic environmental conditions including moisture, temperature and radiative effects, and are tested to eventual failure.

3 Failure Analysis

In carbon fiber-reinforced laminates, different specimen configurations and different loading and environmental conditions lead to different failure modes. The dependency of the failure modes on the many possible parameter combinations of typical laminates is not well understood. In fact, it may be said that the investigation, classification and interpretation of failure modes is still in its infancy. Failure analysis techniques currently existing are: microscopic examination of fracture phenomena; non-destructive evaluation of macroscopic types of damage; and stress analysis of failed parts.

3.1 Microscopic Examinations

Failure modes observable on the microscopic scale include fiber pullout, fiber breakage, matrix micro-cracking, fiber-matrix debonding, and matrix deformations in the form of serrations and cleavage. Their subsequent progression may eventually cause fracture of the laminate. Fractographic techniques based on scanning electron beam microscopy can be applied to identify the fracture origin and the direction of propagation, as well as to analyse material parameters affecting the fracture process such as constituent properties, laminate configurations, fiber-matrix interface properties, loading and environmental conditions. A reliable characterization of the fracture phenomena may ultimately assist in establishing the cause of failure and thus suggest suitable modifications of the composite system.

3.2 Non-Destructive Evaluation

Non-destructive evaluation of all test specimens is mandatory in order to assure the absence of initial damage, and to detect and track various kinds of macroscopic defects. As no single procedure can satisfy this requirement, a combination of mutually supportive procedures must be employed. The techniques utilized by the Institute for Structural Mechanics include: high-precision ultrasonic test facilities with highly-vaporated focussing transducers and a narrow-band transmitter with variable pulse frequency, low-energy X-ray equipment with high lateral resolution, acoustic emission analysis and grid-reflection techniques for in-situ observations of the test specimens [8, 1, 4].

3.3 Stress Analysis

A reasonably accurate analysis of the three-dimensional state of stress around a discontinuity by a standard finite-element approach requires a large number of degrees of freedom so that the solutions may become prohibitively expensive, especially when the tracking of the damage progression involves an interative treatment [16]. This recognition has led to the development of a new analysis approach which, although conceptually not novel, combines an unusual number of features in an economically organized computer program. Its basic component is a triangular hybrid shell element comprising bending and membrane action as well as normal and shearing stress capabilities on its upper and lower surfaces. By stacking several of these elements above one another, a multidirectional laminate can be modelled in great detail with a reasonably small number of degrees of freedom. A special condensation scheme is utilized capable of producing multilayer shell elements and substructures. Failure progression rules are appended to the finite element equation system in such a way that the tracing of damage progression will not require repeated triangularization of the global stiffness matrix [5].

4 Mechanisms of Fracture

In contrast to metals, where fracture under static or fatigue loads results from the nucleation and growth of a single dominant flaw, the fracture of fiber-reinforced composites is characterized by the initiation and progression of multiple failures of different modes such as matrix cracks, interfacial debonding, fiber breaks and delaminations between adjacent plies of the laminates. The kinds of occurring failures, their distribution, time sequence and possible interactions depend on many parameters such as the properties of the fiber/matrix system, the stacking order and curing process, the influence of the environment, etc. The problem is further complicated by different failure modes under static and dynamic load applications [12], and by the possibility of fatigue failure in the compressive as well as in the tensile load regime.

Close observation of unnotched specimens tested under cyclic tension loads indicates that the progression of events follows a more or less distinct pattern. The first discernable damage usually are matrix cracks at certain intervals in the crossplies of the laminates. With increasing cycles more matrix cracks are developed which, at

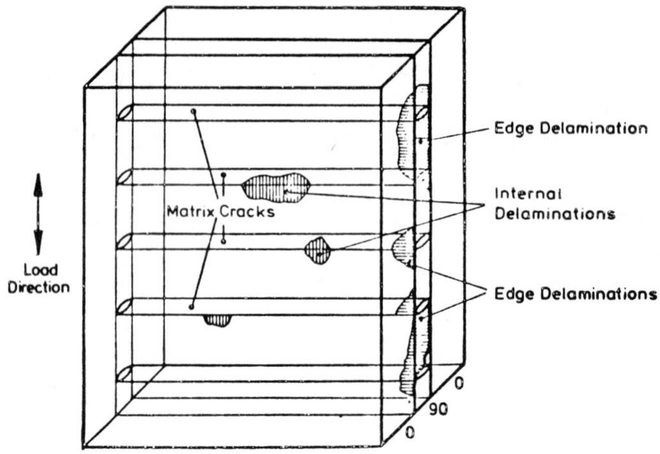

Fig. 1. Matrix cracks and incipient delaminations

the interfaces with neighboring plies, tend to turn and to form small delaminations both inside and, especially, at the free edges of the laminate as indicated in Fig. 1. Additional delaminations may emanate from the locations of fiber bundle breaks.

With respect to compressive loads, critical conditions may arise in the presence of delaminations induced by lack of adhesion because of faulty manufacturing, or because of impact damage. Under sufficiently high compressive stresses the reduced bending stiffness of the separated sections will introduce local buckling and thereby a state of stress at the periphery of the delamination which tends to advance the crack front. Continuation of load cycling then leads to damage growth followed by massive separation and subsequent specimen failure.

Even in the case of very closely controlled tension-tension tests performed under identical parametric conditions, the scatter of the fatigue lives of the specimens is usually very significant. A possible explanation is that, at sufficiently high stress levels, the random distribution of micro-cracks, debonds, fiber breaks, etc., becomes denser with increasing cycles. Toward the end of the specimen life, adjacent failure modes tend to interconnect and form failure paths which, because of their stochastic nature, differ from specimen to specimen and lead to discrepant life spans of the test specimens. A similar argument may apply to the fatigue performance of specimens subject to compressive loads where, of course, different kinds of failure modes interact differently but produce similarly scattered test results.

Accompanying the gradually increasing damage state is a reduction of the overall laminate stiffness. Depending on the stacking order, this reduction may be quite pronounced as shown in Fig. 2 for a matrix-controlled laminate, or more subtle in the case of fiber-controlled laminates. In both instances, however, a particularly critical combination of local failure modes develops which, finally, leads to a rapid deterioration of the stiffness in only a few additional cycles, and to what is commonly called "sudden death" of the specimen [13]. Several of the dominant failure modes are addressed in more detail below.

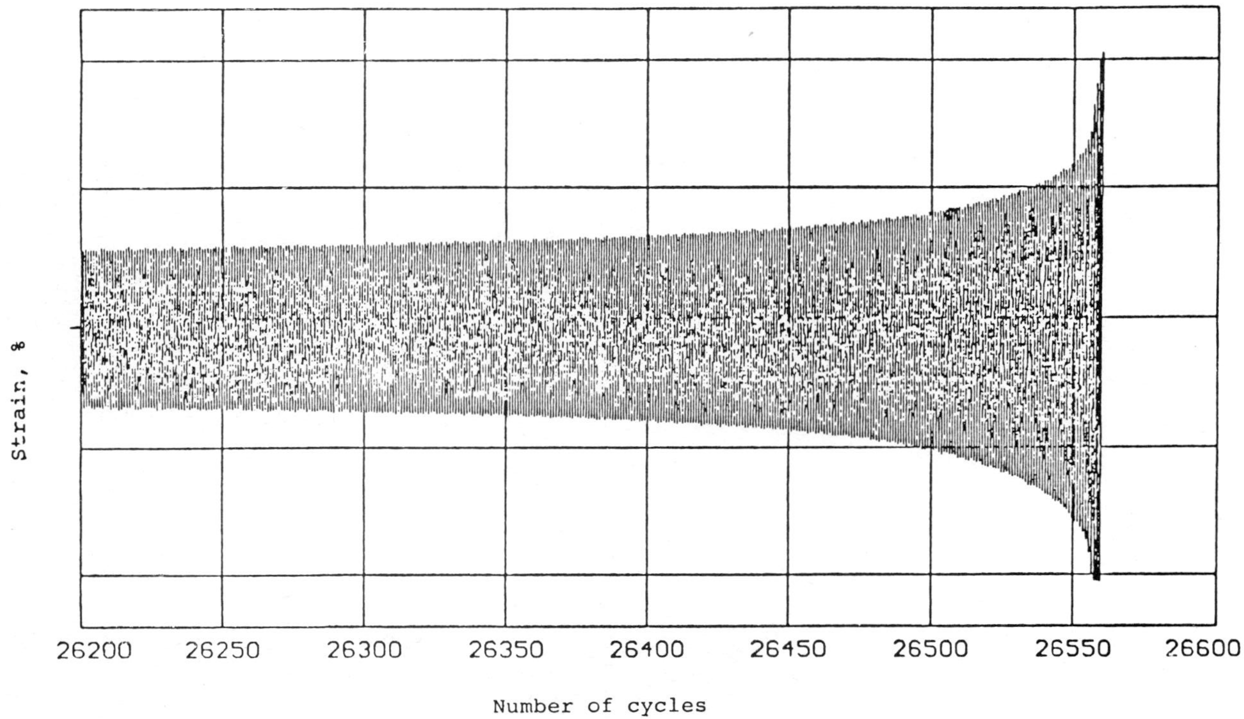

Fig. 2. Stiffness decay shortly before failure

4.1 Matrix Cracks

The normally used epoxy resins have only limited tensile strain capabilities. This condition is aggravated by the introduction of tensile prestresses in the matrix resulting from the discrepant thermal elongations of the fibers and the matrix during the curing process. Consequently, the formation of cracks can be expected rather early in those plies of a laminate which are mechanically or thermally stressed beyond the critical strain values of the resin. Repeated exposures to high mechanical loads or to low temperatures will increase the state of damage [7]. Figure 3 shows that in cross-plied specimens subjected to thermal cycles between +100 °C and −155 °C the cracks turn at the interfaces of adjacent plies and form local delaminations which tend to grow with increasing numbers of thermal cycles. It stands to reason that a similar effect occurs in specimens subjected to sufficiently high mechanical load cycles. Delaminations of this kind occur most severely along the free edges but are

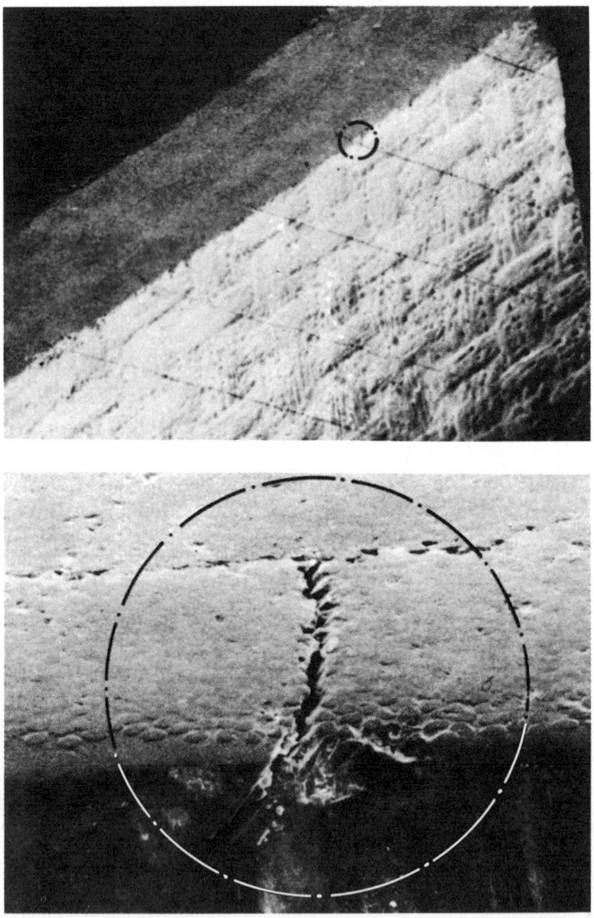

Fig. 3. Damage development due to thermal cycling (T3T F178 material after 1170 cycles)

found in the interior of the test specimens as well. Considering that the strain value associated with the first crack in any one ply is often equated to the limit-load carrying capability of the structure, the urgent need for more ductile resin systems is self-evident.

4.2 Edge Delaminations

The free edges of multidirectional laminates are especially susceptible to the formation of cracks and delaminations because, under imposed axial strains, the enforced compatibility of the lateral contractions of the individual plies introduces interlaminar shearing stresses, as well as normal stresses, in the thickness direction. The resulting state of stress depends on the stacking order and on the load direction, Fig. 4 summarizes the results of a detailed finite-element analysis for laminates stacked $[0, \pm 45.90]_s$ and $[90, \pm 45.0]_s$. The numerical values of the linearly derived stresses account neither for curing prestresses nor for viscoelastic relaxations; however, their potentially dangerous trends can be reversed by altering the ply sequence [17]. The question arises to what extent the actual strength of the laminates is affected by such ply rearrangements, considering that the commonly used lamination theory does not account for these internal states of stress.

Carefully conducted static tension tests confirm, indeed, a superiority of the $[90, \pm 45.0]_s$-laminate over the other by approximately 10%. Apart from the reversed crossply locations, the developing matrix crack patterns were comparable in both cases without noticeable evidence of edge delaminations.

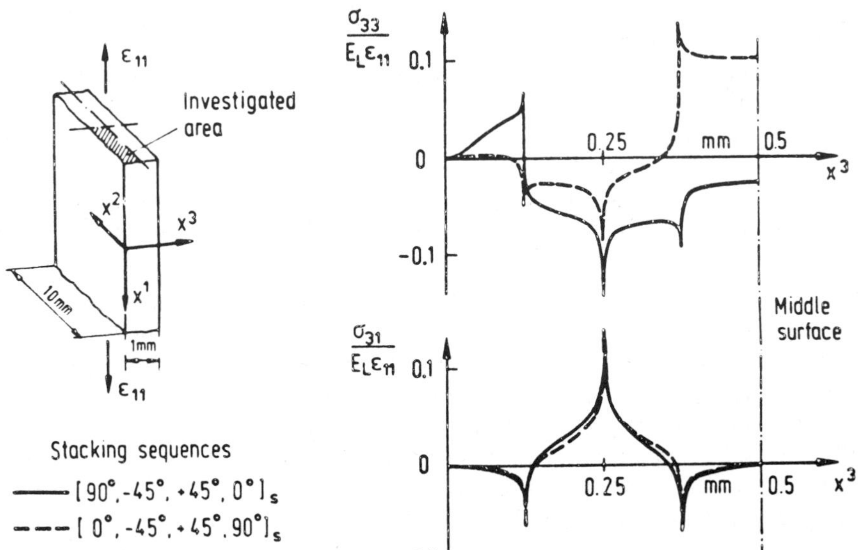

Fig. 4. Effect of stacking sequence on stress distributions in the vicinity of free edges ($x^2 \sim 0$)

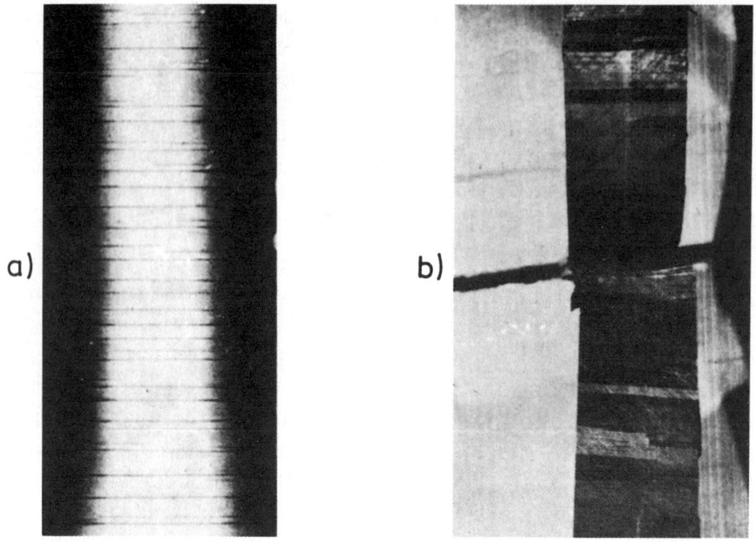

Fig. 5a, b. Progression of edge delaminations

A very different behavior of the two laminates was observed during fatigue loading with R = 0.1 and an upper stress level of 75% of their respective static strengths. As expected, matrix cracking and edge delaminations occurred relatively early in the $[0, \pm 45, 90]_s$-specimens. Monitored by contrast-enhanced radiography, as shown in Fig. 5a, the delaminations proceeded from both sides toward the center of the specimen until, shortly before failure, they were separated only by the narrow strip recognizable in Fig. 5b. The $[90, \pm 45, 0]_s$-specimens, in contrast, exhibited relatively minor evidence of damage. A comparison of the crack patterns in the free edges of the two types of specimens, shortly before failure, is shown in Fig. 6. The onset of delaminations is recognizable between those plies where, according to Fig. 4, the internal peeling stresses are most severe.

In spite of the different fracture behavior, the number of cycles to total failure was comparable for both types of laminates because the two separated halves of the $[0, \pm 45, 90]_s$-laminates were capable of maintaining the 75% static strength level independently. Delaminations of this severity in tension/compression- or purely compression-loaded specimens, of course, would reduce the fatigue life drastically due to buckling of the separated sections. The point here is that an evaluation of test results without an accompanying failure analysis may lead to erroneous conclusions in regard to the performance characteristics of laminates.

In actual applications the boundaries of structural components, such as panels, are often reinforced or less intensely stressed. The issue of edge effects, nevertheless, is of high significance as it also applies to the free boundaries of holes, notches or cutouts. More importantly, as the design allowables are normally defined on the basis of narrow test specimens, a superficial interpretation of test results may entail overly conservative design values.

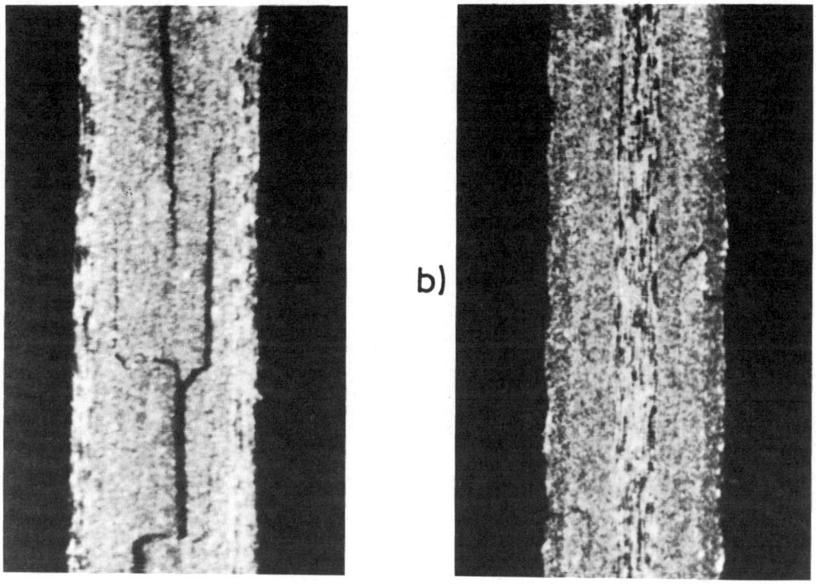

Fig. 6a, b. Severity of edge delaminations

4.3 Central Delaminations

Apart from the free edges of a laminate, delaminations may occur in the central regions initiating from matrixcrack formations, or because of local lack of adhesion between adjacent plies caused either by a processing fault or by impact damage during service.

The introduction of delaminations in test specimens is possible, prior to curing, by the inclusion of an outgassing agent or by the embedment of a very thin teflon disk or, after curing, by controlled impacting. The teflon disk approach has the advantage of providing a well-defined location and geometry and has been adopted for the majority of the studies. Numerous tests have proven that the somewhat blunted crack front at the perimeter of the disk does not significantly retard the eventual growth of the delamination.

Figure 7 shows ultrasonic records of the progressive growth of a delamination under a gradually increasing compressive load in a standard test specimen. The 0.1 mm thick teflon disk was placed between the 90°-plies at the midplane of the specimen. Up to approximately 85% of ultimate load the delamination is seen to be stable and to then grow in the direction of the fibers of the neighboring plies. The same kind of test specimen loaded in tension exhibits a very different behavior. Figure 8 shows that, while the central delamination remains unchanged up to failure of the specimen, edge delaminations occur along the free boundaries at approximately 80% of ultimate load which subsequently gradually increase. Figure 9, finally, illustrates the response of a test specimen under cyclic load with $R = -1$ and a stress level of about 50% of its ultimate strength. After 20,000 cycles the first

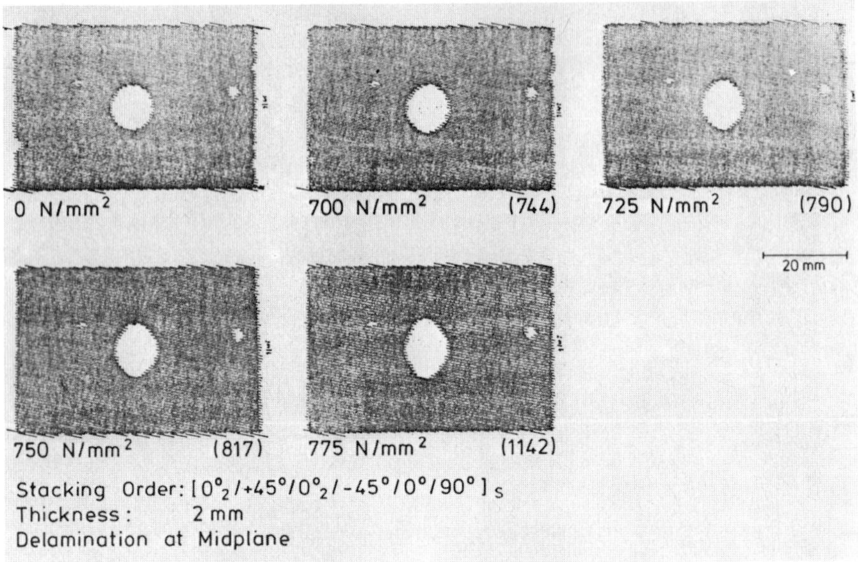

Fig. 7. Growth of central delamination under static compression load

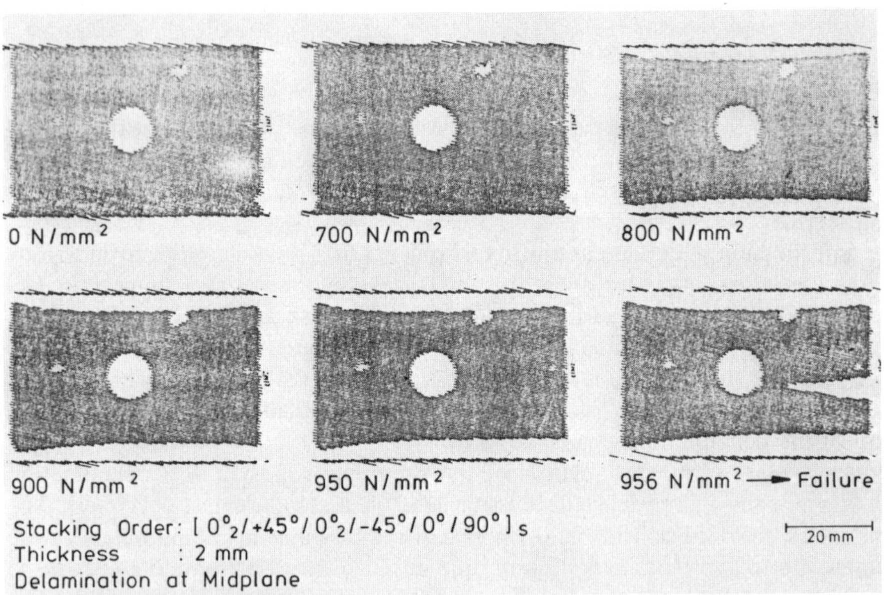

Fig. 8. Growth of central delamination under static tension

evidence of central delamination growth and onset of edge delaminations is noticeable, which gradually increase in severity until failure after 140,000 cycles [8, 12, 13]. Figure 9 also indicates the formation and random distribution of an

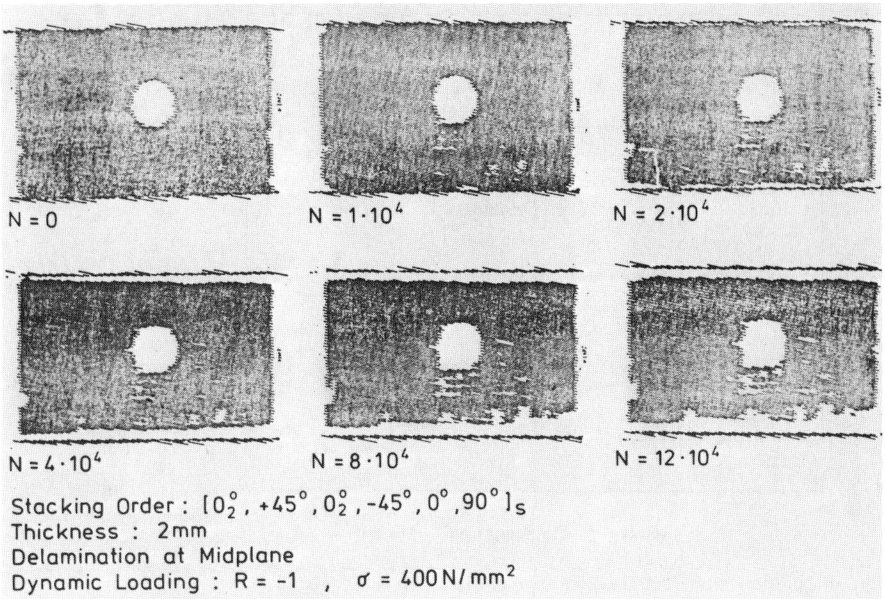

Fig. 9. Growth of central delamination under cyclic load

increasing number of local delaminations, not to be found under static loads and consistent with the argument offered previously.

Variations of the diameter of the delaminations or of their location within their stacking order, obviously, will produce different results. A major goal of the test program is to identify tolerance levels for delaminations below which no growth occurs, and to predict the rate of growth of delaminations above such tolerance levels by empirical/analytical procedures. Figure 10 summarizes some of the accumulated test data which indicate that the size of the delaminations is less significant than their location. A ready explanation is the increasing tendency of the thinner of the separated sections to locally buckle and to thereby aggravate the state of stress at the perimeter of the delamination [10].

4.4 Fiber Breaks and Fiber Debonds

In highly stressed laminates fibers may rupture prematurely, individually or in small bundles, because of their imperfect shapes or imperfect alignment. Especially in fiber-controlled stacking orders the resulting redistribution of stresses in the vicinity of such discontinuities may initiate local failure modes which affect the fracture behavior of the laminate. Figure 11 depicts the observation of the break of a fairly massive fiber bundle on the specimen surface by the grid-reflection technique as well as an enlargement of the associated spalling and cracking of the affected zone.

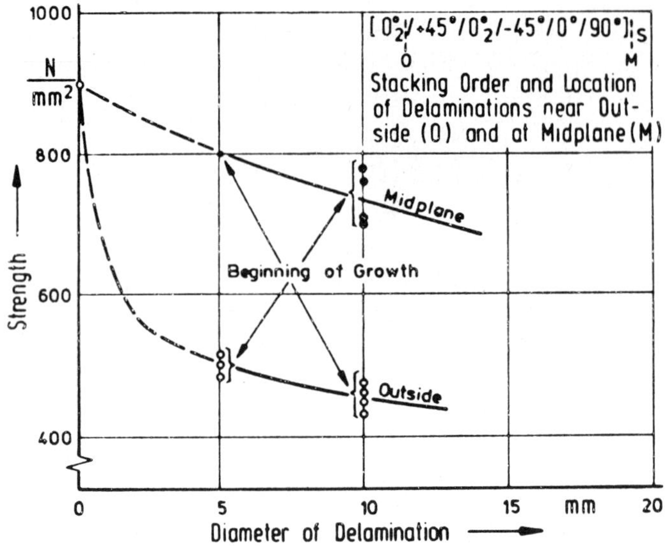

Fig. 10. Influence of delamination size and location on growth under static compression load

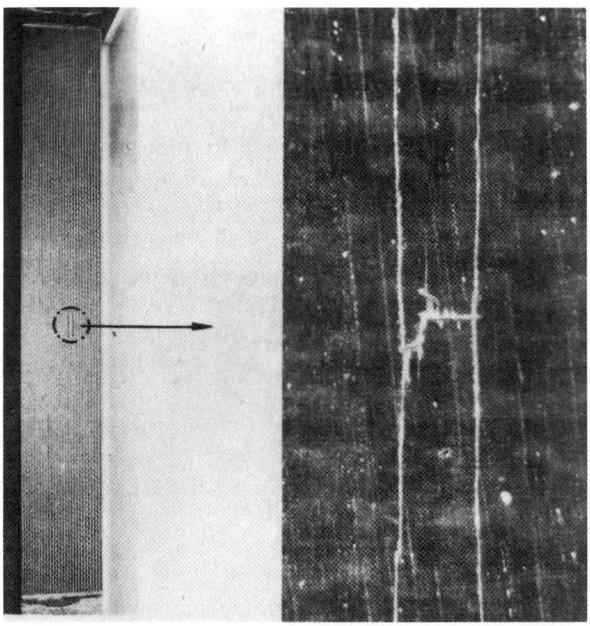

Fig. 11. Observation of fiber bundle break by grid reflection technique

The loss of adhesion between fiber and matrix is often referred to debonding. The cause may be a locally faulty surface treatment of the fibers, or the gradual deterioration of an initially good bond by mechanically or thermally induced fatigue. An

Fig. 12a, b. Fracture surfaces of $[\pm 45]_{2s}$-specimens. **a)** prior to cycling, **b)** after 3480 thermal cycles

example of the latter is given in Fig. 12, showing the fracture surfaces of tension-loaded ±45-specimens before and after several thermal cycling. While prior to cycling the many specks of resin adhering to the fiber surfaces indicate a reasonably strong bond, the much smoother surfaces after cycling seem to signal the loss of it. It may be expected that similar degradations take place under mechanical loads. As a consequence of progessive debonding, a gradual deterioration of the laminate stiffness might be expected.

5 Environmental Effects

The influence of the environment on the strength and stiffness of epoxy resins is a well-known phenomenon. Figure 13 shows, typically, the strength dependence of [0, ± 45, 0]-laminates (Fiberite 934/T300) on both temperature and moisture [6, 3]. The degrading trends are reversible upon drying of the laminate and/or by lowering of the temperature. Potential damage can be inflicted, however, by thermal shocking or by sustained exposure to high temperatures. Figure 14, for example, depicts the gradual loss of weight of epoxy laminates exposed to 100 °C and 120 °C for up to 25,000 h. The tests were performed under atmospheric conditions and produced no visible changes of the surface properties. Additional tests with neat resins in and out of vacuum traced the weight loss to an oxidative process accompanied by significant stiffness degradations.

A different kind of degradation occurs after prolonged exposure of epoxy laminates to ultraviolet radiation. A slight increase in strength, probably caused by postcuring of the resin, is followed by a gradual decrease as a consequence of the erosive deterioration of the laminate surfaces. As a point of interest it may be mentioned that epoxy laminates subjected to 3×10 rads of electron beam irradiation, in vacuum, did not suffer noticeable strength or stiffness degradations in strength of stiffness, although the laminate surfaces exhibited a slight reddish tint [7].

Apart from laminate strength and stiffness, elevated temperatures as well as the presence of moisture affect significantly the strain capability of the epoxy resin. Figure 15 shows the matrix crack formation along the free edges of $[0,90]_s$-laminates tension-loaded to failure under different temperature and moisture conditions. In

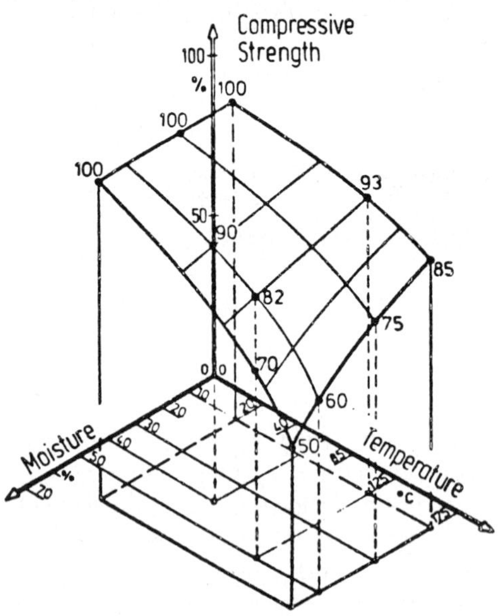

Fig. 13. Typical degradation of compressive strength due to temperature and moisture

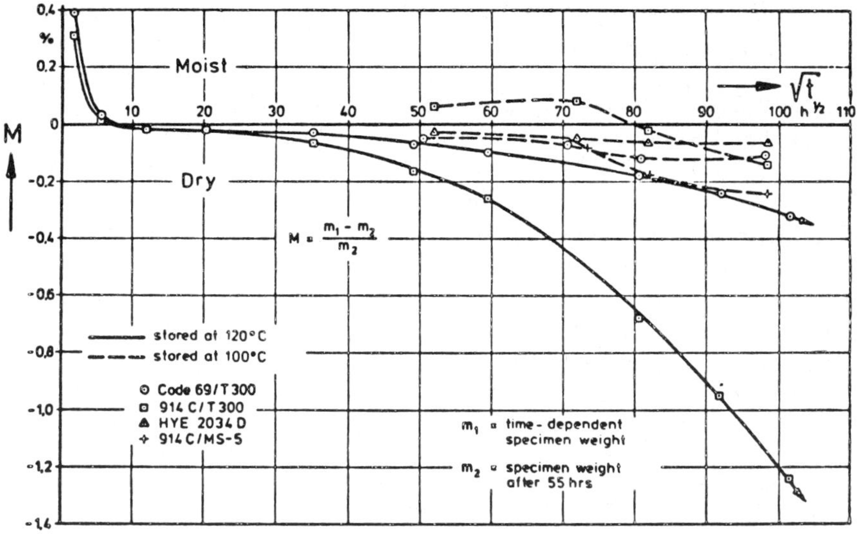

Fig. 14. Weight loss during sustained exposure at elevated temperatures

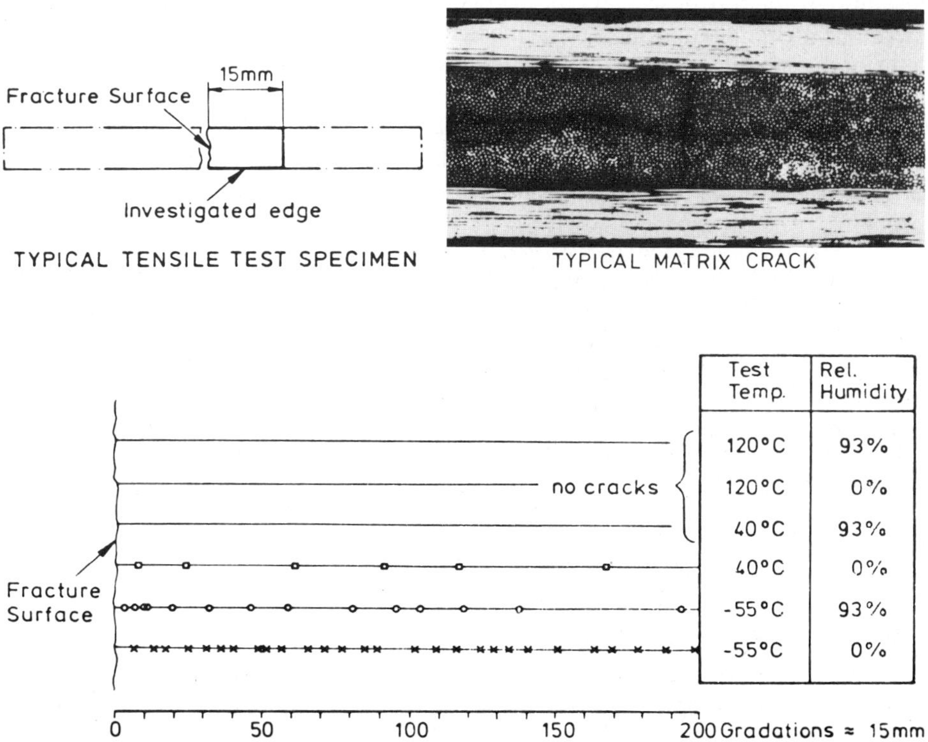

Fig. 15. Influence of temperature and moisture on matrix crack formation

the dry state, the diminishing number of cracks at increasing temperature signals a higher degree of ductility, while in the moist state the onset of cracking is retarded by the relieving superposition of the swelling strains on the curing strains. From the damage-mechanical point of view it is essential that such considerations enter into the analysis of failure modes.

6 Limits of Fracture Mechanics

The complexity of the failure modes of fiber-reinforced laminates makes an adoption or modification of established fracture-mechanical principles difficult or impossible. Exceptions, perhaps, are such simple cases as cracks parallel to the fibers of unidirectional laminates or delaminations between adjacent plies.

Even the classical problems of a drilled hole or an elongated notch prove to be elusive. Several models based on fracture mechanism principles have been introduced for their pragmatic treatment, all based on the assumption that failure will occur when the crack tip damage reaches a critical value. This concept of critical damage zone size represents the complex crack tip damage as an "effective" crack length and stipulates that the damage growth can be modelled as a self-similar crack extension. All of these macroscopic fracture models are semi-empirical and require, for each application, a series of tests in order to correlate the model with the response of the test article. As they do not address, but rather by-pass, the micromechanical complexities in the crack extension process, the attempted generalizations of the models, in regard to, e.g., stacking order or laminate dimensions, have not been fruitful.

A simple challenge to the validity of classical concepts is presented in Fig. 16, showing a laminate with a transverse notch and unidirectionally oriented fibers [10]. Fracture-mechanical reasoning might idealize this laminate into an orthotropic plate

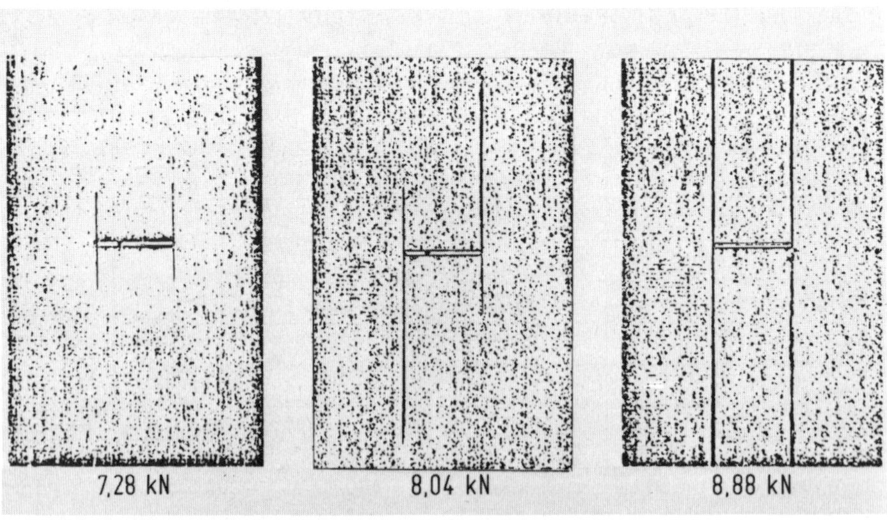

Fig. 16. Crack propagation in centrally notched $[0]_{16}$-specimen

with smeared homogeneity and, under static tensile load, expect a crack extension in the direction of the notch. In actuality the laminate behaves very differently. Emanating from the notch tips, cracks developing normal to the notch direction progress to the point of complete separation of the laminate with no growth in the notch direction at all.

7 Conclusion

Evidently, the issue of damage mechanics is of inordinate complexity and little tangible progress has been made so far. The remaining volume of work seems to be overwhelming and the question is valid whether the effort is worth the gain. The DFVLR Institute for Structural Mechanics has adopted an affirmative position since, without the mastery of damage mechanics, the potential of composite construction cannot be fully exploited. Considering the highly competitive international market, the consequences of failing would be so grave that a relaxed stance in this matter can hardly be afforded.

8 References

1. Awerbuch, J.: (1982), Monitoring damage progression in CFRP by acoustic emission. *In (2),* 47–88
2. Bergmann, H. W.: (1982), Damage mechanics of fiber-reinforced composite materials. *European Space Agency TT 758.* (Translation of DFVLR-Mitt. 81–25; seven individual reports.)
3. Bergmann, H. W., Nitsch, P.: (1980), Predictability of moisture absorption in graphite/epoxy sandwich panels. *AGARD-CP-288*
4. Block, J.: (1982), Monitoring of defect progression by acoustic emission. *In (9),* 3–1 to 3–11
5. Eggers, H.: (1982), Layer model for the calculation of stresses in defect zones. *In (2),* 174–209
6. Gädke, M.: (1983), Kennwertbestimmung für faserverstärkte Werkstoffe. *DFVLR-Mitt. 83–16,* 9–99
7. Hartung, W.: Effects of simulated space environment on the properties of CFRP. Submitted for publication in Composites Technology Review
8. Hillger, W., Schütze, R.: (1982), Nondestructive testing of CFRP laminates. *In (2),* 9–46
9. Jube, G.: (1983), Characterization, analysis and significance of defects in composite materials. *AGARD-CP-355;* 21 individual reports
10. Kirschke, L.: (1983), Schadensmechanismen fehlerbehafteter CFK-laminate. *DFVLR-Mitt. 83–16,* 145–189
11. Kress, G. R.: (1983), Fatigue response of notched graphite/epoxy laminates. M. Sc. thesis, VPI
12. Prinz, R.: (1983), Growth of delaminations under fatigue loading. *In (9),* 5–1 to 5–27
13. Prinz, R., Goetting, H. Ch., Schmidt, K.: (1983), Experimental and analytical study of strength degradation during fatigue of graphite/epoxy laminates. *ICAF-Doc. No. 1336,* 2.3/1 to 2.3/33
14. Prinz, R.: (1983), Analyse delaminierter Bruchflächen statisch und schwingbeanspruchter Probestäbe aus CFK mit multidirektionalem Schichtaufbau. *DFVLR-IB 131–83/01*
15. Prinz, R.: (1982), Damage Propagation in CFRP under cyclic loading. *In (2),* 128–172
16. Rohwer, K.: (1982), On the determination of edge stresses in layered composites. *Nuclear Engineering and Design 70,* 57–65
17. Rohwer, K.: (1982), Stresses and deformations in laminated test specimens of CFRP. *DFVLR-Forschungsbericht 82–15*

Chapter 5
Application of CFRPs in Aerospace

Applications in Aerospace, Especially in the USA

D. F. ADAMS

The introduction of carbon fibers in the mid-1960's, in the United States and elsewhere, almost simultaneously with boron fibers, provided the aerospace industry with composite materials which could compete in stiffness-critical applications with aluminum alloys. Prior to that time, the composite materials available, viz, glass fiber-reinforced polymers, were only competitive in strength-critical structures. For example, pressure vessels and rocket motor cases were major aerospace applications. Most airframe structures are stiffness-critical. The high modulus carbon fibers provided the required stiffness.

Early applications of CFRP were in not-critical-to-flight components, since adequate confidence in this new material's performance had not been established. Components commonly selected were horizontal and vertical stabilizers, flaps, landing gear doors, etc. Laboratory and flight testing quickly proved the superior performance of these components, most of which were very conservatively designed so that success was almost guaranteed. Even though almost all of these components were designed and fabricated on a direct substitution basis, i.e., using the same procedures as for the metal components they were to replace, which was an inefficient mode for these materials, the weight savings achieved were impressive. Weight savings of 15 to 25 percent were common. Costs, however, were typically high, perhaps twice that of conventional aluminum components.

As time has passed, material costs have dropped considerably, material performance has increased, fabrication processes have improved, and confidence has been gained. Thus, today the large airframe manufacturers in the United States have extensive facilities devoted exclusively to composite materials design and fabrication, and are able to produce CFRP parts at costs equal to or less than comparable aluminum or titanium components, while still achieving the weight savings previously quoted. As a result, U.S. military fighter aircraft now in production contain significant amounts of CFRP, particularly in stressed skin structures, and CFRP is being introduced into commercial transport aircraft also.

There is also particular emphasis on the use of CFRP in light civil aircraft, an area somewhat overlooked in earlier times. The Lear Fan Turboprop aircraft, an essentially all CFRP airframe, has received wide publicity as a bold step forward. However, other equally significant developments are now in progress by other airframe manufacturers.

Undoubtedly, the day of the all-composite aircraft is coming.

Experiences in FRG in Aerospace Application

K. BRUNSCH

1 Introduction

Although carbonfibres are wellknown technical products since T. A. Edison, it took them a long while to become a candidate for structural application. In the time before CFC glassfibre components had demonstrated quite a lot of favourable properties, but there was no braod application in aerospace due to the limitations in stiffness.

In the time after the Royal-Aircraft Establishment patent for carbonfibre production had be granted (after 1964) carbon fibres more and more became a strong competitor with the metals used in aerospace. Meanwhile the increasing use of CFC has spured the development of new light alloys to such an extend, that with components made of such alloys weight savings of in excess of 10% can be expected. This report tries to give an overview of the experience gathered with CFC structures for aerospace application mainly in Germany.

2 Materials

CFC always comprises two to three constituents: The fibres, the matrix and the fibre-matrix interface. Therefore experience with CFC material will be reported with reference to the above constituents. Fibre related experience could be gathered with a considerable high volume application of the high strength type fibres for aircraft components. The experiences with high- and ultrahigh modulus fibres mainly result from space activities. Table 1 gives an overview of the elastic properties of the standard UD-laminates made out of those fibres.

With all fibre types a consistant high quality could be observed. Due to the higher quantity used of T300 and XAS-fibres excessive information is available. Within one decade these fibres demonstrated a steady, rather moderate increase in strength and decrease of scatter, Table 2. Only two unfavourable events must be reported, both related more to the interface. Some while ago the surface treatment was executed too intensively, improving ILSS considerable but also dramatically increasing notch sensitivity. This is overcome today.

Table 1. Elastic properties of the UD-layer

60 Vol.%		M40A		75S	T300/914C	XAS/914C
		RT	77K	RT	RT	RT
$E_{\parallel}$	kN/mm^2	225.5	240	298	133	133
$E_{\perp}$	kN/mm^2	7.4	11.45	3.68	9.3	9.3
$G_{\#}$	kN/mm^2	4.3	5.77	3.06	4.6	4.6
$\nu_{\parallel\perp}$		0.26	0.26	0.36	0.28	0.28

Table 2. Improvement of CFC UD-tapes, HT fibres

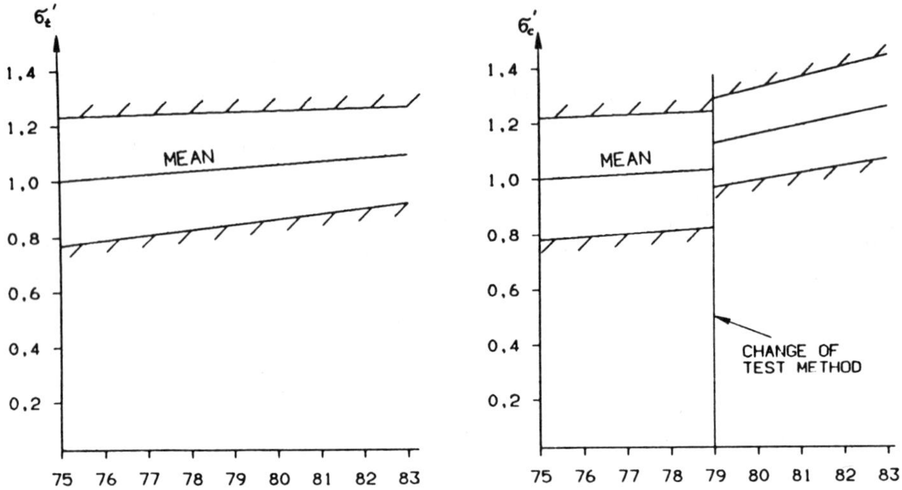

Recently a not notified change in sizing/finish seems to have been made by a fibre supplier. This influenced significantly the results of DSC testing. Shelf live might be reduced.

Investigation of this incident is going on. Like the fibres the resin systems also have developed to a high quality level.

With the 120 °C curing systems no problems have been reported for many years. With the 170 °C curing systems the adverse influence of flow controlling additives on wet glass transition temperature causes some problems.

Table 3. Material strength properties 60% fibre volume, RT

		T300/914C	XAS/914C	M40A/CY209	75S/CY209
Tension					
N/mm^2	$\sigma_{\parallel}$	1613	1580	1422	1572
	s	46		3.8	35
	V	2.9	5.6	0.3	2.2
N/mm^2	$\sigma_{\perp}$	36	36	30	
	s	5.5			1.4
	V	15.3	8	2.4	10
Compr.					
N/mm^2	$\sigma_{\parallel}$	1812	1714	760	526
	s	72.9		51	78
	V	4	5.5	6.7	15
N/mm^2	$\sigma_{\perp}$	218	206	94	72
	s	12.8			2.6
	V	5.9	6.2	4.0	3.6
τ_{ILSS} N/mm^2		109	98	30	38

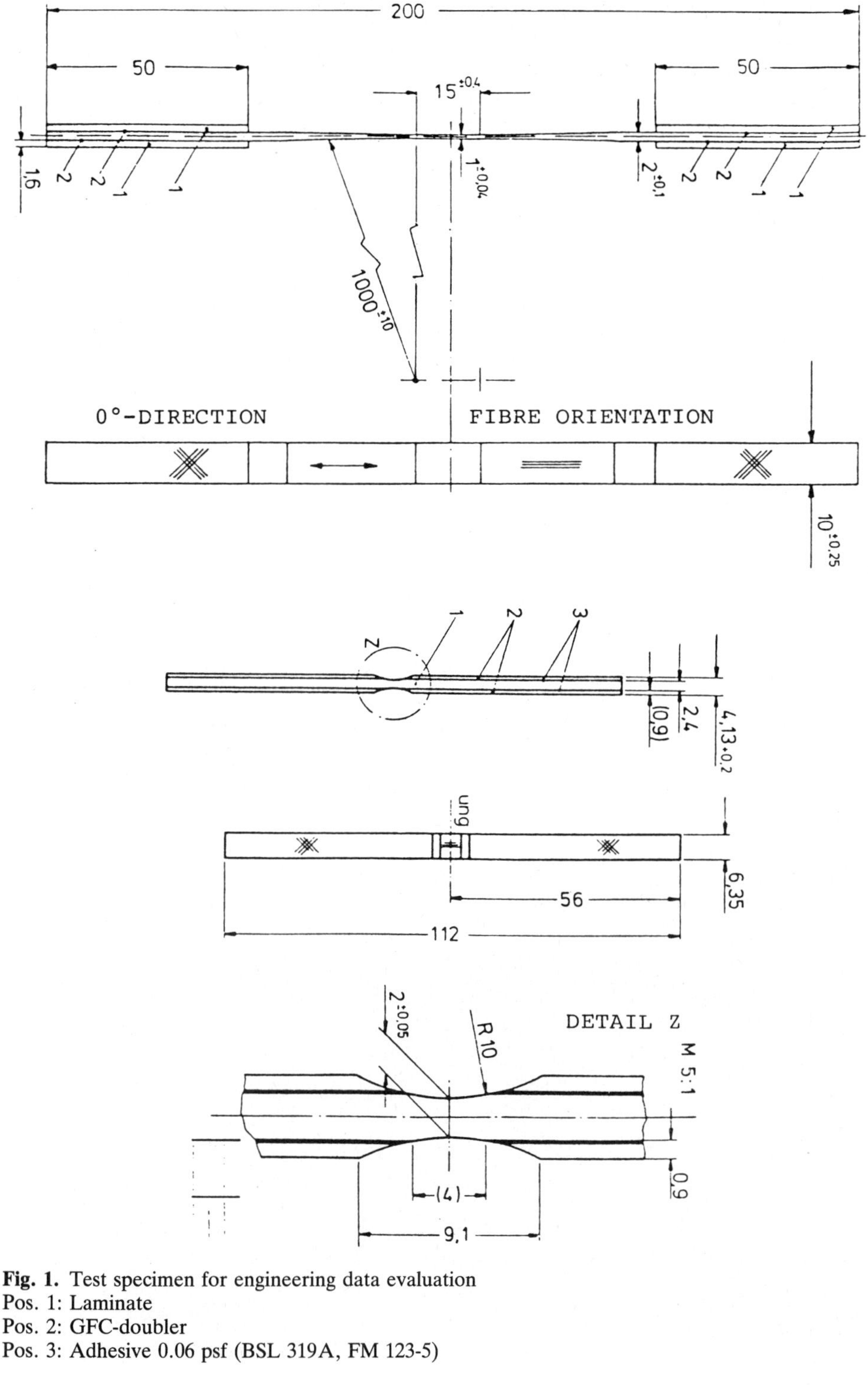

Fig. 1. Test specimen for engineering data evaluation
Pos. 1: Laminate
Pos. 2: GFC-doubler
Pos. 3: Adhesive 0.06 psf (BSL 319A, FM 123-5)

There is a good chance for improved systems to be on the market soon. Again in the early times it had to be learned that influence of additive especially on solvent (MEK) resistance must be considered.

With some resin systems to be used for wet processes like filament winding legal problems have been imposed recently. Due to much more stringent health- and/or environmental requirements some space qualified systems will not be available any longer. Expensive qualification of new systems will be necessary.

For such qualifications a lot of material testing has to be performed. There are two categories of material testing: Acceptance testing and testing for material engineering data evaluation.

Engineering data evaluation must be done with rather expensive test-specimen. For tensile properties evaluation the so called "BAC" specimen proofed to be the best. For compressive strength testing a modified "Celanese" test is recommended. Test-specimen see Fig. 1. Some of the data evaluated using aside of others the test-specimen mentioned before are contained in Table 3.

3 Experience with Space Application

Any material successfully to be used must be selected with respect both to mechanical properties and "design to cost". Carbon fibres, especially the high modulus types are expensive compared to metal. Therefore space applications, which yield the maximum profit of high strength and modulus to density ratio are their domain.

Todays most powerfull civil satellites in orbit are those of the INTELSAT V series. Approximately 25% of the dry satellite mass is made out of Carbon-Fibre-Composites. The structural components are solar arrays, antennas, the feedersystem and the antenna support-structure.

Fig. 2. Solar array

Table 4. Requirements of space structures

START AND LAUNCH PHASE

SURVIVING OF LOADS FROM

- SINUSOIDAL VIBRATION (LONGITUDINAL DIRECTION)
- ACOUSTIC VIBRATION
- STATIC LOADS FROM START ACCELER ATIONS
- 1ST RESONANT FREQUENCY ≥ 30 Hz

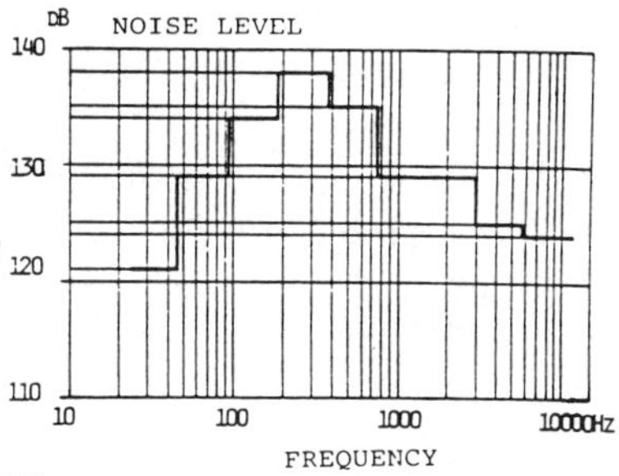

DEPLOYMENT EVENT (GENERATOR)

LATCH-UP SHOCK FROM DEPLOYMENT FINAL ANGULAR VELOCITY 60°/SEC.:

100 NM BENDING MOMENT AT 130 N SHEAR FORCE AT PANEL CORNER PART

IN ORBIT

- LOADS FROM 400 N BOOSTER THRUST
- LOADS FROM ATTITUDE CONTROL MANEUVERS
- LOADS FROM EXTREME TEMPERATURES -180°C ≤ T ≤ 100°C
- NO SIGNIFICANT CHANGES OF GEOMETRY AT WAVE GUIDES AND ANTENNAS
- PRECAUTIONS FOR VENTING AND ELECTROSTATIC DISCHARGE

OUTGASSING RESISTANCE:	TWL ≤ 1%	(TOTAL WEIGHT LOSS)
	VCM ≤ 0,1%	(VOLATILE CONDENSIBLE MATERIAL)

The material used for the solar arrays is Thornel 75S fibre and Ciba CY209/HT972 resin. The ultrahigh modulus fibre was selected to provide sufficiently high natural frequencies of the panels at minimum weight and low coefficient of thermal expansion, similar to that of the solar cells. Design for proper frequencies is required because of the load spectras at start and launch (Table 4). The resin CY209 system was selected first because it is qualified to meet the less than 1% TWL (Total Weight Loss) and the less than 0.1% VCM (Volatile Condensible Materials) requirement and second for filament winding applicability.

By use of these materials the mass per usable m^2 of solar array structure could be reduced from 5 kg for the aluminum design to 1.3 kg for that with CFC.

The solar array consists of 3 panels and the central joke. It is the deployable type. The joke consists of rectangular beams, the panels are CFC skin-Al-honeycomb design.

For all laminate production filament winding was used. Experience gathered is

- the required fibre orientation can be maintained within less than ± 3°. This is required, because of the maximum of flexural and shear stiffness being very narrow for 0/90° and ± 45° high modulus fibre laminates
- ultrahigh modulus fabrics are not available, fabrics are woven to standards. Filament winding technology provides flexibility, the theoretically predicted mass of fibres per unit of area can be applied exactly. The face sheet layer can be cut of the mandrel without problems. The large radius in the dome areas doesn't cause problems

- impregnation is without influence on alignment
- laminates are "quasi symmetrical", minimum thickness is as low as 0.1 mm
- no logistic problems like with prepregs exist – storage time, laminate thickness and so on

The rectangular beams shown in Fig. 3 also completely are produced by filament winding. The ud-chords are wound on narrow drums and cut off.

With the INTELSAT V panels the thickness of the face sheets is 0.16 mm nominally. The cell width of the honeycombs being 10 mm, the support is local only.

The flexural loading of the panels results in compression in the skins which might fail with instability i.e. with wrinkling. For calculation of minimum wrinkling load stands:

$$\sigma_{wr} \leqq 0.816 \cdot \sqrt{\frac{E_c t_f E_f}{t_c}}$$

f = face
c = core

Due to the large ratio of cell width to face thickness local instability – intercellular buckling was expected. For design verification about 150 coupons have been tested.

With these the principal validity of the above equation was found, the factor had to be modified from 0.816 to about 0.30 (Fig. 4).

The second area of unique CFC application for space are antennas. Here the mass saving potential is gratefully acknowledged, but match winning is dimensional stability independent of temperature. To demonstrate this property an antenna of 90 cm Ø was tested from − 140 °C to + 120 °C for 70 cycles. The maximum deformation normally to the antenna dish was measured to be 0.3 mm at one point only. Average dislocation was 0.1 to 0.2 mm.

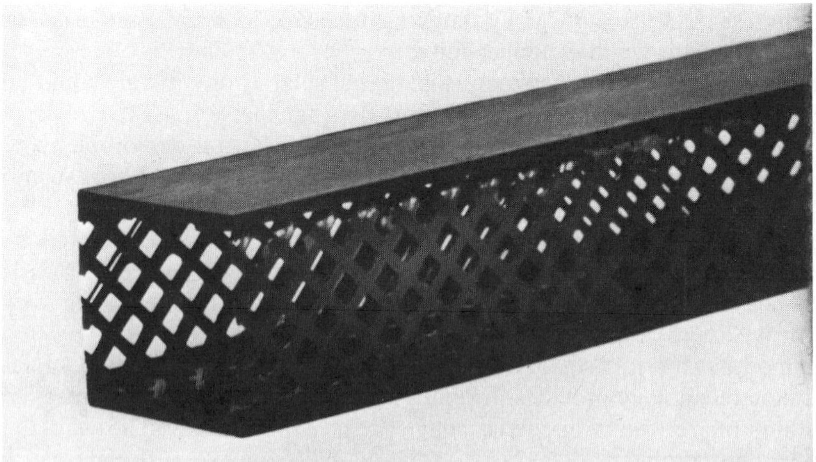

Fig. 3. Filament wound rectangular beam

Reflectivity also was found to be extremely good. Comparative measurements High-Modulus-CFC versus aluminum showed, that in the 1.3 to 1.7 and in the 2.8 to 3.2 GHz frequency, the CFC was as good as the metal. Up to frequencies ~ 14 GHz pure CFC shows reflectivity factors of up to 98%.

The lessons learned with CFC space applications are: HM-CFC structural components feature outstanding good behaviour both mechanically and electrically. Application for central structural components is limited by poor heat-transfer capability. Ongoing R + D efforts will increase effectiveness of structures.

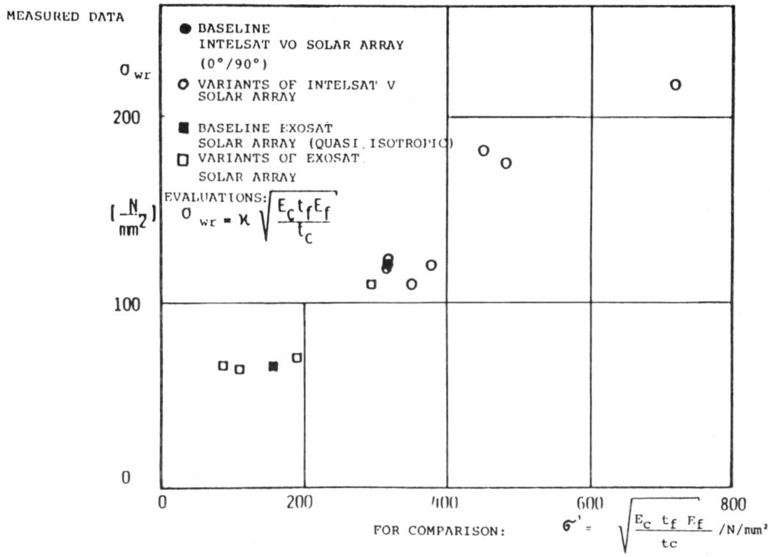

Fig. 4. Wrinkling test results (from compression loads)

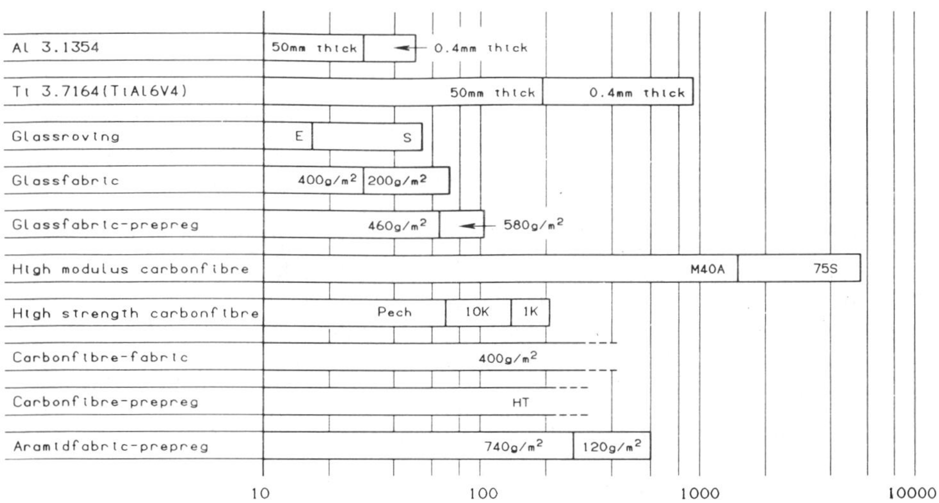

Fig. 5. Material cost comparison (DM/qm^3)

4 Experience with Transportation Aircraft

When the utility of CFC components became evident most of major aircraft companies started programs to build up carriers confidence in the new material. Small structures like the Boeing 737 spoiler were developed and brought into service. These service evaluations showed durability like with metals, ease of repair and did train service personal to handle properly such components.

The programs mentioned above were started when the prognosis for carbon fibre prices were very encouraging. Unfortunately carbon fibre prices are still rather high compared with metals (Fig. 5) and the economic situation of most of the carriers does not let them pay additional money for weight saving.

At MBB's commercial AC division a CFC component will get the go ahead only if it costs equal or less than a comparable one out of metal. This can be achieved only by carefull evaluation of appropriate technology which must be taken in account at the very begin of the design and by dramatic reduction in number of parts and fasteners. A good example for such a technology/design is the CFC rudder of the DC10. This was produced in a one shoot technology using Thermal Expansion Moulding. Silicone rubber bags are a standard tool today with the known disadvantage to stand for a limited number of cycles only. With the development of the airbus vertical stabilizer it was aimed to avoid the costly repeated production of rubber tools. The use of the thermal expansion of aluminum only combined with vacuum bag technology is cost effective. Series production for the airbus family was decided by Airbus Industry.

To the authors knowledge, the airbus vertical stabilizer is the largest civil transportation CFC component to go in series production. Some of the most interesting features will be reported.

Cost effectiveness mainly results from the technology used for the integrally stiffened about 3 by 11 m skin panels, which are produced with the so-called modular

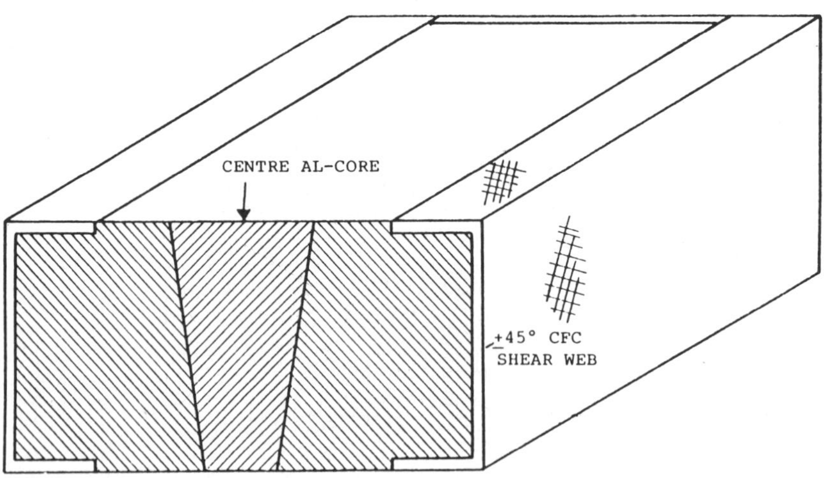

Fig. 6. Crossection of modul with shear web

concept. CFC fabric skins are layed onto the mould, which also is made out of CFC. At the same time ±45° fabric tapes are wrapped around the aluminum "moduli" (see Fig. 6). Once all the moduli are covered they are assembled on a jig. There is considerable compression chordwise, but non spanwise. UD-tapes are layed on that side, which later is layed on the skin. For that purpose the jig is rotated 180° and the mould positioned underneath. For ease of that the mould is air cushion suspended. After the "moduli" have been layed onto the mould and fixed by a frame running all around the mould the inside UD-tapes of the flanges are laminated.

The vacuum bag is then tightened to the mould and the complete lay up cocured. The result of that is shown in Fig. 7.

The resulting CFC stabilizer box is predicted to be approximately 5% less in cost and 17% less in weight compared to the metal version. The predicted cost reduction is based on the assumption, that the wrapping of the moduli, the assembly of the moduli and also the cleaning of the moduli is performed by robots or robot like machines.

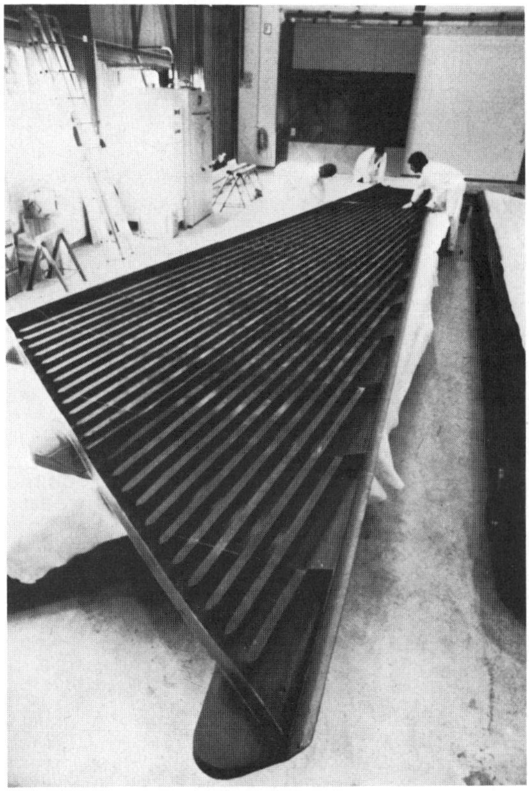

Fig. 7. Integrally stiffened panel

5 Experience with Military Aircraft

In civil application like with space- and transportation aircraft-components design to cost normally is decision making. With military equipment superiority consideration rank first. This is the reason for advanced composites being used for military aircrafts since they became available. Application of CFC components started with the history making boron epoxy horizontal stabilizer of the F14.

Todays advanced composite structures concentrate on use of the by far less costly carbon fibres, the share of composites in structural mass break down increases continuously (Fig. 8). Control surfaces are state of the art, wetwings (Harrier) and front fuselage components are in series production. To come to that state a lot of work needs to be done. Typical sequence of such a development is described using the CFC-Tornado taileron as an example. Aerodynamic shape and also loads were known from the metal version (Fig. 9). Preliminary design made use of so-called carpet plots. These carpet plots – the tensile RT strength of the 0°/90°/45°-laminate

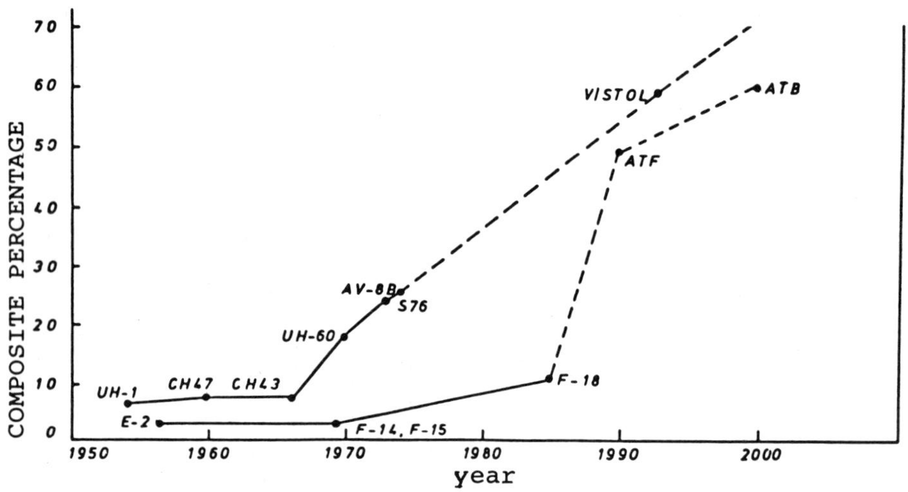

Fig. 8. Composites used with existent and planned for future aircraft

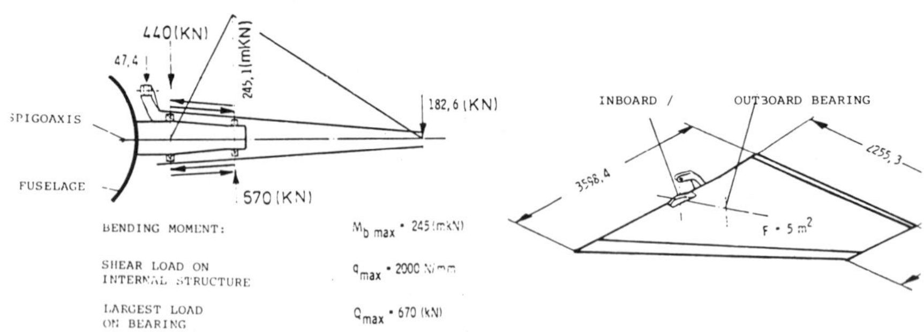

Fig. 9. Tornado taileron loads and dimensions

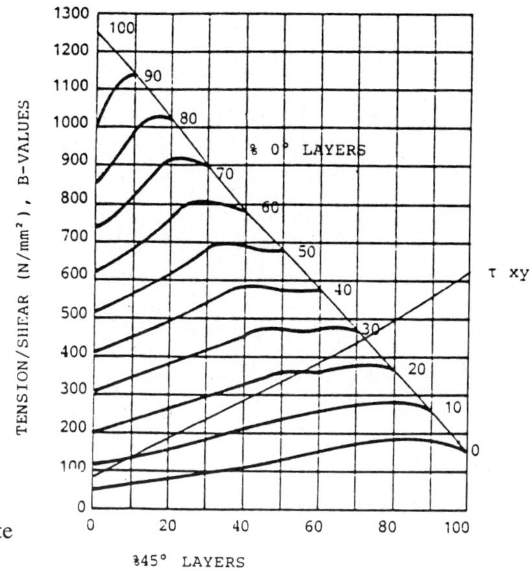

Fig. 10. Tensile strength 0°/90°/45° laminate family, RT

family of T300/914C is shown as an example (Fig. 10) – are calculated by computer programs. The input data are to be evaluated by carefully testing the "standard" 60% volume of fibre unidirectional tape. Carpet plots for example of bearing strength must be based of course on tests with laminates.

In the preliminary design two versions of the highly loaded internal structure have been investigated – an all CFC and a mixed CFC-metal design. Although some novel structural CFC components and the related technology had been developed successfully, the CFC-metal design was selected (Fig. 11) for lower risque.

To finalize the design "Nastran" computation was used. Both skin laminating plan with reference to ply sizes and orientations and honeycomb density have been optimized based on that calculation.

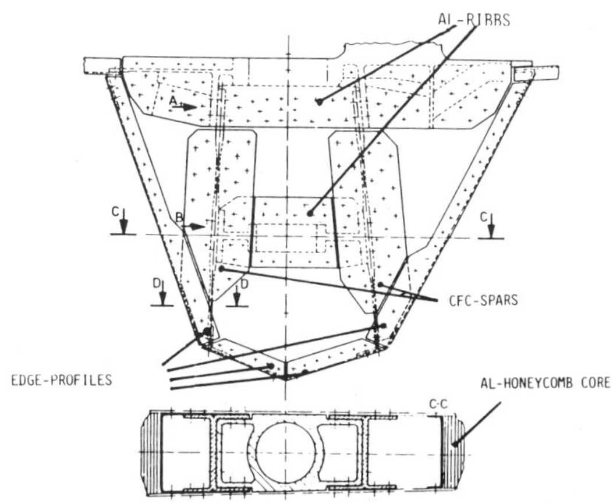

Fig. 11. Taileron internal structures

Fig. 12. Taileron in bonding fixtures

Taileron have been produced without major problems and are in flight and fatigue test now (Fig. 12).

Out of the total mass of the taileron 36% are CFC only. The weight saving compared to metal version is 17%.

Approximately the same weight saving was calculated for a technology research CFC fighter cockpit. The technology research cockpit is about 3.5 m length by 1.4 m width and 0.9 m height. It weighs 290 kg with the reinforcements to allow assembly with the test fixtures and should be 272 kg without. The low weight low cost design consists of two integrally stiffened shells (Fig. 13), the cocured stiffeners are I-shaped. For ease of production the skins are laminated on male mandrels (Fig. 14) and then transferred to female galvano nickel moulds. The stiffeners are preformed on silicon rubber tools and inserted in the mould together with the rubber cores. Skin panels have been produced without problems, one completed cockpit (Fig. 15) will be tested electrically and structurally.

Lessons learned:

- Access to experience with military Aircraft CFC components is limited. Information was obtained from the AFML, that there was no adverse influence on composite structure, that had be flown to skin temperature in excess of the wet glass transition temperatures of the composite. It must be noted, that flights of such a speed always are of short duration only.
- To provide battle damage tolerance the strain at ultimate design load shall not exceed 0.35% to 0.4%.
- Complex one shot cured components must be designed with allowance for minor defects. Especially cocuring of skins and I-beams always is associated with the risque of flaws in the direction of the I-beams. Those flaws tend to be at the corners where the shearwebs attach to the skin.

Fig. 13. Integrally stiffened fuselage shell

Fig. 14. Male laying tool

Fig. 15. Cockpit CFC structure assembly

6 Helicopter Experience

Since the great success of the BO105 glass fibre composite rotorblades everybody uses composite rotorblades. The programs to replace the metal blades of the US-Forces CH46 and CH47 helicopters total approximately 200 Mio. US $.

In spite of the great number of composite rotorblades being either in service production or under development the CFC experience is limited because of the large GFK share.

With the BK117 (MBB-KHI) mainrotorblade some experience with carbon fibre-fabric is being made. Initially these blades were built with glassfibres only. Testing inflight the outboard region of the blade was found to be to low in torsional stiffness. About two meters of glassfabric of the skin were replaced by carbon fibre fabric (Fig. 16) and since that the helicopter is improved dramatically in static and dynamic stability. In the "wet" production process of these blades the mixed fabric skin-glass- and carbon fibre fabrics does make no problems. MBB does expect to have the excellent long time serviceability for the carbon-skin-blades which is known for its

Fig. 16. Rotorblade with CFC-tip in mould

glass fibre-blades. Adverse influence of the electric conductivity of the carbon skin is of low probability, simulated lightning testing with the unprotected CFC skin did not cause catastrophic damage (Fig. 17).

Fig. 17. Lightning damage on rotorblade

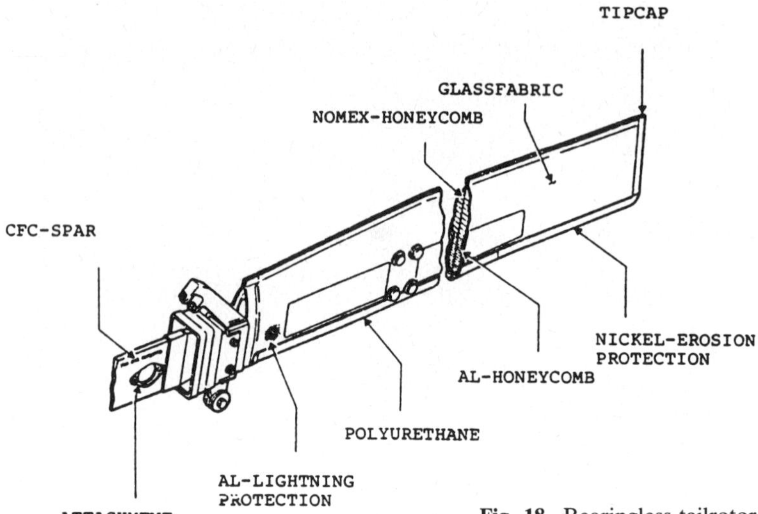

Fig. 18. Bearingless tailrotor

Whilst MBB did use CFC to increase torsional stiffeness, Sikorsky did use it for the opposite objective. The "flex beam" of the bearingless S76 tailrotor (Fig. 18) is entirely made out of high strength CFC. The design of a rotorblade always must avoid coincidence of natural bending frequencies of the rotating beam rotorblade with exciting frequencies. The natural frequencies both in flapwise and chordwise direction must be relatively high, thus defining the required flexural stiffness (EJ) distribution over R.

With the low flexural stiffness of GFC the Js and with them the areas have to be considerably larger than with CFC. As the shear modulus of GFC and CFC-UD is

Fig. 19. Metal version of drive shaft assembly

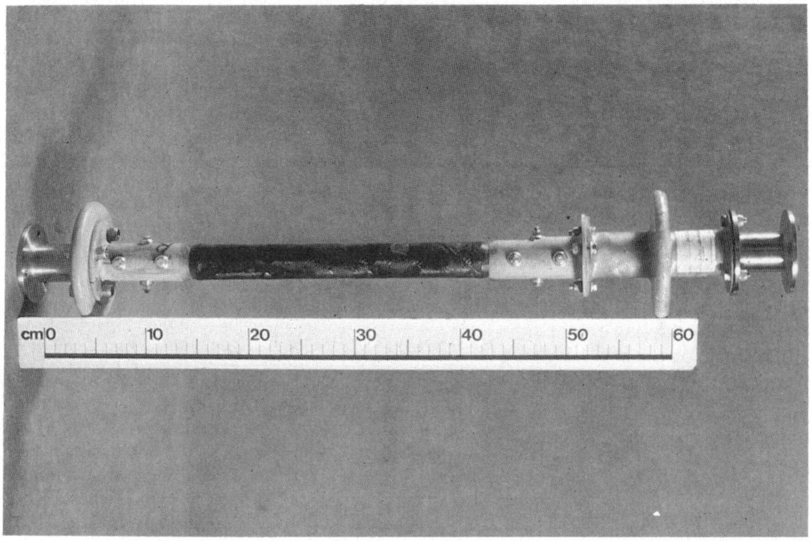

Fig. 20. Composite drive shaft assembly

about the same, the glass fibre version would be much higher in torsional stiffness, the structure would not withstand the repeated deformation for pitching.

The Sikorsky S76 flex beam CFC tailrotor is in service now since 1978. No problems with the CFC flex beam have been reported within many hundredthousands of flight hours. A different area for profitable use of CFC are drive shafts.

This has been demonstrated by the development of an experimental tailrotor drive system for the BO105. The metal version – stainless steel tube, titanium fitting and stainless steel leave springs (Fig. 19) was replaced by a filament wound HM and HT CFC tube and a glass fibre composite coupling. The composite version (Fig. 20) was successfull in static and dynamic test, but most important is that it was superior weight and costwise (Table 5).

With reference to cost must be added, that for logistic reason the number of bearings of the metal version was maintained although the increased flexural frequency of the CFC version would have allowed reduction.

Despite the advantages described above the composite drive shaft system is not in series production. This is because of the miss match of coefficient of thermal expansion of the aluminum tail boom in series production and the CFC tubes. For future MBB helicopters the use of advanced composites both for tailboom and -drive system is foreseen.

The experience with CFC applied for fuselage components is limited only. With respect to the exterior surface, the share of composites is rather large – about 40% for BO105, 50% for BK117 and 70% for the Agusta A129. Most components with these percentages are secondary structures, made out of glass- and aramide fibres, nearby no carbon fibre. A major break through is to be expected once the All Composite Airframe Programs have proven the promised weight and cost reductions. Both Bell and Sikorsky will have an "ACAP" fuselage completed in 1984. Weight saving is predicted to be 22%, cost saving should be 17% compared to the

Table 5. Drive system mass- and cost comparison

Part	Mass (g)			Number of parts		
	Metal	Composite	Mass-reduct.	Metal	Composite	Parts-reduct.
Long tailrotor drive shaft (l = 3940) with a coupling on one side and a flange on the other and additional 4 bearings	7780	3730	4050	89	54	36
Long tailrotor drive shaft (l = 3904) with a coupling on one side and a flange on the other without bearings	6980	2930	4050	72	46	26
Short tailrotor drive shaft (l = 731) with a coupling on each end of the rod	1930	860	1070	101	37	64

Basis: Parts for 200 helicopters

Component	Total costs (%)		Material costs (%)		Manufacturing costs (%)	
	Metal	Compos.	Metal	Compos.	Metal	Compos.
1 drive shaft incl. 2 flanges without bearings	46	69.3	33.8	29.9	12.2	39.4
2 couplings	52.2	20.3	30.0	8.0	22.2	12.3
assembly	1.8	0.4	0.5	0.4	1.3	–
Total	100.0	90.0	54.3	38.3	35.7	51.7

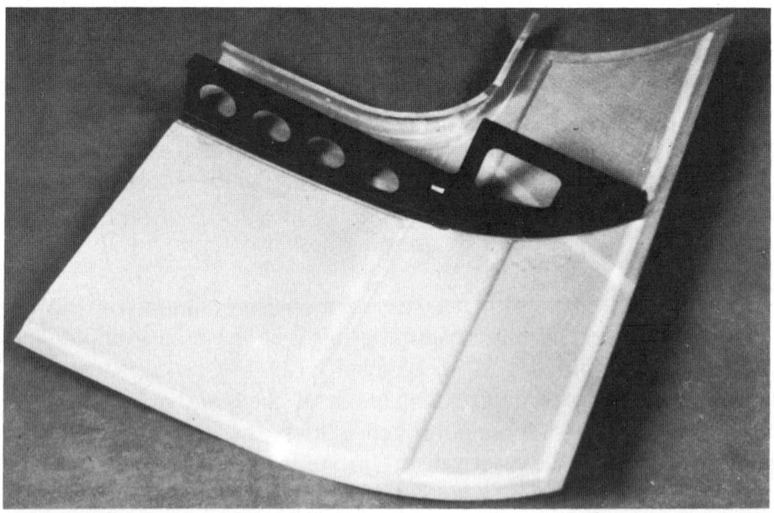

Fig. 21. Composite helicopter airframe section, one shot cured

comparable metal version. A typical fuselage section made out of both CFC and AFC is shown in Fig. 21. This section designed for BO105 size was used to investigate technologies and resulting weight and cost saving potential. Results of these investigation indicate the objectives of the ACAP to be realistic.

7 Technology

For many years the nearby only available technology – aside of filament winding – was hand-lay-up. This did include cutting of the plies, removal of release and supporting film and laying of the plies. Today a lot of machinery is offered and great care is recommended to avoid unprofitable investments. Based on MBB production requirements experience some examples are reported, which are more generally composite than CFC only related.

Materials used mainly are glass- and aramide-fibre-prepreg, i.e. broad good, which is used in secondary structures with one or two layers only. For this application a "Gerber-Cutter" (reprocicating knife) is used to cut the plies. Most of the remaining work is done by hand.

The rather expensive "Gerber" will not be used to cut the plies to be wrapped on the "moduli". The reason is, that the "Gerber" fixes the broad good to the cutting table by vacuum and therefore does not allow to remove supporting and release film in advance of cutting. For series production the plies for the "moduli" will be cut by an NC-controlled waterjet. Waterjet technology allows the automated removal of all auxiliar materials prior to cutting.

In the R + D phase MBB's helicopter division uses steel rule dies to cut plies for the CFC plates of a rotor hub. The plies are identic in geometry but vary in fibre orientation. The steel rule dies work effectively only with the foils on the prepreg. For mass production the cutting technology must be reconsidered due to the lack of a reliable automated technology to remove foils of the plies.

The application of a laser cutting device for serial CFC cutting is – at least now – not planned at MBB.

With laying techniques the situation is similar. Although an "Ingersoll" tape laying machine is installed at the civil AC division it will not be used for the airbus vertical stabilizer. Handlaying of the C-fabric-prepreg is more cost-effective than tape laying of 4 layers to get the same "symmetric" 0°/90°-laminate. The tape laying machine is used for components with great amounts of UD-tapes required.

The military AC division is installing a tape laying machine coupled with a cutting service. As the cutting device operates much faster than the laying machine, ply-families will be laid to balance machine capabilities. With respect to the small military AC market in Germany, ply transfer by hand to the mould appears to be superior in cost effectiveness.

If cutting is performed by means of steel rule dies the die can be used for ply transportation also. This is considered for the fabrication of the rotorhub plates, but first the problem of removing the foils must be solved. Laying of the rotorhub plies is no problem because of the press curing.

The most challenging task in mechanisation of CFC structures production is laying of the plies, especially on double curved surfaces. This area is investigated intensively and with high cost investments, mainly in the USA.

The results achieved up to this day are such, that MBB in the moment considers hand-lay-up of the precut plies to be cheaper than automated laying. This consideration is based on 200 kg or less CFC prepreg to be consumed per day.

Lessons learned with CFC technology are:

- The small volume production of aerospace CFC structures requires a well balanced mix of automated and manual processing to be costeffective
- waterjet cutting of prepreg does not reduce material mechanical properties
- progresses in auxiliary materials can decide on technology and/or design to be used.

8 Quality Control Experience

Composite Components have been developed and produced already before prepregs were on the market. Wet processes still are used, the quality assurance "tools" both for wet and prepreg technology are available.

For the "wet" technologies like filament winding resin and hardener are tested both physically and chemically at income inspection. Typical tests are

for resin:
- density
- epoxid equivalent
- viscosity
- colouring

for hardener:
- amin/anhydrid value
- composition (HPLC)
- melting point (TMA)
- latent heat

As to the testing of the resin of a prepreg there is some controversity in the moment. The requirements range from mechanical testing only to complete physico chemical analysis.

An overview of the physico-chemical methods is given in Table 6. The standard tests are to assure material quality at minimum cost. With that in mind a very costeffective test-specimen for prepreg-UD-tape has been developed. A single layer of prepreg-tape is covered with a thin polyamid foil and then cured. The cured tape is cut to strips of 10 mm width and tension tested to failure. Tensile modulus and ultimate strength are measured.

This simple test combined with the information of fibre-mass/unit of area provides excellent global information on fibre tensile properties and/or fibre distribution and/or fibre alignment.

With the mass production of glass fibre composite rotorblades it had to be learned, that the ILSS does provide limited information on surface adhesion only. In

Table 6. Physico-chemical methods for quality assurance of epoxy-CF-prepreg systems

Method	PCI	Chem. charact.	Specification	Acceptance control
HPLC (HPTLC)	− (fingerpr.)	+	−++	+
GPC		+	−	
GC		+	−?	
IR	+ (fingerpr.)	+	−+	+
NMR		+	−	
AAS		+	−	
DSC	+	+	−+	+
DMA		+	−	
TMA		+	−	
TGA		+	−	
Viscosity	0	+	−+	+
Vanhograph	0	+	−	
Element. analysis		+	−+	
Torsional pendulum DMA (lamin.)		+	−+	+
Standard tests		+	−+	+

Use of methods if available: Methods for specification are specific for each individual system. They can be selected on the basis of chemical characterization in accordance with the supplier. + must; 0 can; − possible

glass fibre acceptance testing a peel test was introduced. A similar test for carbon fibres is being developed.

The single layer test-specimen is used for carbon fibre-fabrics also. For test result interpretation, the specimen is treated as a $[0, 90]_s$-laminate.

The fibre itself, which is dominating the composite properties is tested in laminated form only. For wet processes UD-laminates are produced for evaluation of tensile strength and modulus, ILSS and may be peel strength also.

Within process testing a well balanced compromise must be found to assure quality to maximum reliability at reasonable cost. But prior to that, all the equipment intended for use must be controlled starting with the environment of shops, tool-dimensions, heatflow and temperature distribution of tools and so on. All parameters easily to be reported shall be. For example any scale used is to be linked to a printer. If there are major differences in thickness of components and/or moulds, the readings of many probes for temperature are to be documented.

Also the materials have to be selected with respect to in process quality assurance or inspectability.

Again for example: To reduce the risque of poor fibre orientation with C-fabrics, a "Kevlar" tracer is to be used. In addition to that so-called critical processes have to be identified and special/additional tests/test-specimens have to be specified.

For the time after the component is cured and controlled for dimensions NDT is to be applied. Acoustic transmission testing is in series application with most of the CFC-components produced today.

Thermography, Holography and X-ray technology are well-known also. Computerized processing of the acoustic or X-ray signals improves detecting capabilities.

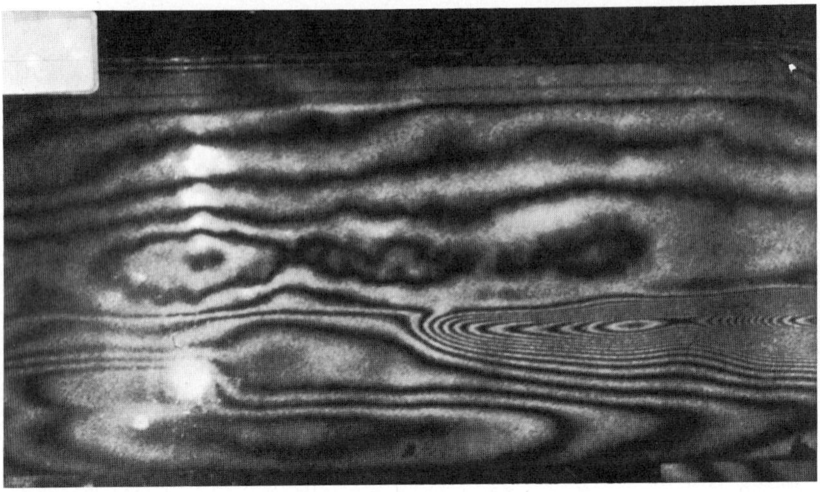

Fig. 22. Holographic image of CFC rotorblade with skin debond

In the development phase computer tomography can be very usefull too. It helps to produce very reliable information e.g. of flaw size which then can be compared with acoustic test reading intended to be used in series production.

The "NDT" to be used depends on the individual structure and on the defects looked for. The holographic image of a great flaw between CFC skin and PVC-foam core of a rotorblade is shown in Fig. 22.

For space structures proof test loading appears to come back again. This can be accepted only with respect to the relative small number of components and the relative high prices payed for space structures.

9 Conclusion

CFC is a well adapted aerospace material today. It has spured metal technology to considerable improvements (Al-Li alloys) and is being improved itself to the maintain lead among the aerospace materials. Ability to handle CFC is an earnest-requirement to be competitive in the international aerospace market.

Experiences in UK and France in Aerospace Application

B. J. Hayes

1 History

It is believed that the first consciously produced, high performance, semi synthetic composites were manufactured from unidirectional prepreg at Duxford, near Cambridge, England in 1941–1942. They were produced from a phenolic resole resin and flax fibre and were developed as a safeguard against the non availability of aluminium. These were very advanced materials at that time and were never put into full production. Figure 1 shows the assembly of a Spitfire Cockpit from flax phenolic composite laminate known as "GORDON AEROLITE".

Some 25 years later high performance carbon fibre was becoming available for evaluation and by the late 1960's the U.K. had three production plants in operation, all licensees of the process developed at the Royal Aircraft Establishment, Farnborough, near London.

By far the largest production capability was at the Rolls Royce plant. This had been constructed to yield up to 60 tonnes per year; principally aimed at providing carbon fibre for their own project the RB 211 engine fan blade.

This fan blade programme was also very advanced for its time but unfortunately failed due to inadequate impact performance. Many trial blades were produced, including the rather smaller Conway blade. Figure 2 shows Rolls Royce trial CFRP Conway fan blade.

Other projects were less ambitious but for one reason or another were not put into full series production, although all were technically successfull, for instance the Harrier ferry wing tip demonstrating early complex ply orientation. Figure 3 shows as another example a helicopter torque tube produced from prepreg, and Fig. 4 an

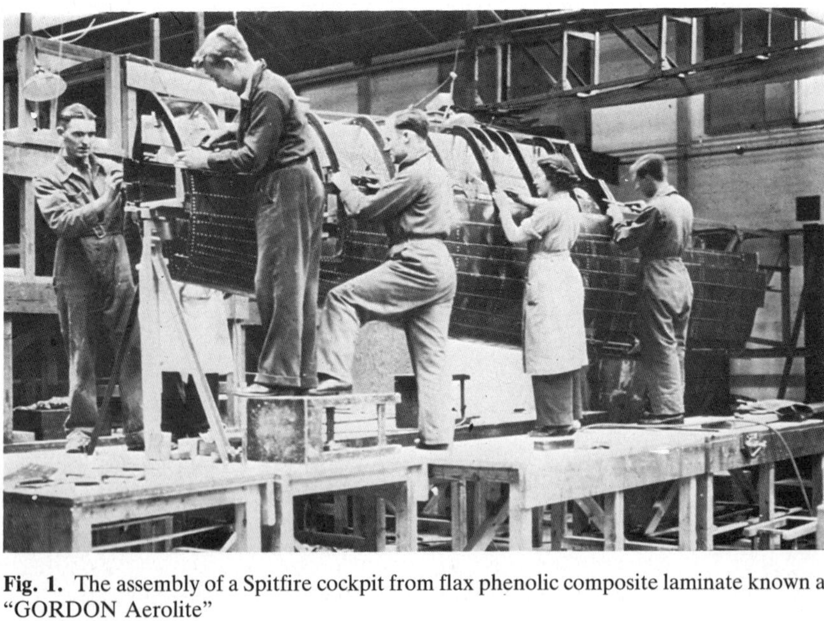

Fig. 1. The assembly of a Spitfire cockpit from flax phenolic composite laminate known as "GORDON Aerolite"

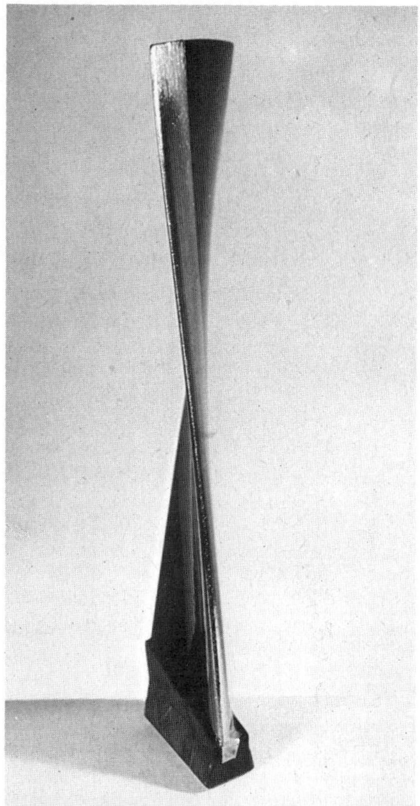

Fig. 2. ROLLS ROYCE trial CFRP Conway Fan blade

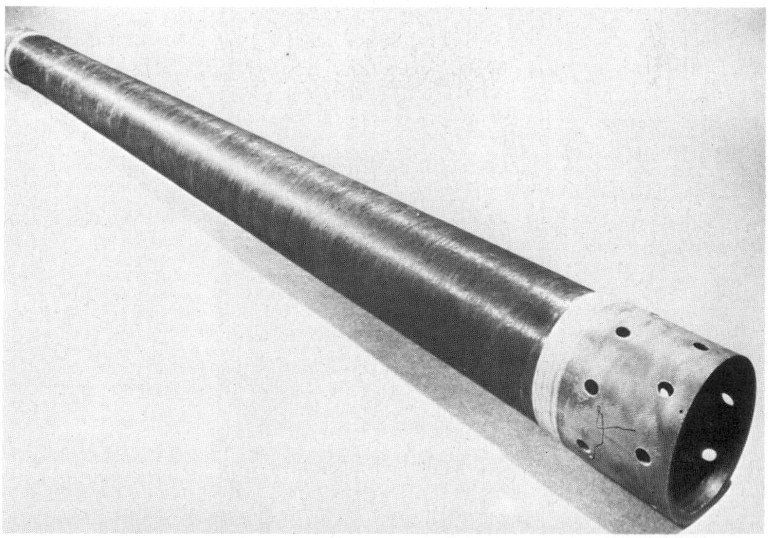

Fig. 3. Helicopter torque tube produced from prepreg

Fig. 4. An early satellite demonstrator

Fig. 5. VC 10 centre rudder section

early satellite demonstrator. This was one of the first carbon fibre structures employing aluminium honeycomb for increased stiffness. The structure was fully fatigue and strength tested and easily met all design requirements, although its construction was time consuming and expensive.

2 Civil Demonstrators

After the demise of the composite RB 211 fan blade project the prospects for CFRP looked doubtful and for some two to three years activity was at a very low level.

From 1972 work increased again and there was a gradual progression from making small non critical parts in carbon fibre which were already designed for metals (access doors etc.) to making larger structures in which a portion of a metal part was replaced by composite. Most of these structures were given practical flying trials.

An example of this type is the VC 10 centre rudder section (see Fig. 5) showing partial replacement of metal with CFRP. Three of these were produced and two are still in service.

A further example is the BAC 1–11 engine CFRP stub wing panel. 5 sets were produced from woven carbon fibre FIBREDUX 914 prepreg and directly co-cured with REDUX 319 A film adhesive to AEROWEB NOMEX honeycomb core. All five sets are successfully in service with British Airways.

Gradually this approach gave way as confidence was gained to redesigning some aspects of existing metal parts for composites and then to designing, producing and testing a much wider range of items as initial composite concepts.

Fig. 6. CFRP Aileron – in production since 1975

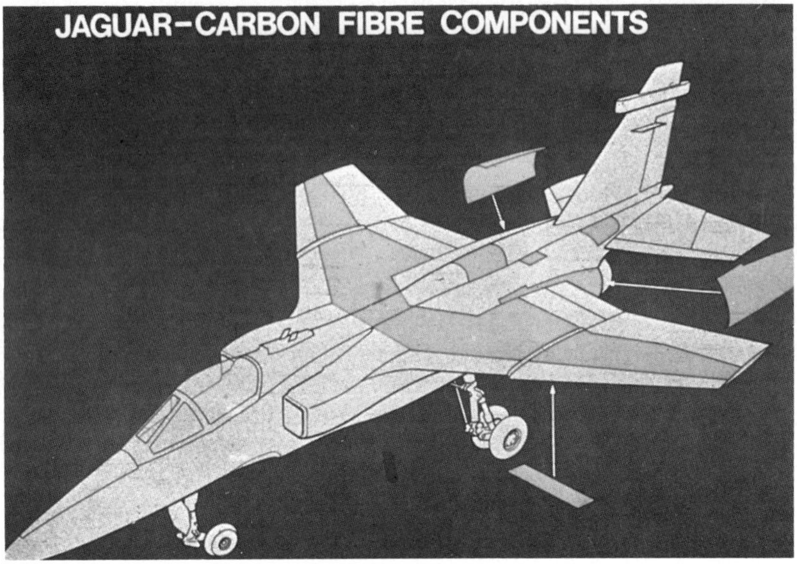

Fig. 7. Schematic representation of the parts manufactured for evaluation in CFRP on JAGUAR

Fig. 8. JAGUAR spine panels demonstrate aluminium honeycomb core

3 Military Demonstrator and Early Production

As various parts were made in CFRP and proved to be satisfactory so some of them were put into limited production. The French Aerospace industry was probably rather advanced in this approach and have a good success record in their early production use of CFRP. Figure 6 shows a CFRP aileron in production since 1975.

In the U.K. more and more parts were produced and tested for the JAGUAR and TORNADO aircraft. Figure 7 gives a schematic representation of the parts manufactured for evaluation in CFRP on JAGUAR. JAGUAR shroud panels show complex skin shape with honeycomb insets. They were produced entirely from unidirectional tape. The JAGUAR spine panels (see Fig. 8) demonstrate aluminium honeycomb core. Many of these have been made and flown with good service experience.

The size of one JAGUAR wing skin can be recognized from Fig. 9.

Also the TORNADO has had demonstrator parts made from CFRP for instance the taileron, and the front fuselage.

Fig. 9. JAGUAR wing skin

Fig. 10. MIRAGE radio bay cover

Meanwhile AMDBA in France have been demonstrating their skill in the production of some outstanding integrally stiffned panels moulded in one step for fin skins amongst many other parts (see autoclave moulding in [1] and Fig. 7 in [2]). Figure 10 demonstrates another example, namely the Mirage radio bay cover.

Fig. 11. Pair of painted doors ready for installation

4 Early Civil Production

Rolls Royce having suffered the early set-back of the fan blade problem had continued to study the weight saving possibility of carbon fibre and took the decision to subcontract the manufacture of the main RB 211 engine pod doors in a carbon/honeycomb sandwich construction to British Aerospace. Some fifty sets of these were made and they have been flying now for many years on Lockheed L 1011 aircraft with complete success. The first of these were the largest carbon composite structures flying in regular commercial operation; weight saving was 40 kg per set. Figure 11 shows a pair of painted doors ready for installation.

Another successful application has been the use of carbon skinned NOMEX honeycomb floor panels for many civil airlines. These represent the minimum weight construction for this use but are rather expensive and less impact resistant than the equivalent glass skinned panels. Nevertheless, they have flown many millions of miles with few problems.

5 Helicopters

Whilst much work was being carried out on this range of airframe and subsidiary components the helicopter industry had a ready made application for composites – rotor blades.

Early rivetted metal blades had a very short life, suffering very badly from fatigue and mechanical damage problems. The introduction of metal bonding greatly improved this situation but the advent of composites, initially as glass but more recently as hybrids, has improved the situation still further. Modern composite blades possess very long lives and are outstanding in fatigue resistance. Recent

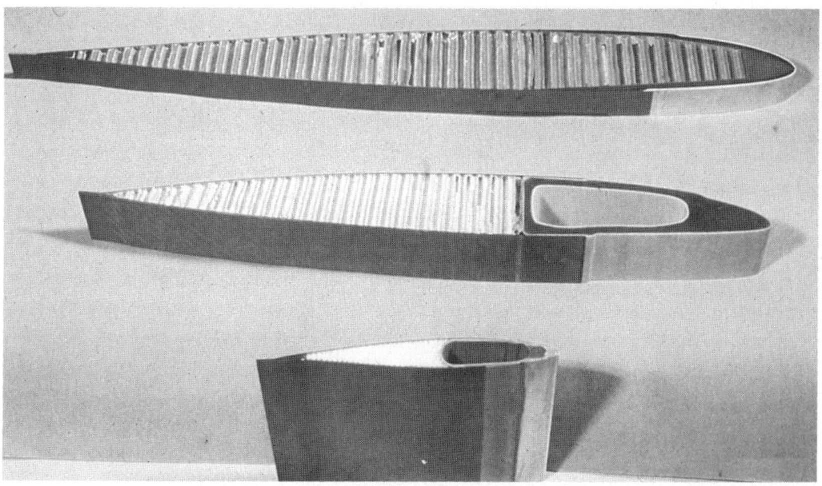

Fig. 12. Sections of PUMA rotoblade (top), SEA KING Retrofit (centre) and LYNX tail-rotor where the NOMEX honeycomb is replaced by a high temperature resistant foam

programmes have produced some very interesting and successful composite blade designs.

The PUMA blade consists largely of low density NOMEX honeycomb skinned directly with ±45° high modulus graphite fibre impregnated with high toughness self-adhesive FIBREDUX 920 resins.
The SEA KING composite retrofit blade contains a unidirectional glass spar overlaid with woven glass and a carbon fibre trailing edge skin. The NOMEX honeycomb is coated with a specially compatible epoxy resin.

Figure 12 shows the sections of PUMA rotoblade (top), SEA KING retrofit (centre) and LYNX tail-rotor where the NOMEX honeycomb is replaced by a high temperature resistant foam.

As well as rotor blades more and more structural fuselage and ancillary parts are being put into production in helicopters to reduce weight and improved performance.

6 Ancilliary Applications

The use of CFRP has not been restricted to standard airframes and components. Many other ancilliary applications and structures have been pursued. Figure 13

Fig. 13. SYLDA (Systeme Lanceur Double Ariane)

Fig. 14. New technology propellor blades

shows SYLDA (Systeme Lanceur Double Ariane) which is essentially a large protective container for satellites carried by Ariane. It is one of the largest European carbon fibre structures measuring some four metres in height and three metres in diameter. The weight saving over the equivalent metallic structure was 43%.

Another important application of carbon reinforced matrix composites in brakes for aircraft has been shown in [2]. The A 310 brake assembly has a thermal capacity three times that of the equivalent steel structure, exhibits low wear and high friction characteristics and saves a total of 400 kg per aircraft.

Figure 14 depicts the latest technology carbon/glass propeller blades manufactured by DOWTY ROTOL using resin injection techniques. These blades are in series production and will shortly be used in the new SAAB-FAIRCHILD SD 340 commuter plane.

7 Modern Civil Production

The latest civil aircraft composite parts are being designed specifically for these materials. An interesting example is the fixed trailing edge structure for the AIRBUS A 300–600. Initially it was designed in metal and an early attempt to convert it to composite construction resulted in a part that contained many mechanical fasteners and was heavier and more expensive than its metal counterpart.

British Aerospace – Weybridge Division then redesigned it as a composite part from the outset and the result was a structure essentially free of mechanical fasteners which is lighter and cheaper than the metal one. A good way to ensure the spread of composite usage.

The construction is of two carbon fabric/NOMEX honeycomb skins bonded either side of carbon fabric moulded ribs. This part is now in series production – to be followed by an equivalent structure for the A 310 as well as many other new composite items.

Finally, consideration must be given to the future generation of aircraft both civil and military which will contain much larger quantities of carbon in their construction.

American designed but built in Norther Ireland is the very "avant garde" LEARFAN 2100. Constructed principally from carbon fibre it has suffered from being the "first in the field" and all the problems that usually entails. Nevertheless it is hoped to obtain FAA Certificate in 1985 followed by full production. The unpainted LEARFAN 2100 over Nevada is shown in [3].

8 Future Prospects

There are at least three military aircraft now being designed in Europe which have much higher levels of carbon fibre composite in them than anything before on that side of the Atlantic Ocean.

These are the SAAB GRIPPEN, for which the first sets of wings are being manufactured by British Aerospace (38% CFRP), the DASSAULT ACX with approximately 40% CFRP, and the joint Anglo/German Agile Combat Aircraft with 38% CFRP.

It is now likely that use of carbon fibre composites in Europe will show a fairly rapid increase in the Aerospace sector. This being due to the replacement of metals rather than any great increase in aircraft numbers or volume.

Nevertheless, the threat of a challenge by lithium/aluminium alloy must be taken seriously. A big opportunity now rests with the fibre and resin manufactures to produce systems which can be used to greater design strains. It has been suggested that if carbon composites can be designed to a safe strain limit of 0.6% instead of the current 0.35–0.4% then they will find much wider use in aerospace and their performance could well be significantly beyond the range of the new aluminium alloy challenge. Whether this objective can be achieved remains to be seen.

9 References

1. B. J. HAYES: "The Fabrication Processes for High Performance Composites", see page 143ff. this volume
2. K. BRUNSCH: "Experiences in FRG in aerospace application", pages 214 and 215 this volume
3. E. FITZER: Introductory lecture on: "Technical Status and Future Prospects of Carbon Fibres and their Application in Composites with Polymer Matrix" (CFRPs), see page 33 this volume

Chapter 6

Further Applications of CFRPs (Non Aerospace)

Industrial, Marine and Recreational Applications of Carbon Fibres and Their Composites

I. L. Kalnin

1 Introduction

These applications are generally considered together for non-technical reasons. First, in selling the material as well as in making market assessments, one must address a diversity of manufacturers, instead of dealing with the relatively few high-technology aerospace and automotive companies, which are in the forefront of materials' R & D. Secondly, the applications in the title sectors are generally much less weight-critical and much more cost sensitive. As a matter of fact, non-aerospace applications did not exist until the price of carbon fiber (CF) dropped below $ 100/Kg in the early seventies. The current U.S.A. prices of the structural CF tow or yarn range from 40–66 $/Kg, depending on the size of the tow and the performance requirements. Thus, the CF is very expensive relative to Fiberglass reinforcement, less so relative to the aramid fibers (e. g. Kevlar), and inexpensive with respect to metal or ceramic fibers, such as boron or silicon carbide, which cost about 400–500 $/Kg.

The estimates of total quantities of carbon fiber (CF) utilized in the U.S.A. for the above applications in 1982 and 1983 are given in Table 1. It is seen, first, that the above markets together consume nearly as much CF as the aerospace and transportation. Also, contrary to the situation in the seventies, the industrial usage now appears to grow at least as fast as the recreational. The marine applications of CF as far as known, are presently very limited, both in scope and in volume.

Clearly, in order to be utilized, the CF performance has to justify its cost. Summarized in Table 2 are the CF properties that impart outstanding characteristics to their composites. It should be noted that the first six properties can be measured directly on the CF, while the other four depend not only on the properties of the CF, but also on the matrix material with which the CF is composited, on the fiber orientation in the composite, the strength of fiber-matrix bonding, the stresses in the composite, and the nature of the ambient in which the composite is operated. Most often, it is not a single property that makes the CF composite cost effective, but a combination of two or more of these. For instance, a combination of low density, high stiffness, and good static as well as cyclic stress resistance is necessary in

Table 1. Estimated carbon fiber usage for 1982 and 1983[a]

Market Sector	1983 Amount			1982 Amount		
	10^6 lbs.	Metric Tons	%	10^6 lbs.	Metric Tons	%
Aircraft, Space Vehicles	1.30	585	50	1.0	455	49
Industrial, Marine, Miscellaneous	0.63	285	24	0.47	215	23
Sports, Recreation	0.52	235	20	0.45	205	22
Sample handouts, Fabrication losses	0.15	65	6	0.13	60	6
Total	2.6	1170	100	2.05	935	100

[a] Adapted from the market surveys by Composite Market Reports, Inc., San Diego, Calif.

Table 2. Outstanding carbon fiber and their composite properties

Property	Typical values or comment
– Low density	1.7–2.0 Kg/dm^3
– High tensile stiffness (modulus)	200–400 × 10^9 Pa (GPA)
– High tensile strength	2.5–3.5 GPA (10^9 Pa)
– Very small axial thermal expansion coefficient	-2×10^{-7} K^{-1}
– High electrical conductivity	5–20 × 10^4 ohm^{-1} m^{-1} (in comparison to non-metals or semi-conductors)
– High thermal conductivity	6–140 Wm^{-1} K^{-1}
– Long service life at relatively high constant or cyclic strain and stresses	Commonly known as "creep and fatigue resistances"
– Corrosion resistance	CF is attacked only by strong oxidizers at elevated temperatures.
– Biocompatibility	
– Low friction coefficient and good wear resistance.	Particularly for the higher modulus CF

Table 3. Commonly used room temperature matrix resins

Name of Resin	Composition
Epoxide, epoxy	di-, tri- or tetra-glycidylethers of various substituted phenols or polyphenols.
Polyester (unsaturated)	Condensates of dialcohols with unsaturated dibasic carboxylic acids or anhydrides plus cross-linkable vinyl monomers.
Phenolic	Condensates of polymerized phenol-formaldehyde.
Polyolefin	Polymerized ethylene or prophylene.
Polyester (saturated)	Condensates of dialcohols with various, mostly saturated dibasic carboxylic acids or anhydrides

dynamic applications; or high stiffness combined with good electrical conductivity and corrosion resistance is needed in electro-chemical or chemical process applications.

In addition, an increasingly popular approach to the lowering of material costs is to combine some CF with a larger amount of a cheaper reinforcing fiber, especially fiberglass, to generate what is known as hybrid fiber reinforcement. This becomes particularly attractive, if the CF reinforcement is placed at the sites in which the outstanding CF properties are most fully utilized. Such a placement, however, requires familiarity not only with the component material properties, but also the knowledge of the mechanical behavior of such hybrid composites. Obviously, a high degree of design capability and engineering experience is called for.

Also, a CF composite cannot perform adequately unless the fibers are uniformly distributed, properly oriented, and well bonded to the matrix. For CF composites

Table 4. High Temperature matrix resins

Polymer Type	Designation or Trade Name	Initial Polymerizable Components
Amide-imide copolymer	Torlon	Trimellitic anhydride; aromatic diamines
Polyimide	PMR-15	Esters of benzophenone tetra-carboxylic acid (BTDE) and of norbornene dicarboxylic acid (NE); aromatic diamines.
Polyimide	LARC-160	Esters of BTDE and NE; aromatic polyamines.
Polyimide	Thermid 600	Benzophenone tetracarboxylic dianhydride; bis(3-aminophenoxy)benzene; metaaminophenylacetylene
Polyetherether Ketone	PEEK	Dihydroquinone; dichlorobenzophenone

the most common matrix materials are thermosetting resins, such as epoxy-based or polyester-based ones. For use near room temperature the less expensive thermoplastics (e. g. Nylon, polypropylene) may be used. The trend, however, appears to be toward CF composites which can withstand service temperatures over 200 °C, and in the recent years polymeric matrix materials, which can be used at 300 °C for up to 1,000 h, have been developed to meet the ever-increasing aerospace demands. Even though there are hardly any industrial applications requiring service at, say 200 °C, at present, there may well be future applications as the use volume increases and the high temperature resin prices drop. Tables 3 and 4 list the most commonly used room temperature and high-temperature, respectively, resins used as matrices in the CF composites. A brief description of the major types of matrix resins can be found in a paper by JUDD et al. [1].

Finally, for temperatures over 400 °C, inorganic borosilicate glasses, or low melting metal (Al, Mg) alloys, or carbonized polymers or pitches have been developed as matrix materials. As far as known, no significant applications for these materials exist as yet in the industrial sector although many R & D prototypes are being described in the literature.

2 Industrial Applications

2.1 Structures and Parts

2.1.1 Moving Parts

In the early seventies much attention was given to the construction of small CF reinforced plastic parts to take advantage of not only the high stiffness/density ratio (= specific stiffness), but also of the better fatigue resistance and vibration damping characteristics in comparison to previously used materials, especially wood. At that time a number of textile machinery parts, like picker sticks, heddle frames, rapier sticks (in shuttle-type weaving looms) or needle bars (in knitting machines) were demonstrated as an advance. Indeed, picker sticks made of pultruded graphite reportedly allowed a modest, 10%, increase in loom speed and extended the service life from about 6 months to 3 years [2]. However, the textile industry has been slow to accept these improvements largely due to the economic conditions not justifying investments in new technology. Other apparently more successful parts made with CF reinforcement have been rolls in paper making and copying machines and carbons in cigarette packaging machines. CF has also been claimed as a reinforcing filler in plastic bushings and bearings providing improved creep and wear resistance [3].

More recently, some moving internal combustion engine parts (e. g. pistons) have been made using CF reinforced high temperature plastics in order to assess the feasibility of a plastic-based, lightweight engine [4]. In applications like this, the corrosion and wear resistance of the CF, combined with the strength and stiffness is clearly a cost effective asset. Whether the resin matrix is capable of surviving the rigorous ambient remains to be seen. Finally, it is notable that some high speed moving bodies, like rotors, fan blades, flywheels, originally reinforced with CF, are now being built with aramid or fiberglass or hybrid reinforcement as a more cost effective approach.

2.1.2 Stationary Structures and Parts

These usually become cost effective by combination of high stiffness (low deflection), a near-zero coefficient of thermal expansion (CTE), and corrosion resistance. The reported applications relate to high quality machine tool supports [5] or instrument parts, such as optical benches, which are usable without the need of precise temperature control, telescope frames, and precision micrometer shafts [6]. Other structural items found in the literature are electrical transmissions line tower frames, antenna supports, hollow telephone poles, masts, or booms [7]. Other supports used in conjunction with x-ray equipment are mentioned under the Medical Applications.

2.1.3 Structures and Parts for Chemical and Corrosion Resistance

It has been reported that the CF by itself is quite resistant to the chemical attack by strong (50%) hot aqueous acids or alkalis; even more so than the bulk carbons [8]. Whenever it is composited, however, it is the resin matrix which is critical to the chemical or corrosion resistance, and different resin systems produce widely different sensitivities to an ambient medium. Data on the chemical resistance of the conventional plastics (epoxies, polyesters, polyolefins, etc.) are readily available, but this is often not the case with the newer, high temperature resins (polyimides, polyether-etherketones). CF reinforced polyphenylene sulfide (PPS) composite has been molded commercially into valves to be used in piping systems which contain corrosive acids or bases at temperatures in the range of -40 to $+150\,°C$ and pressures up to 1.4 MPa (200 psi) [9]. Earlier applications have been focused on improving the corrosion resistance of chemical reactors and storage containers by cladding with CF-rich resin linings [10].

Matrix-free "activated" CF has been fabricated by special processes to yield very high absorptive capacities, exceeding in certain cases even those of charcoal. These fibers are being made commercially into aerosol filters to remove air pollutants or toxic contaminants [11, 12].

3 Medical Applications

These applications are promising because of the excellent compatibility of carbon with human tissues in addition to the high CF strength, stiffness and low degree of x-ray absorption of elemental carbon. Since patients can suffer ill effects from large radiation doses, it is a requirement that medical equipment minimize the patient's exposure to x-rays. Any material located between the radiographed patient and the x-ray film will absorb x-rays. The greater the absorption of the material, the greater the x-ray dose that must be used on the patient in order to obtain a satisfactory x-ray picture. The usual items placed between the patient and the film include film casettes and holders and patient supporting structures. In order to yield sharp radiographs, these must exhibit minimum deflection under load, yet be as transparent to the x-rays as possible. Both the CF and Kevlar reinforced plastics and their hybrids are well suited for the above structures, such as cantilevered support frames, x-ray table tops, C.A.T. SCAN couches, film holders. These are being produced by a number of manufacturers both in the U.S.A. and abroad.

The high CF stiffness is also useful for artificial limbs. In the U.S.A. the Veterans Administration has funded the development of some prototype components, such as heel and ankle assemblies of an artificial leg, but more remains to be done to make it operational [13]. Easier to develop were lightweight external aids such as wheelchairs, braces, and crutches for handicapped and orthopedic patients. For the latter, the fiberglass reinforced plastic casts are often replacing the heavy Plaster of Paris ones, and the use of CF to fashion even lighter but more rigid casts is being mentioned.

Surgical implants containing CF offer a great promise in improving the patient's well-being because of the excellent compatibility between the carbonaceous material and blood or body tissues. CF reinforced carbon pins have been made and used successfully for bone adjustments and in heart valves [14]. Numerous prosthetic applications such as fixation of artificial limbs to the stumps of the amputated ones, joint replacement by CF/carbon composites (especially hip joint arthroplasty), dental implants, bone plates in osteosynthesis, have been mentioned in both the medical and non-medical literature. Patents claiming CF reinforced body implant compositions exist [15, 16] but it is hard to say whether these applications have advanced beyond the research stage.

Finally, filters for improved compatibility with human blood have been developed from bundles of CF, coated with Pyrolytic carbon [17].

4 Electrical Applications

Continuous CF is nearly as good a conductor as the pitch-bonded bulk-graphite used as the conventional electrode material. Thus, the CF Four is in principle equivalent to a multistrand conductor wire, and CF fabricated into paper or fabric can serve as planar conductors. The cost effectiveness of the electrical uses is based on the combination of electrical conductivity with chemical and wear resistances, favorable electrochemistry or temperature independent conductance.

The earliest application of CF was for heater elements, both as space heaters at low temperatures or as furnace heaters at high temperatures. More recently, CF-containing flexible composite sheet heaters have been introduced [18]. Also, the CF is used as filler for structural plastics in order to make them sufficiently conductive to serve as electromagnetic interference (EMI) shielding materials. The advantage of CF is that the quantity of filler needed to achieve a given electrical conductivity level may be up to 100 times lower than the corresponding quantity of a carbon black powder [19]. In addition, CF filled plastics may serve as static charge eliminators.

Process electrodes made of or containing CF have found several applications. The most widely known electrode consists of CF bundles or fabrics arranged in patented configurations [20] and is being used primarily for cathodic reduction of low concentrations of residual metals from spent electrolyte solutions. A wide applicability to various kinds of mining or electroplating industry waste effluents is envisaged. Flexible CF filled sheets have also been proposed as electrodes in such applications [21]. The CF in its various forms is also being used as an electrode for the experimental high temperature (325–350 °C) sodium-sulfur batteries in which the electrode performance usually deteriorates with time [22–24]. In fuel cells felted CF sheets have been found useful as the electrode support plates [25]. In addition, many other CF containing electrode variants are being researched, especially for small high performance battery applications [26].

Combination of high electroconductivity and low wear rate of the high modulus graphitic fiber can be utilized in sliding current collectors ("brushes"). Both all-fiber brushes [27] as well as brushes fabricated from CF filled metal (Cu, Ag) have been evaluated [28, 29].

The CF has also been used as a means for attaining greatly enhanced electrical conductivities, either by intercalation [30] or by vapor plating with a superconductive compound, such as Nb_3Sn [31]. In the former case the conductivity has been found to approach that of a highly conductive metal, such as copper [32].

5 Marine Applications

So far, there are not many marine applications, presumably because of the uncertainty of long term service under a large variety of conditions under water. The Department of Navy is sponsoring the development of two experimental graphite-epoxy structural elements patterned after real components of a hydrofoil patrol boat – a section of a box beam, about 1.2 m long, and a hydrofoil control flap (2 m × 0.5 m in size). These parts are presently undergoing a variety of simulated service tests [33]. In the commercial sector some racing canoes and kayaks are being stiffened longitudinally with a layup of CF/epoxy tapes, in the otherwise fiberglass-polyester body in order to improve their competitive performance. Lately, however, the CF is being replaced by the less costly aramid, and this trend may continue as long as the part does not have to bear sizable compressive or shear loads.

6 Recreational Applications

6.1 Sports Equipment

Here is a very interesting situation where the products are not cost sensitive, but are purchased either for prestige or improvements which matter only to the pros and semipros of that sport. Several of these are large volume applications, even though the amount of CF in each item may be relatively small, typically 0.1 Kg or less. The first of these products was the golf club with a "graphite" fiber containing shaft. The high specific softness of a properly designed shaft resulted in a lighter shaft, allowing to put more weight in the club head and thus attain a longer driving distance owing to the enhanced momentum transfer from the heavier head to the ball. The consumption of CF for golf shafts made a spectacular jump from maybe a few hundred Kg in 1972 to ca. 20,000 Kg in 1973 continuing roughly at that level until 1979. Then it resumed growth, because the cost savings resulting from more efficient prepregging and club assembly processes, increased the sales all over the world. Presently, the consumption of U.S. CF in the golf shafts is estimated at ca. 35,000 Kg/year.

A few years later a similar boom started in fishing rods and tennis racquets. The demand for the former grew at the fastest rate because these CF-epoxy rods are substantially lighter than those made of other materials and owing to the inherent CF stiffness will give an improved casting accuracy and better control of the lure. Presently most of the recreational CF goes into fishing rods, with tennis rackets running a close second, and golf shafts in third place.

Other sports equipment items in limited production are selectively reinforced surfboards, hockey stick blades, bows, oars and paddles, arrow shafts, bicycle frames, molded racquets for squash games. A discussion of the dominant patents pertaining to their manufacture is given in reference [7]. A survey of the history and consumption of the CF for the sporting products was published recently [34]. The total present consumption of CF in sporting goods is now estimated at about 130,000 Kg/year.

6.2 Other Sports

There are a few speciality sports areas that incorporate advanced design features and use CF-epoxy composites. One such is the competitive racing car in which the CF-epoxy composite is used to lower the chasis weight while maintaining strength and stiffness. Presently, the most successful is the McLaren Formula I racing car which won both Grand Prix races in the U.S. in the summer of 1982 [35]. A newly designed two-seat racing car with CF-epoxy chassis, called Mustang GTP, was disclosed by the Ford Motor Co. in May, 1983 [4].

7 Loudspeakers and Musical Instruments

High modulus carbon fibers are also being used to make high quality loudspeakers, because the speed of sound on the surface of a molded CF-filled plastic speaker cone is several times that on the surface of a regular speaker, allowing the use of a shallower cone which is either less conspicuous or can be fitted into otherwise inaccessible areas. These speakers are made in Japan and are being continually improved [36]. Violins and guitars have been designed and built employing CF-epoxy soundboards that are claimed to be more reliable than those of wood, stay tuned, and offer improved sound radiation qualities over those of wood. A number of such soundboards or assemblies are covered by patents [37–39]. One of these patented guitars was reported to sell for $ 2,500; which would seem to keep it away from the fingers of most amateur guitarists.

8 Summary

It is seen that the CF applications are extremely diverse and may not involve cost effectiveness, especially in the areas of competitive sports. Nevertheless, the CF containing materials will alleviate some world wide problems by providing

1. energy savings through the use of lightweight parts,
2. material savings resulting from extended service life due to reduced chemical attack and corrosion, and
3. improved quality of life by means of long lasting biocompatible body parts and accessories and by decreasing the environmental contamination due to certain toxic wastes.

9 References

1. Judd, N. C. W., Wright, W. W.: "Which Resin for Reinforced Plastics", Reinf. Plast. (London) (1978) *22* (2), 39–51
2. Devault, J. P.: "Commercial Applications for Graphite Reinforced Composites", SAMPE Quarterly (1974), 10–16
3. Harrison, M. B., Benion, R.: "Low Friction Bearing Materials" U.S. Pat. 3,741,855, June 26, 1973
4. ACM Monthly Newsletter, No. 139. Composites Market Reports, Inc., San Diego, Ca. May 1983
5. Knight, Jr., C. E. et al.: "Graphite Fiber Reinforced Structure for Supporting Machine Tools", U.S. Pat. 4,072,084; February 7, 1978
6. Devault, J. B., Parks, B.: "Overview of Commercial Applications of High Modulus Composite Materials", Natl. SAMPE Symposium, SAMPE (1974) *19,* 255
7. Sittig, M., ed.: "Carbon and Graphite Fibers", Noyes Data Co., Park Ridge, N.J. (1980), 369
8. Judd, N. C. W.: "The Chemical Resistance of Carbon Fibres and Carbon Fibre/polyester Composite", Plast. Polym. Conf. Suppl. No. 5, The Plastics Institute, London (1971), 258–265
9. Kaiser, R.: "Technology Assessment of Advanced Composite Materials, Phase I", No. PB 283,416, ASRA Information/Resources Center, National Science Foundation, Washington, D.C. April, (1978) 148
10. Pritchard, G.: "Chemical Reactivity of CF Reinforced Composite Materials", Poly. Plast. Technol. (1975), *5,* 55–81
11. Bailey, A., et al.: "Active Carbon Filters", Brit. Pat. 1,376,888; December 11, 1974

12. SPURNY, K. R.: "Developments in Carbon Fibre Filters", Filtration & Separation (1981) *18,* 67–8
13. Reference 9, Page 150
14. JENKINS, G. M., DECARVALHO, F. X.: "Biomedical Applications of Carbon Fibre Reinforced Carbon in Implanted Prostheses", Carbon (1977) *15,* 33–7
15. FARLING, G. M.: "Human Body Implant of Graphitic Carbon Fiber Reinforced Ultrahigh Molecular Weight Polyethylene", U.S. Pat. 4,055,862, November 1, 1977
16. HOMSY, C. A.: "A Method for Preparing a Porous Implantable Material from Polytetrafluoroethylene and Carbon Fibers", U.S. Pat. 4,129,470; December 12, 1978
17. BOKROS, J. C.: "Blood Filter Using Glassy Carbon Fibers", U.S. Pat. 3,972,818, August 3, 1976
18. Takiron Co.: "Polyacrylate – carbon black paper Electrically Conductive Composite Films," Jpn. Kokai Tokkyo Koho Jp 82,100,058; June 22, 1982
19. BIGG, D. M. et al.: "Plastic Composites for EMI Shielding Applications", Polymer Composites (1983) *4,* 40–6
20. FLEET, B., DASGUPTA, S.: "Carbon Fiber Electrode", U.S. Pats. No. 4,046,663; September 6, 1977; No. 4,108,754 and 4,108,757, August 22, 1978
21. BECKLEY, D. A.: "Continuous Filament Graphite Composite Electrodes", U.S. Pat. 4,369,104; January 18, 1983
22. BREITER, M. W.: "Contact Between Metal Can and Carbon/Graphite Fibers in Na/S Cells", U.S. Pat. 4,053,689, October 11, 1977
23. JOO, L. A.: "Sulfur/graphite Fiber Electrode for Na-S Batteries", U.S. Pat. 4,127,634, November 28, 1978
24. ROBINSON, G. et al.: "Cathode Electrode Structures for Na-S Batteries", Brit. 2,095,027; September 22, 1982
25. DECRESCENTE, M. A. et al.: "Fibrillar Carbon Fuel Cell Electrode Substrates", U.S. Pat. 4,064,207; December 20, 1977
26. TAKASHI, N. et al.: "Lightweight, Stable, & Rechargeable Battery with an Activated Fibre Electrode", J. Chem. Soc. Chem. Comm. (1982) 1158–9
27. MENEGAY, D. J.: "Electromotive Brushes Produced From Mesophase Pitch Fibers", U.S. Pat. 4,140,832, February 20, 1979
28. Toray Industries, "Carbon Filament Reinforced Copper Alloys for Sliding Electrical Current Collectors", Jp. Kokai Tokkyo Koho Jp. 57,198,232; December 4, 1982
29. MCNAB, I. R. et al.: "High Current Density CF Brush Experiments in Humid Air & Helium", Elec. Contacts (1979) *25,* 159–163
30. BESENHARD, J. O.: "Permanently Increasing Conductivity of Graphite Fibre Products", DBR Pat. DT 2828-1824, January 10, 1980
31. THOMAS, D. B.: "Superconductors", U.S. Pat. 3,594,226; July 20, 1971
32. CHIEU, T. C., et al.: "Raman Studies of Benzene-Derived Graphite Fibers", Phys. Rev. B (1982) *26,* 5867–77
33. WATTS, A. A., et al.: "Commercial Opportunities for Advanced Composites", ASTM Special Tech. Publ. No. 704, Am. Soc. for Testing & Materials, Phila. (1980), 107–8
34. BURG, M.: "The Market for Graphite in Sports Equipment", National SAMPE Symposium, SAMPE, Azusa, Calif. (1982) 27, 649–58
35. "Formula I Racing Uses the Advantages of Graphite Fibers", Industrial Res. & Dev., July 1982, P. 90–2
36. Matsushita Electric Industrial Co., "Speaker Diaphragms", Jp. Kokai Tokkyo Koho JP 57,155,897; September 27, 1982
37. RICKARD, J. H.: "Reinforced Stringed Musical Instrument Neck", U.S. Pat. 4,084,476; April 18, 1978
38. TURNER, W. A.: "Graphite Composite Neck for Stringed Musical Instruments", U.S. Pat. 4,145,948; March 27, 1979
39. HAINES, D. W.: "Construction Material for Stringed Musical Instruments", U.S. Pat. 4,364,990; December 21, 1982

Experiences of Non-Aerospace Application of CFRPs in Europe

B. J. Hayes

1 Review on Applications and Fabrication Techniques

Unlike the United States and the Western Pacific countries, the use of carbon fibre in Europe for sporting goods applications has been rather small in volume terms. This is probably due to large scale, low unit price, production outside Europe resulting in the import of inexpensive basic fishing rod blanks, racquet frames and golf shafts. Nevertheless, European manufacturers compete in all the prominent application areas and have pioneered some new ideas as well.

More detailed attention has been given to the use of carbon in composites in the transport industry where many of the Formula I racing cars, endurance cars and economy cars have incorporated these materials with great success.

Highly sophisticated design and fabrication approaches have resulted in the production of torque tubes and drive shafts which are under practical road and aerospace evaluation today and which are confidently expected to be in series production for some applications in the near future.

A wide range of small to medium size components for general industrial applications are produced by injection moulding of carbon fibre reinforced thermoplastics. Most use is in the area where great stiffness is required, coupled with low weight, especially where rapid acceleration and deceleration is essential, e.g. in textile weaving equipment.

A potential major volume development is in the field of robot construction where again high specific stiffness is a frequent requirement. Here, as in many other cases, the use of carbon in a honeycomb sandwich to improve this property still further is a very important consideration.

The majority of these non aerospace applications are accomplished by manufacturing from prepreg in both unidirectional or woven fabric form. The principal exceptions to this being the filament wound tubes and the injection mouldings. Very little wet resin hand lay-up impregnation is used because it is time consuming, less accurate in fibre/resin ratio and is generally less satisfactory from the operator's viewpoint (compare also [1]).

For the majority of non aerospace uses the main requirements for prepreg systems are:

a) cure at 120 °C
b) Long out-life at ambient temperature
c) easy handling in use
d) good mechanical properties up to 100 °C maximum
e) excellent toughness
f) processable by press, vac-bag, autoclave, oven/shrink-wrap tape

Woven prepregs are available in a variety of weave styles for different handling and draping properties in widths up to approximately 1.2 m.

Unidirectional prepreg is produced as a standard of 300 mm width although wider and narrower versions are available. Indeed, very narrow tapes or even individual prepregged tows are becoming increasingly popular for winding applications where they replace liquid resin and dry fibre.

Among the many applications in Europe the following are important, either in terms of volume or interest.

2 Special Applications in Sports Goods

2.1 *Fishing Rods*

A great variety of types are manufactured from all-carbon or carbon/glass hybrids. The principal advantage is reduction in weight for equivalent stiffness.

2.2 *Table Tennis Bats*

The use in table tennis bats is a small volume application only. The use of carbon fibre skins on a foam core gives rise to great stiffness at low weight resulting in less arm fatigue during long matches and rallies. The rubber faces still control the ball contact properties (see Fig. 1).

2.3 *Golf Shafts*

Golf shafts are made of carbon prepregs. This application is still a significant prepreg user. More recent developments include the manufacture of carbon reinforced heads as well.

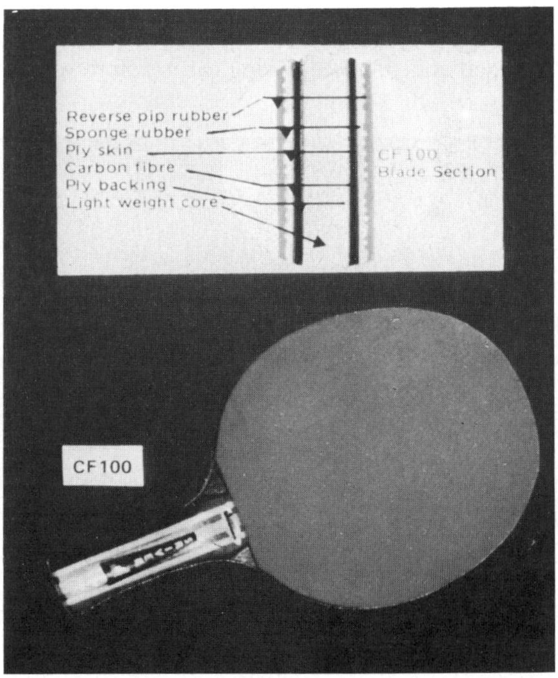

Fig. 1. Table tennis bat and construction

2.4 Bicycle Frames

Lightweight racing cycles with carbon fibre main frame tubes weighing only 8.5 kg have been produced. The carbon tubes are bonded into the aluminium alloy castings. These bicycles are outstanding for racing and are now becoming available to the general public. It is expected that as the price reduces carbon fibre will be used in general touring cycles as well because the weight saving is about the same as the current touring loads carried.

2.5 Ski Poles

Ski poles for cross-country skiing are widely used. Low weight and correct stiffness make the ski pole an ideal and very successful application for carbon. Many tons of fibre have been sold for this application. Fabrication here is usually by wet winding.

Downhill Skis

A number of ski manufacturers have used carbon fibre in the production of skis – usually as a laminate on top of the main wood, foam or honeycomb core.

2.6 Tennis Racquets

Many types of racquets for tennis, badminton and squash amongst others have been designed and produced using carbon fibre reinforcement (see Fig. 2).

Fig. 2. Selection of carbon tennis racquets

The techniques used vary very widely from the incorporation of a simple CFRP laminate into standard wood construction to resin injection into carbon fibre braid supported on low melting metal alloy formers.

As an application there are real advantages to be gained in the use of carbon, although care must be taken not to make racquets too stiff, resulting ultimately in elbow or wrist injury to the user.

3 Musical Instruments

3.1 Guitars

Carbon fibre faced guitars giving very high quality sound at relatively low cost have been made. However, musical instruments manufacture still needs design optimization to fully utilise carbon, but the prospects look good in some areas.

3.2 Violin Bows

Use of carbon prepreg has resulted in substantial reduction in weight for the same stiffness and still yields excellent bowing characteristics. Significant amounts of prepreg have been used already.

4 Sailplanes

Whilst this is an aerospace application in its true sense, most sailplanes and gliders are used for sporting purposes. It is common practice to use some carbon fibre selectively to stiffen the otherwise predominantly glass structure.

5 Racing Cars

The use of composites in racing cars has been very successful. Most formula I cars now have composites as part of their structure. They not only result in lowering the body weight of the cars but they have proved very energy absorbent in many crash situations – particularly when used in the form of honeycomb sandwich. The reinforcement used may be carbon, Kevlar or glass in any combination. The LOLA type uses an air intake scoop, main tray, frame and body shell manufactured from glass,

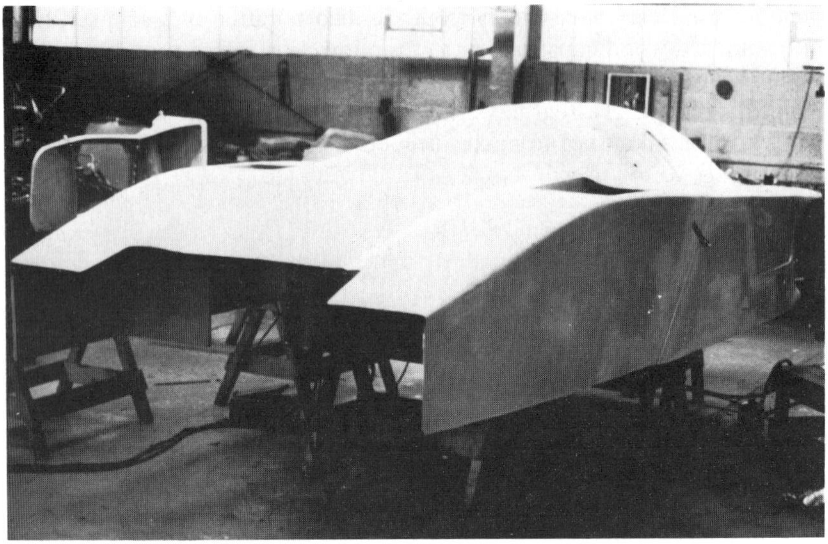

Fig. 3. Lola endurance car body shell

kevlar, carbon and metallic honeycomb to give very low weight and high stiffness for good aerodynamic flutter resistance. Figure 3 shows the Lola endurance car body shell (compare [1]).

6 Drive Shafts

Carbon fibre and hybrid drive shafts, torque tubes and support struts are often manufactured by wet winding techniques, but also increasingly from one of the various types of prepreg tape [2].

The fibre reinforced plastic tube is not only much lighter than its metal counterpart, but it can be designed to be completely vibration free and usually eliminates the expensive intermediate universal joint used in long metal shafts. Examples shown illustrate power transmission shafts for the LEARFAN 2100 (Fig. 4) (compare [1] and [2]).

7 Marine Use

At present the majority of composites used in boat hulls, masts, superstructure, sail boards, surf boards, water skis etc. are manufactured from glass fibre and glass/honeycomb/foam sandwich. Figure 5 shows an example with carbon; a large catamaran hull under construction in France.

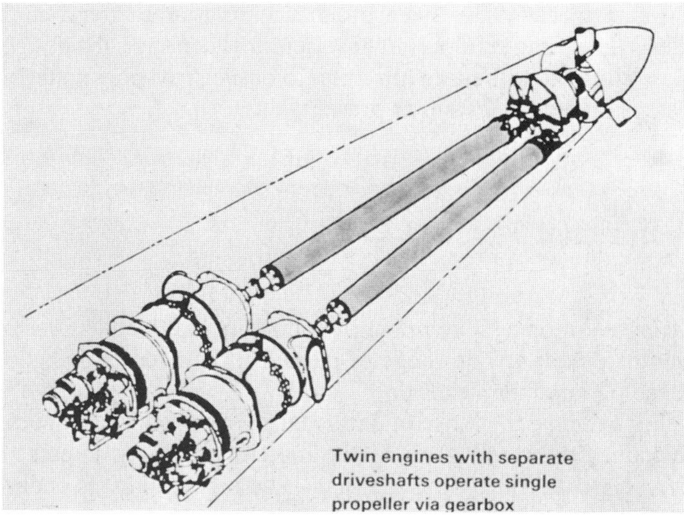

Fig. 4. Diagrammatic situation of Learfan 2100 driveshafts

Fig. 5. Large carbon catamaran hull under construction in France

Many specials have been produced incorporating carbon fibre, particularly into masts, keels and rudders but have seldom been specially designed for such materials. New attempts with more attention to design are being made but their cost effectiveness is open to question at present.

8 Injection Moulded Parts

Using thermoplastic compounds containing up to 30–40% short length carbon fibre reinforcement a wide range of parts can be made by injection moulding including those of very complex shape.

For example the range of parts that can be produced from carbon reinforced nylon include: fishing rod gear, textile marchinery parts, impeller blades, brackets and even cigarette filter tip dispensers. The potential list is endless.

9 References

1. E. FITZER: Technical status and future prospects of carbon fibres and their application in composites with polymer matrix (CFRP's) (page 35 this volume)
2. B. J. HAYES: The fabrication processes for high performance composites; see page 145 this volume

Biomedical Applications

G. W. Hastings

1 Carbon as a Surgical Implant Material

Carbon is a natural selection for a biomaterial, comprising as it does, one of the major components of the human body. However, it was not investigated for use because at the time of increasing interest in surgical implant materials, the technology had not developed to make it in a form suitable for medical use. Metals were the materials of choice for replacement of the load bearing joints in the body and in combination with polyethylene (ultra high molecular weight) seemed to be satisfactory, at least for the age group of patients considered. As deficiencies in metals began to be realised, corrosion and fatigue, and with the demand for replacement surgery in every younger patient newer materials were sought. Another factor was the mismatch in mechanical properties between metals and living bone (Fig. 1). The stiffness of metallic implants leads to interruption in the normal physiological stress loadings and leads to bone resorption. This has led to lower rigidity materials, in particular for fracture fixation devices (Table 1).

In cardiovascular surgery the biocompatibility problem is more difficult due to the complex nature of blood interactions with foreign materials and the consequences of changing the blood flow patterns by use of less compliant materials.

Various ceramic materials have been studied, notably alumina for orthopaedic use and apatite for dental surgery. Clinical experience is still being gained with these and attention has also turned to various forms of carbon. The biocompatibility had been established by work of BENSON [1] and particularly BOKROS [2] who studied blood and soft tissue.

Table 1. Factors in fracture healing

Stability of bone fragment fixation
Normal mechanical strain levels
Biological factors –
blood supply
anatomical position
absence of infection

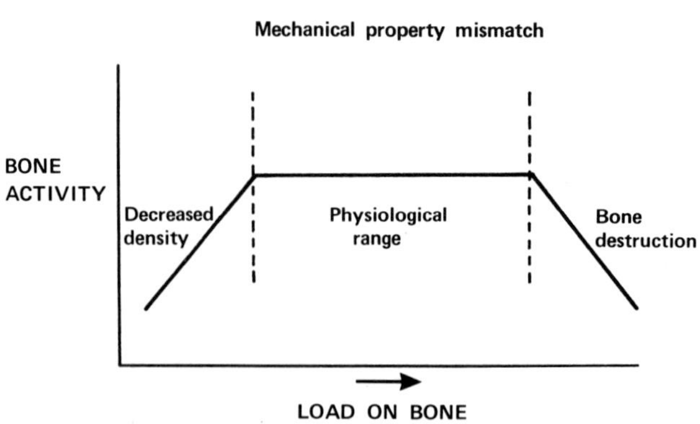

Fig. 1. Only load applied within a narrow physiological range is acceptable for health of living bone

Fig. 2. Heart valve prosthesis with glass-like carbon disc occluder post-implantation

Already LTI isotropic carbon, a pyrolytic form made by vapour deposition of carbon from hydrocarbons on to carbon substrates is used in heart valves and in excess of 200,000 have been inserted in patients. Biocompatibility in terms of non-thrombogenicity is excellent for the material used in the disc occluder in a prosthetic valve, but wear has been observed and failures reported [3] (Fig. 2). Vitreous forms of carbon, prepared from an organic polymer precursor, have been investigated for tooth root implants.

Coatings of carbon on to various substrates have been investigated since this is a means whereby the bulk mechanical properties of the substrate material can be utilised with the compatible surface produced by the coating. Particularly low temperature methods now available (ULTI carbon) mean that plastics and heat sensitive materials can be treated [4]. Dacron knitted vascular prosthesis are one example [5]. Various methods for coating blood contact devices are being developed.

In bulk carbons three types are being studied:

a) high strength isotropic carbon,
b) silicon of carbide/carbon composites and
c) carbon fibre reinforced carbon (cfrc),

generally for orthopaedic application (joint replacements).

This paper is only concerned with the latter, cfrc, being a fibre based product.

2 Carbon Fibre

Since the development of carbon fibre it has seen application in many areas in leisure products and industrial developments, particularly aerospace.

The first medical application of carbon fibre was in the repair of tendons and ligaments [5–9] and more recently for repairing defects in the cartilage of joint surfaces [10]. It appears that the fibre induces formation of collagenous fibrous tissue ordered in the direction of the carbon fibre filaments. The carbon fibre is partially fragmented and considerable deposits are observed in the regional lymph nodes. One of the problems in replacement of knee ligaments by carbon fibre

Table 2. Biological acceptance

Biocompatibility (biological acceptance) is multifactorial:
Mechanical factors
Design considerations
Material integrity
Tissue tolerance

appears to be the laxity which gives a poor end result. Tendon replacement seems to be more satisfactory but the fragmentation may be a cause for concern.

Braided carbon fibre, made up of 32 tows with a braid angle of 43° has been used in medial collateral ligament repair in sheep, the end being inserted under a bone shelf to provide anchorage [11]. This anchorage was found to be firm when implants were later examined and connective tissue surrounded the strands under the shelf with single fibres embedded in bone. Oriented collagen fibres were observed associated with the strands. In a similar experimental series in which the braid was fed through a drill hole in the bone [12] histology confirmed the presence of irregular collagen bundles in the direction of the ligament but less dense than normal. Fragments of fibre were seen intracellularly. The observation was that at one year there remained a granulomatous tissue showing that it is a foreign body reaction but well oriented. Complications reported included intraarticular dissemination of fibre fragments, rupture and failure of tissue ingrowth.

The application of carbon fibres to these replacements does not seem to be a completely satisfactory technique and is not suitable for use near to the skin. The tissue formed around the braid, although partly organised does not seem to be normal tendon or ligament and long-term use is probably not yet indicated.

Biocompatibility is dependent on several factors not simply tissue response as usually understood (Table 2). Mechanical factors are equally as important and therefore present results may depend as much on mechanical design deficiencies. A study on different braiding methods has established the differences and has optimised the design to a 43° braid angle with 32 tows (each 3000 filaments) [13]. In vivo sheep experiments and laboratory tensile testing and fatigue studies were performed, for the latter the braids were attached in human cadaver knees. The importance of an elastic fixation was demonstrated.

Experimental use as a patch for hernia repair is reported using it in the form of a non-woven fabric [14]. A polyacrylonitrile fabric was made and then carbonised at 1000 °C. Dense tissue permeation was observed with inflammatory reaction up to 8 weeks. In contrast to polyester patches, the carbon ones increased in strength with time of implantation.

3 Carbon Fibre Reinforced Composites – General Comments

An important distinction in the properties of carbon composites in general compared with the other forms of carbon is that fibre delamination is a strength determining factor. They have an advantage that a short duration overload will not cause catastrophic failure although some fibres will be broken. As a consequence of delamination effects, strength effectively decreases as the l/d ration (thickness/support distance) decreases, i. e. as the span increases. This imposes design limitations on biomedical uses and may account for the unacceptable level of breakage found when endeavouring to use this material as a bone plate (B. MCKIBBIN, personal communication).

The use of unidirectional or multidirectional composites is an area for discussion. The best strength properties are given by unidirectional reinforcement but only when the loads are applied parallel to the fibre direction. The natural reinforcement patterns of bones in general accord with this concept and certainly the pattern of trabeculae in the head of femur show the lines of reinforcement to correspond to the directions of the principal stresses (Fig. 3). However, bone is a living material which constantly responds via cellular activity to the variations in load patterns. It will also repair damage. Since a replacement material may not be placed in the exact alignment with the stress patterns, since these may change anyway, and since the stress distributions are very complex in vivo, it would seem better to use a multidirectional or cross ply, laminate. This is a compromise and results in some loss of strength but does give assurance against torsional forces as well as those which are purely tensile.

Fibre content will also considerably affect strength and Young's modulus. These increase linearly with increasing fibre content. For biomedical applications it is

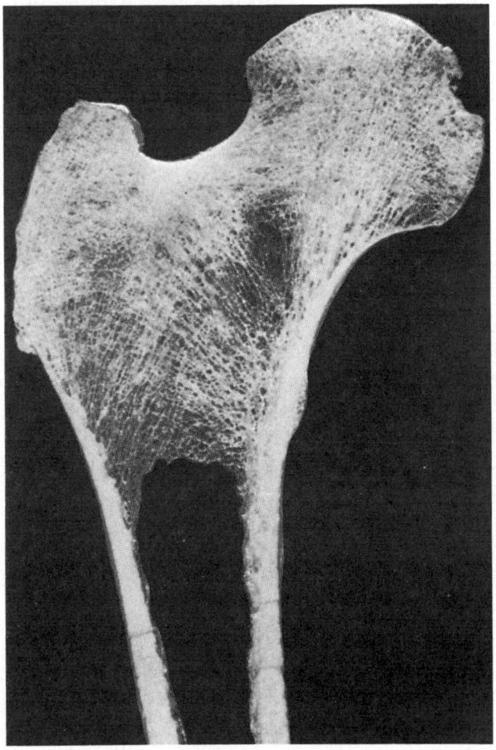

Fig. 3. Structure of human bone – head of femur

Table 3. Stiffness of composite plates used in tibia compared with stainless steel of recommended size

	Elastic Modulus E	Flexural Rigidity E.I.
Solid Section of Plate		
Composite	65×10^9 N/m^2	45.9 Nm^2
Steel	210×10^9 N/m^2	53.7 Nm^2
Section Across Screw Hole		
Composite	–	32.4 Nm^2
Steel	–	31.3 Nm^2

The importance of dimensional factors is shown

probably preferred to opt for maximum possible fibre content and to rely on variations in fibre properties (e. g. low modulus/high strength, high modulus/low strength) to achieve product modification. Particularly in the case of resin based composites, a minimum amount of resin is preferred to optimise properties and aid quality control.

The relationship between Young's modulus of an implant and bone properties is an area of controversy. It is observed in clinical practise that a stiff implant material will lead to loss of bone density and thus of strength of bone material. To avoid this, it is suggested that low modulus materials should be used. However, since stiffness depends on the section of the device, i. e. it is design dependent, modulus reduction alone is not the solution. Some reduction in modulus will be beneficial but not the extent of reduction of strength to a dangerously low level (Table 3).

The use of hybrid composites has not been really explored for biomedical use. Some of our results with carbon-Kevlar hybrids showed that although toughness was enhanced Kevlar was unacceptable for biocompatibility reasons (unpublished work).

One limiting factor in surgical implant design with metal alloys is the inadequate fatigue life shown by the surgical materials. Cyclic loadings in an insufficiently supported load-bearing prosthesis lead to rapid failure by a corrosion enhanced process. Since carbon fibre composites do not show a yield point and the strain at failure is greater than would be expected in normal clinical use they are unlikely to fail in service. Progressive delamination and therefore loss in rigidity is observed and some of the weaker fibres may fail progressively, but catastrophic failure other than by high velocity impact, is not likely (see later).

4 Carbon Fibre Reinforced Carbon

Carbon fibre reinforced carbon made by gas densification of a carbon fibre matrix may be made with unidirectional or multidirectional reinforcement. The properties of a unidirectional material are given in Table 4.

Table 4. Carbon Fibre Reinforced Carbon [19]

Density	g/cm^3	1.7
Flexural Strength	MN/m^2	800
Young's Modulus	10^3 MN/m^2	140
Compression Strength	MN/m^2	800

The high strength, low modulus combination led to the proposal for its use in the stem portion of an upper femoral prosthesis. [15] Various methods for producing cfrc have been described [16].

It is clearly important for biocompatibility to be established and NEUGEBAUER et al. [17] implanted fragments into the femoral medullary canal of rabbits. Some particles were found in other parts of the animal but without foreign body reactions. No bone destruction was seen. There was a small amount of fibrosis and foreign body giant cell reaction and also some new bone formation. Toxic effects and severe inflammatory reactions were not observed.

5 Carbon Fibre Reinforced Plastics (CFRP)

5.1 Carbon Fibre Reinforced Polyethylene

Ultra high molecular weight polyethylene (UHMWPE) is well established for use in prosthetic joint replacements. It does, however, wear and also suffers from a certain amount of creep. Reinforcement with carbon fibre has been proposed. One of the problems is that UHMWPE is difficult to mould due to its high ($>10^6$) molecular weight and it would be difficult to obtain a good composite. Using a slightly lower molecular weight product this has been achieved. This was also investigated using high density polyethylene [18], for greater ease of fabrication. As reported, when mechanical properties were optimised, the wear rate was reduced in comparison with results reported by others.

Creep results indicate that the reinforced polymer may have a more steady state behaviour than the unreinforced.

A carbon fibre reinforced UHMWPE (Poly Two_R-Zimmer USA) has been introduced into clinical practise. In vivo studies on abraded particles introduced into mice indicated a similar tissue response to that observed with unreinforced polymer [19] although it should be noted that the volume of debris produced is important.

It appears that there is a lack of compatibility between fibres and polymer and this requires further investigation.

5.2 Carbon Fibre Reinforced Epoxies (CFRE)

This project was developed from work on water emulsified epoxy resins and the reinforcement of these as potential orthopaedic implant materials [20–22]. However, for load-bearing devices conventional systems are required and attention turned to use of pre-preg to make laminated composites [23]. Type II fibre was used in a heat hardening resin system.

Two aims in the programme were to attempt to reduce the effects of stress protection in bone arising from use of stiff metallic implants and to provide better fatigue life. It is important in fracture treatment to provide stable fixation so that healing can proceed. This occurs naturally by an internal bridging process (callus formation). Rigid metal implants lead to a by-passing of this stage in healing and rely for their function on a minimal gap between the broken ends of bone. The plate is thus supported in a new bone/plate/screws system. If excess movement occurs the plate will fatigue rapidly (a few weeks).

Laboratory fatigue studies in reversed plane bending at 10 Hz were carried out on bone plate designs (8 hole DCP) while sodium chloride solution at 37 °C was circulated round the plates. Stainless steel controls showed a maximum safe load for 2×10^6 cycles minimum lifetime (corresponding 1–2 years in a patient) to be 6 Nm. None of the carbon fibre composite plates broke at 2×10^6 cycles for loads up to 13.8 Nm. Reduction in stiffness was observed and used as a measure of endurance limit and an upper limit of 10.5 Nm was proposed. A stiffness reduction of 10.3% was noted. The endurance limit is approximately 30% better than that for stainless steel and of course, catastrophic failure did not occur.

Following in vivo functional testing in sheep, human trials were begun for tibial fractures and have been reported [24]. Approximately 30 fractures were so treated and in all cases satisfactory healing was obtained. Stainless steel screws were used. The trial has now been extended to forearm fractures for which this is more clearly a recommended treatment and satisfactory stability and healing have been obtained. The level and type of activity in the forearm is quite complex and results may in fact show a relationship to these factors. The healing pattern is distinctively different from that seen in the tibia but this may be an effect of the relative stiffness of the two different systems (Fig. 4).

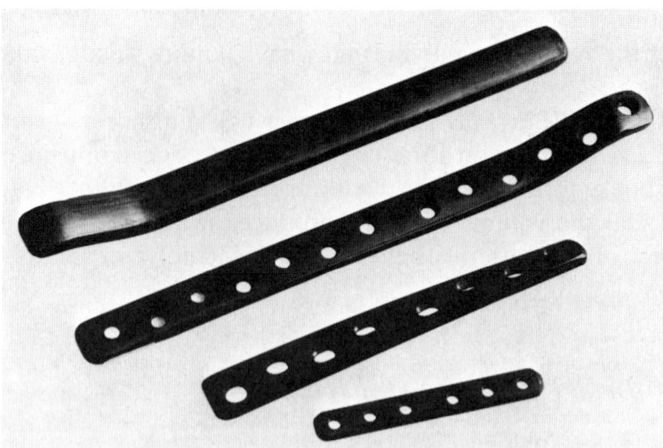

Fig. 4. Some examples of composite bone plates – carbon fibre reinforced epoxy (Stoke Plate)

Calculation shows that the cfrc tibial plates were of comparable stiffness to the recommended steel plates and the main advantage was the improved fatigue properties for a greater possible range of elastic deformation (see Table 3).

Laboratory studies show that the method used by the surgeon can have profound effects on the overall stiffness of the system and may, in fact, mask differences in materials [25].

Strain gauged plates have also been used to determine effects of screw fixation on plate characteristics and recent developments in telemetry should permit direct in vivo evaluation to take place.

A full analysis of results is presently being performed and includes results of healing, tissue reaction, plate properties post implantation in patients [26]. Long term studies are also in hand to study the biological performance. This, it is hoped, will lead to development of joint prosthesis and a programme of design evaluation and mechanical testing is under way and will incorporate our latest developments in ceramic materials.

Polysulphone is another matrix polymer receiving attention [27, 28]. HUNT has extensively studied its potential as a lower modulus implant material [29].

Other groups also report work on composite bone plates [30] using carbon fibre reinforced carbon (see earlier).

5.3 Carbon Fibre Reinforced Acrylics

Acrylic bone cement used to stabilise the position of joint endoprostheses has limitations in long-term durability and work has begun to study the benefit obtained by carbon fibre reinforcement [31]. Mechanical properties are improved and biocompatibility is being evaluated.

A possible use for reinforcement of acrylic dentures has been investigated and as anticipated fatigue life of the material is improved [32], but the scatter in results, for various reasons, makes evaluation of a complete denture difficult [33].

Composite bone plates based on polypropylene and on a diacrylate resin have also been examined [34]. Poor fibre-matrix adhesion led to fibre migration and cell culture studies are not completely satisfactory for the diacrylate material.

6 References

1. BENSON, J.: Progress on biomedical applications of carbon. North American Rockwell, Rocketdyne Report. R-7855, May, 9, 1969
2. BOKROS, J. C.: Carbon Biomedical Devices. 1977. Carbon, 15, 355–371
3. ANDERSON, JAMES M., ANKENEY, J. L., HILTNER, A.: Material and component failure in prosthetic heart valves. In L. R. RUBEN (Ed). Biomaterials in Reconstructive Surgery, 1983, C. V. Mosby, St. Louis, Chapter 48
4. LEAKE, D., MICHEOLI, S., PIZZOFERRATO, A.: Histological evaluation of carbon coated mandibular condyles in G. D. WINTER, D. F. GIBBONS, H. PLENK, Biomaterials 1980, 1982 J. Wiley & Sons, Chichester, 377

5. HAUBOLD, A. D., SHIM, H. S., BOKROS, J. C.: Developments in carbon prosthetics. 1979. Biomat. Med. Dev. Artif. Organs, 7, 263–269
6. FORSTER, I. W., RALIS, Z. A., MCKIBBIN, B., JENKINS, D. H. R.: Biological reaction to carbon fibre implants. The formation and structure of a carbon-induced "Neotendon". 1978. Clin. Orthop. 131, 299–307
7. JENKINS, D. H. R., FORSTER, I. W., MCKIBBIN, B., RALIS, Z. A.: Induction of tendon and ligament formation by carbon implants, 1977, J. Bone Jt. Surg. 59 B, 53–57
8. JENKINS, D. H. R.: The repair of cruciate ligaments with flexible carbon fibre. A longer term study of the induction of new ligaments and of the fate of the implanted carbon. 1978. J. Bone Jt. Surg. 60 B, 520–522
9. JENKINS, D. H. R., MCKIBBIN, B.: The role of flexible carbon fibre implants as tendon and ligament substitutes in clinical practise. A preliminary report. 1980. J. BONE Jt. Surg. 62 B, 497–499
10. MINNS, R. J., MUCKLE, D. S., DONKIN, J. E.: The repair of osteochondral defects in osteoarthritic rabbit knees by the use of carbon fibre. 1982. Biomaterials, 3, 81–86
11. NEUGEBAUER, R., MÄURER, H. J., HÜTTINGER, K. J., BURRI C., CLAEŞ, L., HELBING, G.: Trap door fixation of braided carbon fibre knee joint ligament prosthesis in G. D. WINTER, D. F. GIBBONS, H. PLENK Junr. "Biomaterials 1980". 1982, J. Wiley & Sons, Chichester, 287–293
12. GOUTALLIER, D., BERCOVY, M., BLANQUAERT, D., VOISIN, K., GAUDICHOT, A., ROUVEIX, B.: Cruciate ligament replacement with carbon fibre in A. J. C. LEE, T. ALBREKTSSON, P. I. BRÅNEMARK, Clinical applications of biomaterials, 1982, J. Wiley & Sons, Chichester, 3–10
13. CLAES, L., BURRI, C., NEUGEBAUER, R., WOLTER, D., ROSE, P.: The influence of various carbon fibre braiding techniques and methods of fixation in the extensibility of ligament prostheses. 1983. Biomaterials, 4, 134–138
14. MINNS, R. J., DENTON, M. J., DUNSTONE, G. H., SUNTER, J. P.: An experimental use of a carbon fibre patch as a hernia prosthesis material. 1982. Biomaterials, 3, 199–203
15. BRÜCKMANN, H., HÜTTINGER, K. J.: Carbon as a promising material in endoprosthetics. 1980. Biomaterials, 1, 67–72
16. BRÜCKMANN, H., MÄURER, H. J., HÜTTINGER, K. J. RETTIG, H.: New carbon materials for joint prostheses in G. D. WINTER, D. F. GIBBONS, H. PLENK Junr. Biomaterials 1980. 1982. J. Wiley & Sons. Chichester, 27–31
17. NEUGEBAUER, R., HELBING, G. WOLTER, D., MOHR, W., GISTINGER, G.: The body reaction to carbon fibre particles implanted into the medullary space of rabbits. 1981, Biomaterials, 2, 182–184
18. BRADLEY, J. S., HASTINGS, G. W.: Carbon fibre reinforced plastics for orthopaedic implants in G. W. HASTINGS, D. F. WILLIAMS (Ed). Mechanical properties of Biomaterials, 1980. J. Wiley & Sons. Chichester. Chapt. 30
19. RAE, T., RUSHTON, N.: Biocompatibility of carbon fibres, high density polyethylene and the resultant composite material in Biomedical Polymers. Proc. Conference, 1982. Biol. Eng. Soc. London, 23–29
20. HASTINGS, G. W., WYATT, W. A.: Water emulsifiable epoxy resins. Brit. Patent, 1380108 (1972)
21. HASTINGS, G. W., THAN THUY, N. G.: Biomechanically compatible materials in D. F. WILLIAMS (Ed). Biocompatibility of implant materials. 1976, Sector Pub. London, 134–140
22. HASTINGS, G. W.: Carbon fibre resin composites for surgical implants. 1978. Composites 9, 193–208
23. BRADLEY, J. S., HASTINGS, G. W., JOHNSON-NURSE, C.: Carbon fibre reinforced epoxy as a high strength, low modulus material for internal fixation plates, 1980. Biomaterials, 1, 38–40
24. TAYTON, K., JOHNSON-NURSE, C., MCKIBBIN, B., BRADLEY, J. S., HASTINGS, G. W.: The use of semi-rigid carbon fibre reinforced plastic plates for fixation of human fractures, 1982, J. Bone Jt. Surg. 64 B, 105–111
25. SHELLEY, P.: PhD thesis, 1983. North Stafordshire Polytechnic, Stoke on Trent
26. ALI, M. S., FRENCH, T., HASTINGS, G. W., RAE, T., WYNN JONES, C.: Clinical and laboratory biocompatibility of carbon fibre reinforced epoxies used for less rigid fracture fixation in the forearm (The Stoke plate) presented at Biointeractions, 84. (London 4–6 January, 1984. sponsored by "Biomaterials")

27. SPECTOR, M. et al.: Histological evaluation of porous polysulphone coated femoral prostheses in dogs, 1980. Orthop. Trans. 4, 206–207
28. SPECTOR, M., HARMAN, S. L., ELDRIDGE, J. T., DAVIS, R. J.: Porous polymeric coatings for orthopaedic implants in G. W. HASTINGS, D. F. WILLIAMS (Eds). Mechanical properties of Biomaterials, 1980. J. Wiley & Sons, Chichester. Chapt. 23
29. HUNT, M. S.: Carbon fibre reinforced polysulphone for orthopaedic surgical implants. In Biomedical Polymers Conf. Proc. 1982, Biol. Eng. Soc. London, 89–96
30. FITZER, E., HÜTTNER, W., MANOCHA, L. M., D. W.: Carbon fibre composites as materials for internal bone plates. 1978. Proc. London Intnl. Carbon Graphite Fifth Conference. 454–456
31. SCHULTZ, J. H., GASTMANN, H. D., WOLTER, D.: Bone reaction following intramedullary implantation of carbon fibre reinforced bone cement (Sulfix). Presented to 2nd Europ. Conf. on Biomaterials, Gothenburg, August, 1981. Abstract book p. 63
32. MANLEY, T. R., STONEBANKS, J. A.: A multi-station machine for the fatigue testing of denture base materials. 1980. Biomaterials, 1, 125–128
33. STONEBANKS, J. A., MANLEY, T. R.: A rig for loading dentures with carbon fibre reinforcements. In Biomedical polymers. Conf. Proc. 1982, Biol. Eng. Soc. London. 75–79
34. CHRISTEL, P. S., LERAY, J. L., SEDEL, L., MOREL, E.: Mechanical evaluation and tissue compatibility of materials for composite bone plates in G. W. HASTINGS, D. F. WILLIAMS (Eds). Mechanical properties of biomaterials. 1980. J. Wiley & Sons, Chichester, Chap. 29

New Technologies – The Promise for Future Applications of Composites Throughout the World

D. F. ADAMS

A decade on more ago, the mention of composite materials, and CFRP in particular, implied aerospace applications. Sporting goods eventually became a major market for composites, although often not recognized as such. Currently, the automotive industry holds great promise as a large volume user of composite materials. In near term applications, glass fiber reinforcement will be used primarily, because of cost factors. However, longer term, it is expected that CFRP will be used extensively. The automotive industry is of special interest because of the large volume potential; hundreds of thousands of units are produced by each manufacturer each year. Thus, if even only a few pounds of CFRP per vehicle an incorporated initially, the total usage becomes significant.

Most automobile manufacturers are already using glass reinforcements in many secondary structures, and certain speciality vehicles such as the Chevrolet Corvette in the United States have used fiberglass bodies for some time (the Corvette since 1955). However, the Corvette is a low volume, high performance (and high cost) vehicle, approximately 50,000 units per year being produced. The introduction of the 1984 Pontiac Fiero, a potentially high volume, moderate cost vehicle, with its all-composite body (SMC and RRIM panels), is a more significant development. Even more exciting, however, is the rumor that General Motors is developing an all-composite, filament wound space frame to mount the composite body panels onto. This one-piece structure, expected in the 1990 product line, will replace the present steel space frame, which is fabricated of over 300 individual stamped pieces spot welded together. The fabrication cost reduction implications of the composite structure are obvious. Undoubtedly, similar developments are underway in other countries as well.

In addition to innovative future applications in the worldwide automotive industry, other application areas are also very promising. These include both the farm machinery and machine tool industries. With increased emphasis on large scale automation of farm operations, and high fuel costs worldwide, the increased use of composite materials is becoming increasing attractive. Larger equipment must still be able to move across cultivated fields without sinking, hence reduced weight becomes critical. This reduced weight also implies reduced powerplant sizes, and reduced fuel consumption. Also, lighter components mean fewer laborers to adjust the equipment, reducing labor hours.

Machine tools have historically relied on mass to provide rigidity and reduce high frequency vibrations. However, as production rates have increased, the rapid movements required have become incompatible with massive components. Thus, it can be expected that there will be a major revolution in the machine tool industry in coming years, as radical new designs are introduced. Secondary benefits will include greater dimensional stability (reduced thermal expansion), greater portability of equipment, and reduced shipping costs.

On course, the aerospace and sporting goods industries can be expected to continue to increase their usage of CFRP in the future. But it undoubtedly will be other, larger volume industries which will have an even greater influence on the composite materials industry.

Chapter 7
Carbon Fibres Activities in Brazil, PR China, India and South Korea

Development of Carbon Fibre Technology in Brazil

T. NAGABHUSHANAM

1 Introduction

Through the third national development plan, the Brazilian government has explicated its objectives in the development of the agricultural sector and for independence in terms of energy. Light weight, high strength composite materials are very closely related to these objectives. Composite materials are the products made from resin and high tensile and/or high modulus fibers. Carbon fibers are one of the best known high tensile and high modulus fibers. The Brazilian government through the Centro Técnico Aerospacial (CTA) has initiated the developmental activities in carbon fiber technology.

2 Developmental Activities Before the Sponsorship of UNFSSTD/ UNIDO

CTA initiated the developmental activities in the field of carbon fiber technology in 1976. These activities were mainly on the laboratory scale with an objective to conduct the qualitative experiments on locally available precursor materials for carbon fibers. The activities were initiated in the Division of Materials of CTA by Francisco Carvalho with few of his colleagues. During the early years of development, Prof. E. Fitzer, Director, Institut für Chemische Technik der Universität Karlsruhe, was invited for his expert advice in this field of technology which added to the morale of the group to develop this technology.

During these years, commercial acrylic fibers of "Fibras Sinteticas da Bahía, S. A." (FISIBA) were used to convert into carbon fibers and to arrive at qualitative results. The heat treatment equipments were fabricated locally in Brazil and were used. After 4 years of research activities CTA has accomplished the following:

a) Evaluated the locally available precursor fibers;
b) Developed a process, though not perfect to oxidize and carbonize the commercial precursor fibers available in Brazil;
c) Designed and fabricated critical parts of the furnaces;
d) A few facilities were developed to characterize the precursor and heat treated fibers.

Even though the CTA efforts were commendable in this field of technology, it had to go a long way before achieving fibers with appropriate properties. There was a necessity for a comprehensive program which should include:

i) Modifying procedures for stabilization and carbonization.
ii) Purchasing new equipment for heat treatment and precursor fiber spinning;
iii) Acquiring expertise to guide the group;
iv) Training the counterpart scientists in this technology elsewhere in the world;
v) Exchanging views with the experts in this field.
vi) Redesigning the existing furnaces;
vii) Training the counterparts in the characterization of the fibers.

Hence, a project was drafted by CTA which was approved and signed by CTA/ UNIDO/UNFSSTD on 18/05/81.

3 Developmental Activities After the Sponsorship of UNFSSTD/UNIDO

The main objectives of this project were:

i) To create a scientific nucleus in Brazil (CTA) to produce precursor fibers and carbon fibers;
ii) To optimize the conditions to process acrylic fibers into carbon fibers and to extend to pilot plant scale;
iii) To generate information to produce precursor acrylic fibers and to transfer it to production line, and
iv) To complete minimum infrastructural facilities required for carbon fiber technology development.

With the sponsorship of the project by UNFSSTD/UNIDO the developmental activities were improved due to the visits of the project experts, arrival of project purchased equipment, accessories etc. The detailed developmental activities are described in the following pages.

3.1 Research Activities in Acrylic Polymers and Fibers

3.1.1 Polymerization Studies

Initially the activities were concentrated on the characterization of the commercially available FISIBA fibers. Later, due to the visits of project experts initial polymerization experiments were conducted. But, due to non-availability of reaction vessels of suitable quantities to use for spinning, the polymerization experiments were postponed to a later date.

3.1.2 Studies in Spinning

With the arrival of the laboratory model spinning assembly, the activities started in spinning acrylic fibers using FISIBA commercial acrylic polymers. The line diagram of the spinning equipment is given in Fig. 1. In order to standardize the working

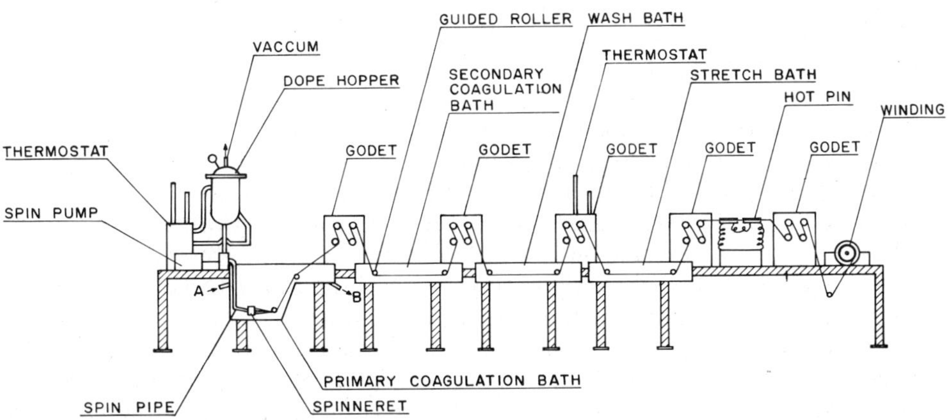

Fig. 1. Spinning assembley

Table 1. Preliminary spinning results

Spinning No.	Stretch ratio	Tensile Strength Kg/mm^2	Modulus Kg/mm^2	Density G/m^3
1	2.35	7.6	236	1.17
2	2.35	19.4	545	1.17
3	2.25	17.7	600	1.18
7	3.3	14.3	393	1.18
8	3.3	43.3	1297	1.16
9	3.3	35.7	990	1.17
10	3.6	39.3	960	1.16
11	3.6	17.6	408	1.15
12	3.6	20.95	426	1.15
13	3.6	17.4	376	1.16
14	3.5	33.7	937	1.15
15	3.5	33.7	937	1.15
16	5	32.2	415	1.18
17	4	21.4	354	1.19
18	4	22.5	320	1.19
19	4	21.8	466	1.18
20	4	22.1	325	1.17
21	4	21.6	335	1.17
22	4	21.9	362	1.17
23	4	20.2	335	1.17

parameters of this equipment the preliminary spinning experiments were commenced using 23% polymer in 50% N.N. dimethyl formamide (DMF). The results of the initial fiber spinning experiments are presented in Table 1.

In 1983 the spinning experiments were conducted both at CTA using lab model spinning equipment and at FISIBA using project supplied pilot plant spinning assembly.

3.1.2.1 Developmental Activities at CTA

Several of the spinning experiments were conducted using lab model spinning equipment. For example:

i) Use of successive coagulation baths;
ii) Variation in coagulation temperature;
iii) Variation in pH of the coagulation bath;
iv) Variation in spin pump speed, spin pump size, and speed of takeup godet etc.

Use of successive coagulation baths gave interesting results showing in improved fiber properties up to 30% DMF concentration in the 2nd coagulation bath (50% DMF being in the primary coagulation bath). Low coagulation temperatures improved the circularity of the fibers; lower pH values in the primary coagulation baths higher density fibers. Some of the results using various spin pump size, and spin pump speed and speed of the first godet are given in Table 2. The best spinning results of 1983 are given in Table 3 using lab model spinning equipment.

Table 2. Spinning results

	Spin pump speed Meters/Minute	Speed of take-up godet 1 Meters/Minute	Maximum Stretch Obtained	Denier of the fiber
Spin pump size 0.4 cm^3/Rev.	3.0	5	7.0	3.990
	2.0	5	7.0	2.470
	1.4	5	6.4	1.828
	1.4	7	6.7	1.560
Spin pump size 0.6 cm^3/Rev.	1.4	5	6.2	1.430
	1.4	6	5.4	1.370
	1.4	7	5.7	1.009
	1.4	7	6.0	1.020
	1.4	7	6.3	0.970
	1.4	8	5.6	0.960

Table 3. Best spinning results of 1983 (Lab. model)

	Stretch ratio	Denier	Tensile strength GPa	Modulus GPa
	5.6	3.79	0.36	5.8
Early 1983	5.6	8.20	0.38	5.4
	5.6	1.92	0.34	7.8
	4.9	1.35	0.28	7.72
	5.7	1.00	0.34	9.24
2nd half of 1983	6.0	1.02	0.28	8.04
	6.3	0.97	0.421	7.73
	5.6	0.94	0.349	9.86

3.1.2.2 Spinning Activities at FISIBA

Meanwhile, the pilot plant spinning assembly supplied by the project arrived in 1983 and was assembled at FISIBA. FISIBA, though a fiber producing industry had no previous wet spinning experience. FISIBA scientists were trained by project experts in understanding the spinning conditions and crucial points. The preliminary spinning results at FISIBA towards the end of 1983 are presented in Table 4. The

Table 4. Preliminary spinning results at pilot plant in FISIBA

Spinning No	Stretch	Final Denier	TS GPa	Modulus GPa	Density g/m^3
1	4	5.77	2.22×10^{-1}	–	–
2	2.99	3.1	1.616×10^{-1}	4.31	1.43
3	3	2.2	2.449×10^{-1}	5.292	1.14
4	4.11	1.9	2.711×10^{-1}	5.542	1.14
5	4.47	1.9	3.163×10^{-1}	6.119	1.13
6	4.94	1.84	3.21×10^{-1}	5.36	1.18

Table 5. Spinning results at FISIBA in 1984

Spinning No	Final Denier	TS GPa × 10^{-1}	Modulus GPa	Density g/m^3
7	1.64	3.53	6.258	1.19
	1.71	8.19	5.75	1.15

experiences in the bench scale experiments at CTA enabled to identify and to improve the properties of fibers in the pilot plant. Also, the scientists trained at CTA by project experts further helped the FISIBA people in commissioning the preliminary spinning experiments. In view of mutual contacts between CTA scientists and FISIBA people, the properties of the fibers at the pilot plant were improved.

Further, in order to conduct the preliminary heat treatment experiments on CTA spun fibers, the fibers were spun with selected best conditions of 1983 in larger

Table 6. Results of polymerization conducted in 1 liter capacity using two monomers, i. e. acrylonitrile (AN) and methylacrylate (MA)

Experiments		AN (%)	MA (%)	Temp. (°C)	Time (hours)	Quant. (grams)	Yield (%)
A-	1	97	3	40	5	104,70	52,35
	2	97	3	35	5	84,91	42,455
	3	97	3	40	5	171,52	85,76
	4	97	3	50	5	171,50	85,75
	5	97	3	60	5	173,94	86,97
	6	97	3	70	5	164,71	82,355
	7	97	3	80	5	172,00	86,00
B-	8	96	4	40	5	71,71	35,855
	9	96	4	60	5	109,77	54,885
	10	96	4	50	5	157,72	78,86
	11	96	4	70	5	185,24	92,62
	12	96	4	80	5	179,11	89,555
C-	13	95	5	40	5	29,95	14,975
	14	95	5	40	5	58,73	29,365
	15	95	5	50	5	26,84	13,42
	16	95	5	70	5	142,40	71,20
	17	95	5	80	5	165,10	82,55
	18	95	5	60	5	83,56	41,78
D-	19	98	2	40	5	53,57	26,785
	20	98	2	50	5		
	21	98	2	60	5		
	22	98	2	70	5		
	23	98	2	80	5		
E-	24	99	1	50	5		
	25	99	1	70	5		
	26	99	1	40	5		
	27	99	1	60	5		
	28	99	1	80	5		

quantities. Single spinnings were conducted for 7–8 h with lab spinning assembly at CTA. The fibers from a number of bobbins were pooled together. To make this, the precursor group constructed an equipment. These fibers were pooled into a tow consisting of at least 1000 filaments and preliminary heat treatment experiments were conducted at CTA. Upon the analysis of preliminary results on the heat treatment of the CTA spun fibers, it was felt necessary to increase the density of the fiber and to further stretch the fibers while spinning. In view of these observations, a set of spinning experiments were designed to increase the density of the fiber and to heat stretch the fiber in spinning. These included the spinning with increased dope concentrations and increase in the path length of coagulation bath etc. These observations have also helped in improving the fiber properties at FISIBA. The subsequent spinning results of FISIBA are given in Table 5.

Table 7. Results of polymerization conducted in 1 liter capacity using three monomers – acrylonitrile (AN), methyl acrylate (MA) and itaconic acid (ITA)

Experiments		AN (%)	MA (%)	ITA (%)	Temp. (°C)	Time (hours)
A-	29	96	3	1	40	5
	30	96	3	1	60	5
	31	96	3	1	70	5
	32	96	3	1	50	5
	33	96	3	1	80	5
B-	34	95	4	1	50	5
	35	95	4	1	60	5
	36	95	4	1	70	5
	37	95	4	1	40	5
	38	95	4	1	80	5
	39	95	4	1	80	5
C-	40	94	5	1	70	5
	41	94	5	1	60	5
	42	94	5	1	80	5
	43	94	5	1	40	5
	44	94	5	1	50	5
D-	45	94	4	2	50	5
	46	94	4	2	40	5
	47	94	4	2	60	5
	48	94	4	2	70	5
	49	94	4	2	80	5
E-	50	94	3	2	40	5
	51	94	3	2	60	5
	52	94	3	2	50	5
	53	94	3	2	70	5
	54	94	3	2	80	5
F-	55	95	3	2	40	5
	56	95	3	2	50	5
	57	95	3	2	60	
	58	95	3	2	70	5
	59	95	3	2	80	5

In 1984 the project supplied polymerization reactors, chemical reagents, initiators and monomers arrived at CTA. Polymerization experiments were conducted in smaller scale initially using two monomers, i. e. acrylonitrile and methylacrylate, and later, using acrylonitrile, methylacrylate and itaconic acid. The smaller scale polymerizations were of 1 liter capacity. These polymerizations were conducted in aqueous medium. The results are given in Tables 6, 7. Also the polymerization experiments were conducted in larger quantities (5–10 Kg capacity). The polymers are now being characterized, and will be used to spin at CTA. Also these polymers will be spun in inorganic salt solutions such as $Zncl_2$ and NaSCN and in nitric acid in addition to DMF.

3.2 Research Activities in Heat Treatment

As it has already been mentioned, the developmental activities were further intensified with the involvement of UNFSSTD/UNIDO. In view of the visits of the invited project experts and interaction of the Brazilian scientists with the internationally known scientists the quality of the research improved. The best heat treatment results of 1981 are given in Table 8 and the line diagram of the heat treatment equipment is given in Fig. 2.

In 1982, several of the parameters in heat treatments were changed due to the advices of experts who visited the project site. The studies were concentrated mainly using Courtauld's SAF precursor.

The best results of heat treatment studies in 1982 are represented in Table 9 (continuous process) and Table 10 (discontinuous process). Also, in order to find out the effect of air flux during the oxidation stage, several experiments were conducted both in the presence and absence of moved air.

In 1983 due to gained experience a few modifications were effected in the design of the existing heat treatment equipment. For easy handling and temperature control the existing stabilization oven was cut into two. Project supplied trios were introduced in between ovens. The modified version of heat treatment equipment is shown in Fig. 3. Meanwhile CTA has built a new building for carbon fiber work.

In 1984, in order to explore new oxidants, studies were conducted using nitrogen dioxide. But it was observed that the stabilization by NO_2 did not improve the physical properties of the carbon fibers.

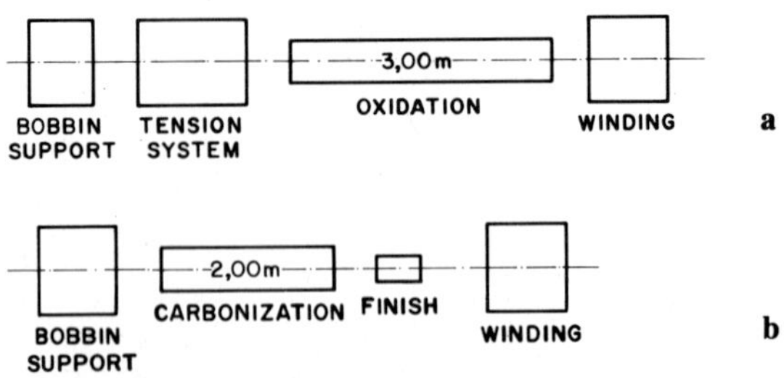

Fig. 2. Heat treatment equipment **a** old systems, **b** system since 1981

Table 8. Best results of 1981 – One STEP oxidation (FISIBA precursor)

	Oxidation				Carbonization			Carbon fibers properties	
	Elong. %	Temp. °C	Res. time min	ATM.	Temp. °C	Res. time min	ATM.	Tensile Strength	Modulus
Continuous process	+7	230	180	AIR	1000	120	Argon	200×10^3 PSi 1,38 GPa	20×10^6 PSi 138 GPa
Discont. process	+6	220	220	Forced air	1000	120	Argon	345×10^3 PSi 2,30 GPa	26×10^6 PSi 180 GPa

Table 9. Best results of 1982 (Courtaulds SAF PAN)

Continuous process Elongation Pre-stretch %	Temperature Pre-stretch °C	Pre-oxid. Temp. °C	Oxidation Temp. °C	Res. time Pre-oxid. min.	Res. time oxid. min.	ATM.	Temp. Carb. [a] °C	Res. time Carb. min.	ATM. Carb.	Tensile Strength	Modulus
+6	190	220	230	60	120	AIR	1100	120	Argon	275×10^3 PSi 1,90 GPa	21×10^6 PSi 145 GPa

[a] Shrinkage = (8–9)%

Table 10. Best results of 1982 (Courtaulds SAF PAN)

Discontinuous process Elongation oxid. %	Temperature oxid. °C	Res. time min.	Atmosphere oxid.	Temperature Carb. [a] °C	Res. time min.	Atmosphere Carb.	Tensile Strength	Modulus
+6	220	200	Forced air	1100	120	Argon	350×10^3 PSi 2,40 GPa	$2{,}8 \times 10^6$ PSi 193 GPa

[a] Shrinkage = (8–9)%

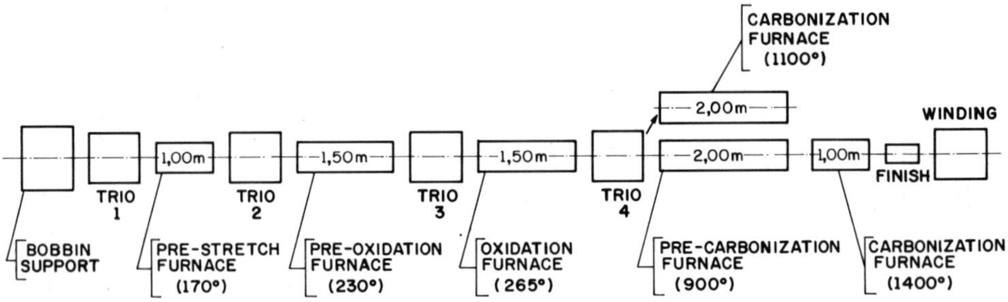

Fig. 3. Heat treatment equipment introduced 1983

3.2.1 *Activities in Oxidized Polyacrylonitrile (PAN-Ox)*

PAN-Ox is a good substitute for Asbestos. It is a heat resistant material and has several industrial applications. Some of the companies which participated in the International Conference on Carbon Fibre Applications at CTA were interested to use oxidized polyacrylonitrile fiber as thermally resistant material. A company "ASBERIT" has tested CTA PAN-Ox material and was pleased to inform CTA that it can be used and they have a demand of 60 ton per year. The physical properties of this PAN-Ox are given in Table 11. As per the requirement of the ASBERIT company, about 3.0 kg of the processed PAN-Ox fiber was sent for testing at ASBERIT Co.

3.3 *Developmental Activities in Surface Treatment*

Surface treatment of carbon fibres is very important to form good composites with resin. The bonding of resin to the fiber depends mainly on the functional groups present on the surface of the carbon fibers. Poor bonding results in poor properties of final composites. The surface treatment to carbon fiber is done by different

Table 11. Physical properties of CTA oxidized fiber

Fiber No.:	CTA PAN Ox-1
Tensile strength:	0.175 GPa
Modulus:	6.577 GPa
Density:	1.498 g/cm^3
Denier:	1.0888
Elongation:	10%
No. of filaments:	18000
Tenacity:	1.3 gm/denier 1.29 gm/dtex
Breaking strength:	1.57 gm
Filament dtex:	1.2088
Heating temp. final:	300 °C, (first 230)
Residence time:	120 min.

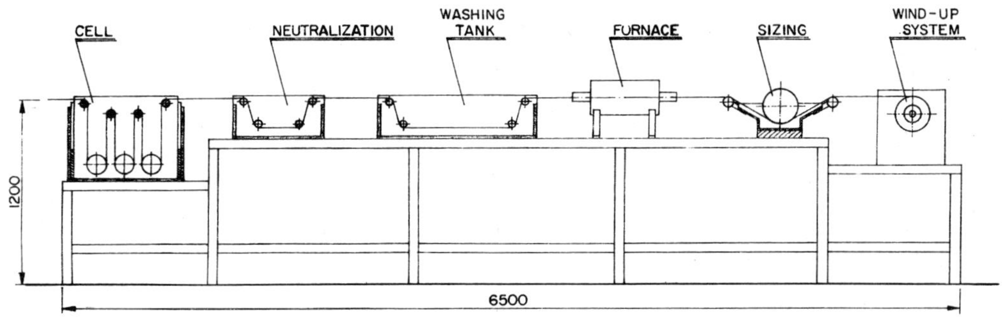

Fig. 4. Surface treatment equipment

methods. For example, electrolytic oxidation, acid treatment, alkali electrolysis, thermal oxidation, etc. At CTA surface treatment by electrolytic process is being developed.

With the formation of a group in surface treatment, in 1983, the activities restarted in this area. Anodic oxidation equipment was assembled by the CTA group. In view of the information and assistance given by project experts, a few surface treatment experiments were conducted using CTA processed carbon fibers. A 20% sulphuric acid was used as the electrolyte. About 125 m of the fiber was surface treated. Unidirectional composites were made using epoxy resin DER 383 (Dow) and curing agent DEH 50 (Dow). These composites were tested. In the subsequent experiments, the graphite electrode used earlier was replaced. Modifications were made using glassy carbon tube made out of polyperfural alcohol resin at CTA. Also subsequently several other modifications were made in the electrolytic cell. Also the experimental conditions such as current density, electrolyte were changed and several of the experiments were conducted. Meanwhile, it is felt that untreated carbon fibers obtained from a commercial firm should be surface treated in order to standardize the surface treatment work. In view of CTA's latest contacts with RK Textiles UK, a sample of the untreated carbon fiber was obtained and surface treated with improved parameters.

The final objective is to introduce a surface treatment equipment in line with the carbonization processing equipment (see Fig. 4). Also, an equipment has been fabricated for the sizing of the surface treated carbon fibers. This will also go on line in the total carbonization equipment. A few batch scale experiments were already performed for sizing using 2% epoxy resin DER 757 (Dow) dissolved in acetone.

4 Present Status of R and D Activities

CTA, having demonstrated its capabilities in the development of carbon fiber technology is now involving COPENE (COMPANHIA PETROQUIMICA do NORDESTE S.A.), the parent organization of FISIBA, for the industrialization of this technology in the country. Discussions are going on in this direction. Should these discussions become successful, the industrialization of carbon fiber technology could be seen in the next few years by COPENE in Brazil.

The Development of Carbon Fibre and its Composites in China

H. Zeng

1 Introduction

Research and applications of glass fibres and their composites, which serve as the first generation of composite materials to be used in engineering structural materials in China, have been started since the mid 1950's. At present, an all-round system for research, design, production and education in the field of glass fibre composites in our country has been established. The glass fibre composites have come to play more and more an important role in various sectors of the national economy in China. As we all know, the glass fibre composites possess high tensile strength qualities, but have not been success in developing high modulus-to-weight ratio qualities (as well as high temperature resistance) which will be required in those application which is sensitive to stiffness-to-weight ratio, or under very severe technical environments. So that, we began our research and development on carbon fibre in early 1960's, although of relatively poor quality initially. Following is a brief account on the work on carbon fibres and their composites in China during recent years.

Carbon fibre has a combination of superior properties including high strength-to-weight ratio, high modulus-to-weight ratio, high temperature resistance, ultralow thermal expansion coefficient, good chemical resistance etc. Various carbon fibres have been judiciously combined with the matrices such as polymers, metals, carbons, ceramics, and cements etc., forming high performance composites. These composites serve as the newest structural materials, thermally insulating materials, antihunt materials and other advanced functional composites with great significance for the development of aircraft and aerospace industries as well as for the modernization of other industrial and commercial applications. For the preparation of carbon fibres in China since the early 1960's, the early precursor fibre was rayon, followed by polyacrylonitrile (PAN), and most recently pitch. However, at the present time, PAN fibre is the mainly used precursor for the carbon fibre fabrication. The acrylic fibre is manufactured by solution polymerization which can be carried out directly in the spinning solvents. In our works the effective solvents of a highly polar nature including organic and inorganic solvents have been studied. We have also investigated the mechanisms of polymerization of acrylonitrile, and of copolymerization including one or more monomers in addition to acrylonitrile, and the wet-spinning process as well. The relationship of the properties with structure and chemical composition of various acrylic fibres was studied.

2 The Treatment of Polyacrylonitril Fibres in Order to Produce Carbon Fibres

According to the fact that the properties of carbon fibre are closely associated with its graphitic crystallite size, and its degree of orientation of the graphitic layers parallel to the fibre axis, which was remarkably dependent upon the nature of precursor fibre and carbon fibre manufacturing processes, the processing of manufacturing PAN-based carbon fibres involves stabilization step, followed by carbonization step, and in the case of graphite fibres, the graphitization step. The stabilization of PAN fibres is a prerequisite to the carbonization, the purpose of which is to make the PAN fibres unmeltable (leading to the formation of ladder polymers) for following carbonization. This process of preoxidation-stabilization (cyclization) is carried out under condition of stretching the fibres to avoid any polymer relaxation, which is most effective to realize the orientation of polymer structure. We have widely studied the mechanisms of the preoxidation-stabilization and of the carbonization (including the thermal degradation and shrinkage behaviour arising from the pyrolysis of PAN fibre, as well as microstructural transformation occured during these reaction steps). In order to explore the optimal conditions for the stabilization of PAN fibres before carbonization, by the use of a number of techniques including infrared spectroscopy, X-ray diffraction, light microscopy, electron microscopy (TEM, SEM, EPMA), electron spectroscopy for chemical analysis (ESCA), thermal analysis (DSC, DTA, TMA, TGA) and mensuration of the mechanical properties etc., we have investigated the air-preoxidation and accelerated-preoxidation with catalysts (for example $SnCl_4$) or by heating in air under radiation of the far-infrared ray. In the meantime, we have also investigated a variety of factors including chemical composition, molecular weight distribution, supermolecular structure and orientation degree, as well as characteristic defects and impurities of precursor PAN fibres, which have effects on the qualities of carbon fibres resulted. The characteristics of the texture-structure and the defects in acrylic fibres are directly related to their spinning process used. As we all know, the knowledge of structure-property relationships is the key to better materials, thus, in order to provide scientific fundamental understanding for improving the qualities of carbon fibre, in recent years a number of basic researches on the carbon fibre structure-property relationships in association with its manufacturing process were extensively carried out in China. At present, our experimentally prepared carbon fibre may reach tensile strength about 300 kg/mm^2 (2.94 Gpa) and tensile modulus about 22,000 kg/mm^2 (216 Gpa), as to say, its mechanical property is approaching high strength level.

Carbon fibre is one of the polymorphism of polymeric carbon. The microstructure and texture of carbon fiber are completely different from amorphous glass fibre. To prepare a fibre-resin composite with good interfacial adhesion (improving its interlaminar shear strength), we began some basic researches on various surface treatments (dry and wet-oxidation) of carbon fibres. These include the investigation of the morphological structure and surface oxygen-containing functional groups on the treated carbon fibres, as well as their effects on the wettability of matrix and the interfacial effects (physical, chemical and physicochemical), which are able to exert an influence on the properties and the mechanism of the fracture of the composite obtained.

3 Application of Carbon Fibres

In company with the increment of the production and application of carbon fibres, as well as with the developing understanding of structure and structure-property relationships of carbon fibre, in recent years the introduction of oxidation techniques made the preparation of highly activated carbon fibres possible, which are characterized by high surface areas (about 1200 m^2/g), resulting mainly from micropore formation, and, so that, having high sorption and desorption capacities as compared with the conventional powdered forms of activated carbon. Activated carbon fibres, yarns and fabrics were shown to effectively sorb gases, vapors, and dissolved solids. Such new high-effective sorbent materials are now becoming available for various specific applications (such as air filters, purification of solutes and solvents, or protective garments against hazardous chemical vapors etc.). Activated carbon fibres in China are presently in research and development stage. Moreover, we have also investigated the utilization of intermediat products obtained during manufacture of carbon fibre as functional materials, for example, preoxidized-stabilized PAN fibres as flame proof fabrics or semiconducting fibres used in corona resistant conditions and, if further treated with various chemical methods, resulting various functional fibres (such as ionexchange fibres or cloth). These researches, as mentioned above, attempt to explore comprehensive utilization of carbon fibres.

Early applications of carbon fibre composites were primarily in the aircraft and aerospace industries. However, for the long-term objective, the high technology industry such as aerospace is only a smaller market for the application of carbon fibre composites. In fact, any application in other industrial sectors which could benefit from a material of higher stiffness-to-weight ratio or higher strength-to-weight ratio is a potential candidate, in which the interest in carbon fibre composite materials is growing day by day. They are rapidly beginning to adapt these aircraft and aerospace technologies to their own application. At present, in China a number of prototype components are being fabricated and tested in the field of carbon fibre composite application for other industries, and extensive research work is being done to meet this long-term objective. At the present stage in our country prominent matrix materials for carbon fibre composites are thermosetting resins such as epoxy resins and phenolic resins. With the aim of developing better carbon fibre-resin composites, we have carried out some basic research on the mechanisms of rubber-toughened epoxy resin systems and of curing, as well as their effects on properties and fracture mechanism of the carbon fibre composites obtained. It should also be mentioned that advanced carbon fibre composites with thermoplastic resin-matrix have a number of advantages as compared with those with thermosetting resin-matrix. In recent years we engaged in research work on the carbon fibre thermoplastic composites. A large research program is focused on the studies of preparation methods, structure-property relationships of carbon fibre (or multiple fibres) reinforced thermoplastic resins (such as PE, PP, polycarbonate, nylon, polysulfone, PTFE, PPS etc.), as well as of their applications. The performance of a fibre-resin composite is neither uniquely dependent on the properties of the fibre nor of the matrix. In fact, they are widely controled by the interfacial interactions and the bonding between the components of the composite. In particular, flexural strength, interlaminar shear strength and the mode of failure, are directly dependent on the peculiar nature of the interfacial interactions. It should be noted when crystalline polymers (such as nylon and PPS) as thermoplastic matrices are used in combination with carbon fibres (or multiple fibres) in composites, all fibres used will exhibit interfacial nucleating effects on the crystallization of polymer-matrix, but the crystalline morphologies induced by different fibres will be different, arising mainly from the difference of surface structure of various fibres. In composites there exists an interfacial effect, which is responsible for the stress gradient in the matrix during the crystallization of crystallizable polymer-matrix. This will exert an influence on the overall composite comprehensive properties and their microstructural features of fracture. Although a good interfacial adhesion improved by surface treatment of carbon fibre (or of matrix-resin used) will enhance the interlaminar shear strength of composites obtained, however, the structural features and tensile strength of matrix-resin used will be a limitation to the interlaminar shear strength of some composites. Therefore, for preparing a carbon fibre composite with outstanding properties, it is also important to select a correct process and proper matrix resin. Carbon fibre composites in China were experimentally used in sporting goods (such as badminton racquets, racing boats, masts for the sailboat, racing model aeroplane), textile machinery parts, chemical plant equipments (such as corrosion resistant sealings), mechanical egineering components (such as wear and frictional components or self lubricating composites), electronic components (such as effuser) etc. These primary miscellaneous applications, as mentioned above, have obtained remarkably technical benefits. However, a major drawback of these industrial applications is the price of carbon fibres. The desired properties are inherent to the carbon fibre composite materials, but their present high price is an obstacle to a further development of miscellaneous large volume utilization. As a result, in recent years by the means of hybrid construction, it has been possible to develop acceptable products, for example, glass fibre to partially replace the more costly carbon fibres. The hybrids with lower cost glass fibre will not only be necessary to offset the higher cost carbon fibres, but available for a better impact strength. With this long-term objective of developing hybrid composites, which should grow in harmony with the optimum performance and cost benefits, we have been conducting some basic researches on the "hybrid effects" or the hybridization qualities of carbon fibre composites with glass fibres. In the meantime, some other new high-performance carbon fibre composites, such as carbon/carbon composites, carbon/metal composites and carbon/ceramic composites etc. are also investigated in China.

4 Summary

A great deal of research work on the mechanics and the engineering structural design of composite materials has been carried out in China. On the basis of research, production and applications of glass fibre composites, in China today, a scientific and technical corps including chemists, physicists, mathematicians, mechanics, design engineers and technologists etc. are working with great effort in the field of carbon fibre composites. According to the concrete condition of our country and learning from world's advanced experience, to develop advanced carbon fibre composites will play more and more an important role in our socialist modernization construction.

Above are the status of our works on carbon fibre and its composites conducted in recent years. Although some of them have come up to certain level, most of our works remain at the beginning stage and lack depth, and many important field of carbon fibre composites still remain gaps. But we have confidence, that, we could push our work to a high level, since now-a-day in our country, science and technology are receiving the best attention, and our researchers are working with enthusiasm. On the other hand, we are also willing to further strengthen scientific and technical exchange and co-operation with international institutions as well as with other friendly countries on the basis of equality and multual benefit. In the meantime, we sincerely welcome valuable comments and suggestions from our colleagues at this conference. Let our scientists join hand to hand to make our contribution to the further development of carbon fibres and its advanced composites, and to the further strengthening of friendly ties between our peoples.

Acknowledgment

Let me take this opportunity, on behalf of my Chinese colleagues, to convey warmest thanks to you, organizers of this international conference on carbon fibre application – United Nations Financing System for Science and Technology for Development (UNFSSTD), United Nations Industrial Development Organization (UNIDO) and our host, Brazilian Government as well as its Centro Técnico Aerospacial (CTA), – for arranging this conference.

The Carbon Fiber Industry in Korea

C. Y. Kim and B. Rhee

1 Introduction

The Korean government has been driving the 5th five year economic development plan since 1982, which includes the development of five major fields of scientific technologies. In order to perform this plan successfully, we have to prepare lots of industrial raw materials. But Korea is lacking in natural sources and must develop some new industrial materials continuously. Therefore, developing the special structural materials is one of these fields, and the carbon-fiber-reinforced composite materials have the most important place.

The carbon fiber is one of the most attractive products studied presently. The lack of research personnel, however, drags the activity.

The public research institutes hire a fair number of researchers who have been trained either in Korea or abroad and are capable of doing research.

The government has started supplying the matching fund for a research in 1982 if a private company requests a public research laboratory to develop a product with a 30% of the total cost.

In Korea, a study on the fabrication technology of CF on a laboratory scale was for the first time presented to the annual meeting of Korean Institute of Chemical Engineers in April, 1971.

Soon after, a research project supported by government (MOST) was reported in 1972, and then, continual research on developments of CF have been performed at Chungnam National University (CNU).

Some of them were cooperated with Prof. E. Fitzer in Karlsruhe University (W. Germany) and one is now going on.

The work during this period resulted not only in accumulation of knowledge and development of fabrication technologies of the carbon fiber and C/C composite materials, but also in development of an interest for such materials among industries, and at last, one industry have invested a fund into co-work team consist of KIMM (Korean Institute of Mechanic and Metallurgy) and CNU to develop carbon fiber from PAN precursor on a pilot plant scale from the beginning in 1983.

This work is more seriously supported by the government organization nowadays and the research is being derived towards both development of fabrication technology of high performance CF and use of CF for CFRP composites.

2 Research Activities

Another study on development of carbon fibers was performed in KAIST (Korea Advanced Institute of Science and Technology), which was funded jointly by the government and a private industry in 1982 although it was not continued in 1983.Results of research activities so far obtained are following:

1. A complete fabrication and surface treatment technology of CF has been developed on a pilot plant scale (0.25 t/month) starting with PAN precursor, and properties of these fibers are competitive to those of commercially available CF.

2. A lot of knowledge and technology for production of CF have been accumulated, and most equipments necessary for the industrial scale fabrication of CF could be designed and made by these co-workers.
3. A great interest in development and application of CF has sprung up to the government, industries, researchers, technicians, and so on.

3 Expected Future Development

There is no industry in Korea to be producing carbon fibers presently although a few of companies are studying seriously to do so. If they do, it may be favorable that the starting material will be PAN, because we have already developed the carbon fiber with PAN precursor on a pilot scale and trying to transfer this technology up to the industrial scale.

However, in spite of such results and interest for CF, there are lots of work to do to ensure a stable industrial production of good quality CF. First of all, the starting PAN precursor should be changed to pitch, which is a by-product of oil refinery and iron industry. By using pitch, cost of CF production will be down and CF quality will be much improved, if it is treated to form a mesophase.

Fortunately, – mentioned above, the going on bilateral laboratory study which is being supported by DFG/GTZ is to develop mesophase pitch from Korean raw pitch, and we are now getting valuable results day by day. Furthermore, about 0.8 million BPSD of clude oil can be refined and almost 10,000 BPSD of petroleum pitch is produced as a by-product in Korea. Also, about 240,000 tons of coal-tar pitch is coming from iron industries in a day.

There is one company which produces carbon fiber prepreg with a capacity of 20,000 square meter per month. Presently the plant is operating in the range of 20 to 50% of the capacity and the carbon fibers as well as the epoxy resins are imported mostly from Japan.

There are four companies, at least, to make use of the prepreg to produce mainly fishing rods and much less tennis rackets. The prepreg making company supplies about 90% of the prepreg used in Korea.

No other industry related to carbon fiber products such as producing composites, weaving carbon fibers, etc. is known yet in Korea.

But, now there are some public research institutes, private sector's research laboratories, and university laboratories, all of these could develop the carbon fiber and its applications if they are fully sponsored by the government or private industries.